T0250604

Lecture Notes in Computer Science 1253

Edited by G. Goos, J. Hartmanis and J. van Leeuwen

Advisory Board: W. Brauer D. Gries J. Stoer

Springer
Berlin
Heidelberg
New York
Barcelona
Budapest
Hong Kong
London
Milan
Paris
Santa Clara
Singapore
Tokyo

G. Bilardi   A. Ferreira
R. Lüling   J. Rolim (Eds.)

Solving Irregularly Structured Problems in Parallel

4th International Symposium, IRREGULAR'97
Paderborn, Germany, June 12-13, 1997
Proceedings

 Springer

Volume Editors

Gianfranco Bilardi
University of Padova, Department of Electronics and Computer Science
I-35131 Padova, Italy
E-mail: bilardi@artemide.dei.unipa.it
and
The University of Illinois at Chicago
Department of Electrical Engineering and Computer Science
851 South Morgan Street, Chicago, Illinois 60607-7053, USA
E-mail: bilardi@eecs.uic.edu

Afonso Ferreira
CNRS, LIP ENS Lyon
46, allée d'Italie, F-69364 Lyon Cedex 07, France
E-mail: afonso.ferreira@ens-lyon.fr

Reinhard Lüling
University of Paderborn, Department of Mathematics and Computer Science
Fürstenallee 11, D-33095 Paderborn, Germany
E-mail: rl@uni-paderborn.de

José Rolim
University of Geneva, Computer Science Center
24, Rue General Dufour, CH-1211 Geneva, Switzerland
E-mail: rolim@cui.unige.ch

Cataloging-in-Publication data applied for

Die Deutsche Bibliothek - CIP-Einheitsaufnahme

Solving irregularly structured problems in parallel : 4th
international symposium ; irregular '97, Paderborn, Germany, June
12 - 13, 1997 ; proceedings / G. Bilardi ... (ed.). - Berlin ; Heidelberg
; New York ; Barcelona ; Budapest ; Hong Kong ; London ; Milan ;
Paris ; Santa Clara ; Singapore ; Tokyo : Springer, 1997
   (Lecture notes in computer science ; Vol. 1253)
   ISBN 3-540-63138-0

CR Subject Classification (1991): F.2, D.1.3, C.1.2, B.2.6, D.4, G.1-2

ISSN 0302-9743
ISBN 3-540-63138-0 Springer-Verlag Berlin Heidelberg New York

This work is subject to copyright. All rights are reserved, whether the whole or part of the material is
concerned, specifically the rights of translation, reprinting, re-use of illustrations, recitation, broadcasting,
reproduction on microfilms or in any other way, and storage in data banks. Duplication of this publication
or parts thereof is permitted only under the provisions of the German Copyright Law of September 9, 1965,
in its current version, and permission for use must always be obtained from Springer -Verlag. Violations are
liable for prosecution under the German Copyright Law.

© Springer-Verlag Berlin Heidelberg 1997
Printed in Germany

Typesetting: Camera-ready by author
SPIN 10549983   06/3142 – 5 4 3 2 1 0   Printed on acid-free paper

Preface

The Fourth International Symposium on Solving Irregularly Structured Problems in Parallel (IRREGULAR'97) was scheduled for June 12 and 13, 1997, in Paderborn, Germany.

IRREGULAR focuses on algorithmic and system aspects arising in the development of efficient parallel solutions to irregularly structured problems. It aims, in particular, at fostering cooperation among practitioners and theoreticians in the field. It is an annual meeting, with previous editions held in Geneva, Lyon, and Santa Barbara. Previous proceedings appeared as LNCS 980 and 1117. IRREGULAR has been promoted and coordinated by the Symposium Chairs,

Afonso Ferreira (ENS Lyon) and José Rolim (Geneva).

This volume contains all contributed papers accepted for presentation at the '97 Symposium, together with the invited lectures by Charles Leiserson (MIT), Burkhard Monien (University of Paderborn), Friedhelm Meyer auf der Heide (University of Paderborn), Alexandru Nicolau (University of California, Irvine), and Jean Roman (University of Bordeaux).

The contributed papers were selected out of several dozen submissions received in response to the call for papers. The program committee consisted of:

Mikhail Atallah (Purdue)	Reinhard Lüling (Paderborn)
Gianfranco Bilardi, Chair	Nelson Maculan (Rio de Janeiro)
(Padova & UIC)	Brigitte Plateau (IMAG Grenoble)
Jacek Blazewicz (Poznan)	John Reif (Duke)
Bruno Codenotti (IMC-CNR Pisa)	Horst Simon (NERSC Berkeley)
Jack J. Dongarra (Tennessee)	Jon Solworth (UI Chicago)
Jean Marc Geib (Lille)	Paul Spirakis (CTI Patras)
Jun Gu (Calgary)	Katherine Yelick (Berkeley)
Friedel Hossfeld (KFA Jülich)	Emilio Zapata (Malaga)
Oscar Ibarra (Santa Barbara)	Albert Y. Zomaya (W. Australia)

The selection of 18 contributed papers was based on originality, quality, and relevance to the symposium. Considerable effort was devoted to the evaluation of the submissions by the program committee and a number of other referees. Extensive feedback was provided to authors as a result, which we hope has proven helpful to them. We wish to thank the program committee and the referees as well as the authors of all submissions.

The final program represents an interesting cross section of research efforts in the area of irregular parallelism. It is organized into six sessions, respectively devoted to discrete algorithms, randomized methods and approximation algorithms, implementations, programming environments, systems and applications, and scheduling and load balancing. We expect the relevance of these areas to grow with the increased spread of parallel computing.

IRREGULAR'97 could not have happened without the dedicated work of the local organization committee, consisting of:

Reinhard Lüling (Chair) and Ralf Diekmann,

from the Paderborn Center for Parallel Computing of the University of Paderborn, whose sponsorship of the Symposium is gratefully acknowledged.

Finally, we wish to thank Sylvie Boyer for invaluable assistance with the preparation of these proceedings.

Padova, April 1997   Gianfranco Bilardi

List of Referees

Richard Beigel

Rudolf Berrendorf

Robert Blumofe

John Bruno

Martha Castañeda

Lenore Cowen

Thomas Decker

Yves Denneulin

Ulrich Detert

Rüdiger Esser

Graham E. Fagg

Paola Favati

Paulo Fernades

Mihai Florin Ionescu

Michael Ger

Eladio Gutierrez Carrasco

Zouhir Hafidi

Maximilian Ibel

Esther Jennings

Renate Knecht

Mauro Leoncini

Bruce Maggs

Lenka Motyckova

Noël M. Nachtigal

Raymond Namyst

Angeles G. Navarro

Machael A. Palis

Marcelo Pasin

Antoine Petitel

Andrea Pietracaprina

Geppino Pucci

Sanguthevar Rajasekaran

Sanjay Ranka

Giovanni Resta

Jean-Louis Roch

Jean-François Roos

Luis F. Romero

Kwan Woo Ryu

Jürgen Schulze

Jens Simon

Georg Stellner

El-ghazali Talbi

Oswaldo Trelles Salazar

Maria A. Trenas Castro

Luca Trevisan

Jeffrey S. Vitter

Walter Unger

Rolf Wanka

Peter Weidner

Matthias Westermann

Tao Yang

Contents

* Invited speaker

Systems and Applications

Parallel Mesh Generation

Dr.-Ing. Lutz Laemmer
Dipl.-Ing. Michael Burghardt
THD, Informatik im Bauwesen
Petersenstr. 13, 64287 Darmstadt, Germany
E-Mail: laemmer@iib.bauwesen.th-darmstadt.de
Tel. +49 6151 16-4190
Fax. +49 6151 16-5552

Abstract. The efficient parallelisation of the finite element method is based on non-overlapping partitioning of the computational domain into an appropriate number of subdaomins. The problem size required for efficient application of parallel solution techniques is considerably large. The problem description in terms of finite element nodes and elements is complicated and difficult to handle with respect to the required main memory and file size. We describe a parallel solution method to perform mesh partitioning without prior mesh generation. The geometric description of the computational domain consists of vertices, edges and faces and boundary conditions, loads and mesh density parameters. The geometric description is recursively partitioned. The domain interfaces are minimized with respect to the number of coupling finite element nodes. Load balance is ensured for graded and locally refined meshes. Applications for two-dimensional models in structural mechanics are demonstrated.

1   Introduction

Mesh generation is a well-known hard problem in finite element method. The finite element method allows modeling of physical phenomena on unstructured, graded meshes in contrast to finite difference method using structured or block-structured computational meshes. Usability and flexibility of finite element method is proved in a vast amount of applications. Usually, the input data for the unstructured meshes result from CAD data.

The need to generate finite element meshes quickly is a common requirement of most computational fields. This is especially true for adaptive finite element processes. Therefore, the need for parallel mesh generation is well justified. Parallel solution techniques are mainly based on partitioning of the computational data and mapping them onto an particular parallel machine. For distributed memory parallel machines geometrically motivated partitioning is very efficient to construct fast solvers. Additionally parallel mesh generation avoids the bottleneck of serial mesh generation arising from storing, partitioning and loading large computational meshes.

The partitioning problem is known to be NP-complete (proof see [21]). That means the problem cannot be solved in polynomial time. Parallelising the partitioning approach reduces significantly the time for partitioning a given mesh. The mapping problem is known to be NP-complete, too. Although, usefull heuristic approaches are available (see [5], [10]).

A large number of research groups works in the field of parallel mesh generation [30], [22], [13], [23], [15], [28], [31]. The usual attempt to generate and partition a given problem in parallel is made in two steps using a so-called background mesh.

Step 1 Solve the partitioning problem for the background mesh.

Step 2 Refine the background mesh to achieve the required mesh density.

Step 1 is usually performed in a recursive manner resulting in a number of idle processors during the first recursion levels. Step 2 is inherently parallel.
The motivation of the development of these algorithms is twofold:

1. The size of the partitioning problem for the background mesh is significantly reduced compared to the original problem size resulting in a significant shorter runtime.

2. The mesh refinement process is parallelized using the (geometric) locality.

The generated partitioning should ensure connected and well-shaped subdomains on every processor and equally distributed numbers of elements or nodes for the non-overlapping domain decomposition approach.

The mapping problem is usually not solved by the approaches in the techniques referenced. But the parallel generation of the mesh is motivated by the elimination of the mesh loading and distribution process. Re-assigning the mesh to a more appropriate mapping is contra productive. Therefore, the generated partitioning should also ensure a near optimal mapping.

The strategies based on background meshes are dual to the multilevel methods developed mainly for the high quality partitioning algorithms based on the spectral bisection method. Origins of this method can be found in the works of Barnes and Hoffman [3] and it was first proposed by Pothen [25] for improving graph ordering for sparse matrix factorisation. The recursive partitioning is well suited for a near optimal mapping of the partitioning to hypercube and mesh topologies.

Although spectral bisection produces best quality decompositions (compare [26], [29], [20]) the complexity of the numerical Eigenvalue problem to be solved is significant. Therefore, multilevel methods reduce the Eigenvalue problem by gathering elements of the fine computational mesh to larger elements. The partitioning is performed for the smaller problem with the larger elements (see [2], [14], [17]). Jones and Plassman [16], Barnard and Simon [1], Roose and Driesche [8], [7], [9] are proposing parallel versions of the algorithm to reduce runtime.

Nevertheless, the quality of the final partitioning of the refined mesh has to be improved by local techniques to minimize cut length and optimize the subdomain shape. Techniques applicable both for serial and parallel improvements are Kernighan-Lin heuristic [18] or Helpful-set strategy [6].

In contrast to the background mesh approach we propose a method for parallel mesh generation working directly on the CAD input data and without a predefined background mesh. Partitioning cuts are introduced as additional geometric constraints to the original data. The quality of the partitioning is directly ensured by straight cut lines. The resulting subdomains are well shaped. Final improvement by a local heuristic is not necessary. The generated partitions are nearly optimal mapped to the underlying hardware topology. The parallel mesh generation algorithm is described in section 2. Results obtained for unstructured not graded and graded meshes are discussed in section 3.

2 Parallel Mesh Generation

In the following subsection we describe the input data for the proposed meshing algorithm. The inertial method is then described briefly. We explain more detailed aspects of the subdomain boundary generation in the following two subsections.

2.1 Input data

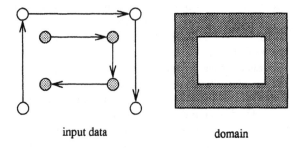

input data   domain

Fig. 1. *Input data of mesh generator*

The two-dimensional geometry is described by linear primitives. These data origin from a CAD description and consists of points, lines and polygons describing the outline and holes of a two-dimensional computational domain, properties assigned to subdomains, polygons (or boundaries of subdomains), lines (segments of polygons) and single points (see Figure 1).

Points are assigned information on the expected finite element size. Expected element size is interpolated between known points. Lines are defined as linear connections of previously defined points. Polygons consist of an ordered set of lines. Polygons are assumed to be closed. One of the polygons describes the outer boundary of the two-dimensional region. Holes are described by special tags to polygons. Additionally, lines and points can reflect lines or points in the computational model.

2.2 Inertial method

The main idea for the partitioning origins from the inertial method, for example [11]. The domain is split into two equally weighted subdomains along the axis of the larger inertial momentum. The quality of the cut is problem dependend. The inertial mehtod is applied to pre-existing meshing the co-ordinate information. Our method is directly based on the partitioning of the CAD geometry. Therefore, cut lines are added to the geometry description of the CAD data and extend the input data of the meshing algorihm.

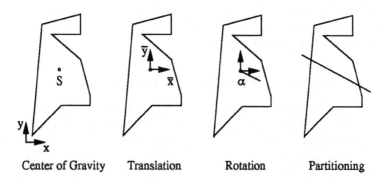

Center of Gravity   Translation   Rotation   Partitioning

Fig. 2. *Recursive bipartition by inertial method*

The mesh density is governed by a pointwise defined density function $f(x, y)$. Usually, such a function defines the requested element size. This element size defines the number of elements in the surrounding of these points. To ensure equally distributed computational load in the mesh generation and in the finite element computation the number of elements should be the same on every processor. Therefore, we construct from the function describing the element size a function describing the computational load.

We are using in the two-dimensional case for an expected element size w the function $f(x, y) = \frac{1}{w^2}$ measuring the element maximum diagonal length or the surrounding circle, respectively. Alternatively, we are using the function $f(x, y) = (\bar{w} - w)^2$ with $\bar{w}$ indicating a mean or maximum element side length. Both functions are well suited for a weighting function applied to the domains describing the problem to be used in the inertial method.

One recursion step of bipartition is performed as follows (see Figure 2). Initially, the center of gravity of the polygonial bounded geometry is calculated by numerical integration with

$$x_s = \frac{\int x \cdot f(x,y)dA}{\int f(x,y)dA}, \quad y_s = \frac{\int y \cdot f(x,y)dA}{\int f(x,y)dA} \tag{1}$$

and the problem is translated into the $\bar{x}, \bar{y}$ co-ordinate system. The inertial momentum are computed by

$$I_{\bar{x}} = \int \bar{y}^2 \cdot f(\bar{x},\bar{y})dA, \quad I_{\bar{y}} = \int \bar{x}^2 \cdot f(\bar{x},\bar{y})dA$$

$$I_{\bar{x}\bar{y}} = -\int \bar{x} \cdot \bar{y} \cdot f(\bar{x},\bar{y})dA \tag{2}$$

giving the principal direction $tan2\alpha = \frac{2I_{xy}}{I_x - I_y}$. The domain is split by the principal axis according to the largest principal inertial moment.

The numerical integration is performed for a triangulation constructed from the existing point set. The existing point set has to describe the domain in a user acceptable way. Usually, triangulation of a given domain is arbitrary. Therefore, the user has to ensure the modeling of the mesh density function is appropriate with the given set of points (see also 3.2).

2.3 Coupling nodes on the subdomain boundaries

The coupling nodes on the subdomain boundaries have to be generated deterministically to avoid additional communication. The problem is twofold. First, the global namespace of the coupling nodes has to be unique. Second, the co-ordinates of the nodes generated on the boundary between two or more subdomains have to be the same to ensure conforming meshes.

The subdomains are recursively generated in parallel. Initially, the problem is loaded onto all processors. All the processors perform the first split. Processors $0, \ldots, p/2 - 1$ are assigned the first part of the domain and processors $p/2, \ldots, p - 1$ are assigned the second part of the domain. Assignment is performed straightforward by the identity mapping, namely subdomain 1 is mapped onto processor 1, subdomain 2 is mapped onto processor 2 and so on.

Resulting coupling nodes on subdomain boundaries are generated as described below in section 2.4 and are numbered consecutively. The generation of boundary nodes is deterministic. The same algorithm is performed on every processor. Therefore the available information on the co-ordinates of the coupling nodes is the same on every processor. Therefore, no communication is needed after the split.

In the second and the following recursion levels first the number of newly generated nodes is counted in every group of processors performing the same split, second the total number of coupling nodes is exchanged between the interested processor groups and third the globally unique names are assigned to the coupling nodes. The collective communication (**MPI_Allreduce**) synchronises the processor groups.

The mesh generation is performed sequentially on every processor without further communication on the last recursion level. The implemented mesh generation algorithm is decribed in [24].

2.4 To guarantee conform meshing on subdomain boundaries

Based on the input data we are using a technique defining piecewise node placements on linear segments. The segment is defined by two nodes with prescribed mesh size values f_1 and f_n and a connecting straight line. The node with the smaller mesh size value f_1 is assumed to be the starting point of the recursive node generation process illustrated in Figure 3.

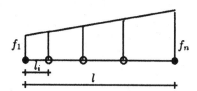

Fig. 3. *Discretisation of a line segment*

The distances of the $n-1$ intermediate nodes on the segment are computed by the recursion

$$l_1 = \beta f_1$$

$$l_i = l_{i-1} + \beta \sum_{j=1}^{i} j \qquad (3)$$

whereby β is computed in a preprocessing step.

Following [19] the length of the line segment to discretize L is calculated from the co-ordinates x_1, y_1 and x_n, y_n of the starting point and the endpoint, respectively. The mean value of the expected element sizes in the starting and end point $\bar{f} = \frac{1}{2}(f_1 + f_n)$ is distributed linearly onto the number of intermediate line segments $n = \lceil \frac{L}{\bar{f}} \rceil$ in proportion to the weighted distances giving $n(n-1)$ fractions of length

$$\Delta l = \frac{2(f_n - f_1)}{n(n-1)} \qquad (4)$$

The total length of the distances is checked by

$$l_{tot} = n f_1 + \Delta l \sum_{i=1}^{n} i(n-i). \qquad (5)$$

and β is determined by $\beta = \frac{\Delta l}{l_{tot}}$ to scale the single distance values l_i to fit into the total segment length L.

3 Results

Results of the proposed method are compared to the traditional sequential approach of mesh generation as described in [24]. Parameters to be compared are the mesh quality, the partitioning quality and the parallel efficiency of the mesh generation. The influence of the some what different mesh on the solution of PDE will be reported in a follow up paper. Two examples are composed here with detailed result data. The first example (m1) has a simlpified geometry without mesh grading. Figure 4 shows the mesh generated on 1 and on 4 processors, respectively. The other example is a graded mesh. Figure 5 shows the mesh generated an 1 and on 8 processors, respectively.

3.1 Mesh quality

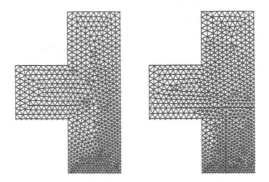

Fig. 4. *Serial and parallel mesh generation example m1*

Meshes generated by the proposed parallel generation algorithm differ from sequentially generated meshes with respect to the additional line segments of the subdomain cuts. These line segments are clearly visible in the parallel generated mesh. they are not part of the sequentially generated mesh. The main advantage of these line segments is the a-priori short cut length. Local improvement of the mesh by flipping single elements between subdomains to shorten the subdomain boundaries is not necessary.

Smoothing the element shapes in the parallel generation process is restricted to the elements of a single subdomain. Nodes can only be moved inside the subdomains. Intersubdomain smoothing is not implemented to avoid communication. Nevertheless, the comparison of the resulting minimum angle distribution for a number of triangulations shows similar results for the sequential and the parallel case, respectively (see Table 1 with results for the generation of Example m2 (compare Figure 5) on different numbers of processors).

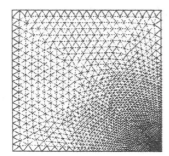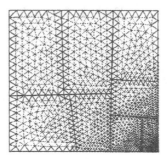

Fig. 5. *Serial and parallel mesh generation example m2*

Table 1. *Minimum angle distribution for Example m2 (Figure 5)*

Number of processors	1	2	4	8
$\bar{\alpha}$	59.73	59.70	59.70	59.66
σ	7.43	7.44	7.74	7.96

3.2 Partitioning quality

Quality of the partitioning is estimated by the following features

- load balance,

- interprocessor communication and

- network complexity or subdomain connectivity.

The attraction of the inertial method results from the perfect solution to the last two features ([11],[14]) . Inertial method is a member of the recursive bisection partitioning class. Therefore, the mapping of the resulting mesh is well investigated and the network complexity is bounded. Straight line cuts produced by the application of the inertial method to the original, unrefined subdomain geometry result in extremely short subdomain boundaries. Minimum interprocessor communication is achieved without additional smoothing and improving a preliminary partitioning like in the sequential approach based on exisiting meshes.

The load balance of the resulting partitioning is more critical. Load balance is achieved by the minimum mean difference from an average load per subdomain in the least square sense. The known sequential implementations of the mesh based inertial method guarantee a perfect result.

For the examples shown in Figures 4 and 5 the resulting mean difference from average load are shown in Table 2. The quality of the proposed method depends on three characteristic features

Table 2. *Mean difference from average load 4*

Number of processors	Example m1	Example m2
2	12.18%	4.12%
4	13.01%	4.56%
8	6.60%	3.45%

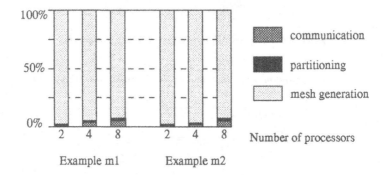

Fig. 6. *Runtime distribution for parallel mesh generation*

– quality and appropriatness of the input data,

– quality (accuracy and smoothness) of the mesh density function and

– the number of the processors.

Errors in the estimation of the load distribution are propagated with the recursion levels. The accuracy of this estimation depends on the accuracy of the numerical integration. The numerical integration is performed for intermediate triangular meshes generated with the given point set. If the given point set does not reflect the users description of the mesh distribution or the triangular mesh is to coarse to transfer sufficient information the estimation of the mesh distribution and the load distribution will be less accurate. For a given problem the loss in accuracy grows with the number of processors.

3.3 Parallel efficiency

Parallel efficiency depends on the time complexity of the mesh generator itself and on the parallel overhead due to additional or not parallelisable parts of the algorithm. Parallelisation overhead introduced by additional communication is not significant for the proposed method. All the communications needed are performed completely synchronised and can be implemented as collective communications of type **ALLREDUCE** within processor groups (compare details of

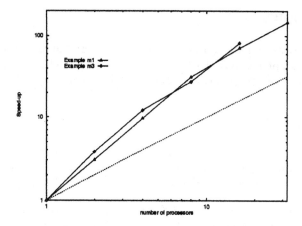

Fig. 7. *Speed-up of parallel mesh generation*

MPI implementations in [12]). These communications are known to be implemented in logarithmic time very efficiently on a large number of multiprocessor architectures (see [27], [4]).

The mesh generation process is perfectly parallel. Therefore, the application scales very well. Due to the overlinear complexity of the meshing algorithm the speedup is superlinear (see Figure 7). Runtime analysis on a Pyrsytec PowerXPlorer-32 (MPC 601 with Parix 1.3 and gcc2.7.1) shows the minimal impact of the additional parallelisation overhead as shown in Figure 6.

The ratio between the time for the mesh generation and the time spent on communication determines the efficiency of the parallel approach. The time for the communication is bound by the number of messages to be exchanged. The number of messages to be exchanged depends on the number of processors, only. It is independent on the problem size. Therefore, the efficiency of the proposed method can be tuned by scaling up the problem size - increasing the number of elements to be generated on every subdomain - for a given processor configuration.

4 Summary

The method presented generates adaptive meshes in parallel. Input data origin from CAD description. The partitioning is based on the geometric description. Triangulation in advance is not necessary. The mesh generation process works on the partitioned geometric description.

Results are comprehensive and promising. Mesh quality is ensured in the parallel approach. The quality of the partitioning is comparable to the mesh based methods. Addional improvement by local heuristics, flipping of elements between subdomains to minimise the interprocessor communication and optimise subdomain shapes is not necessary. The parallel overhead can be neglected. The algorithm is scalable.

Acknowledgments

The work presented is part of a project concerned with parallelisation of structural mechanics applications. This project is currently funded by the Deutsche Forschungsgemeinschaft, Bonn-Bad Godesberg, Germany.

References

1. S. T. Barnard and H. Simon, *A parallel implementation of multilevel recursive spectral bisection for application to adaptive unstructured meshes*, Proceedings of the 7. SIAM conference on Parallel Processing for Scientific Computing **18** (1995), 627–632.
2. S.T. Barnard and H.D. Simon, *A fast multilevel implementation of recursive spectral bisection for partitioning unstructured problems*, Concurrency: Practice and Experience **6** (1994), no. 2, 101–117.
3. E.R. Barnes and A.J. Hoffmann, *Partitioning, spectra and linear programming.*, Tech. Report RC 9511 (No. 42058), IBM T.J. Watson Research Center, 1982.
4. M. Barnett, D.G. Payne, R. van de Geijn, and J. Watts, *Broadcasting on meshes with wormhole routing*, Tech. report, Univ. of Texas at Austin, November 1993.
5. S.H. Bokhari, *On the mapping problem*, IEEE Transactions on Computers **3** (1981), 207–213.
6. R. Diekmann, B. Monien, and R. Preis, *Using helpfull sets to improve graph bisections.*, Tech. Report RF-008-94, Universität Paderborn, June 1994.
7. R.v. Driesche and D. Roose, *An improved spectral bisection algorithm and its application to dynamic load balancing*, Parallel Computing **21** (1995), 29–48.
8. _____, *Dynamic load balancing of iteratively refined grids by an enhanced spectral bisection algorithm*, Workshop Dynamic load balancing on MPP systems, Daresbury, November 1995.
9. _____, *Dynamic load balancing with a spectral bisection algorithm for the constrained graph partitioning problem*, High-Performance Computing and Networking (B. Hertzberger and G. Serazzi, eds.), LNCS 919, Springer, 1995, pp. 392–397.
10. C. Farhat, *On the mapping of massively parallel processors onto finite element graphs*, Computers & Structures **32** (1989), no. 2, 347–353.
11. C. Farhat and M. Lesoinne, *Automatic partitioning of unstructured meshes for the parallel solution of problems in computational mechanics*, International Journal for Numerical Methods in Engineering **36** (1993), 745–764.
12. Message Passing Interface Forum, *Document for a standard message-passing interface*, Tech. Report CS-93-214, University of Tennessee, November 1993, Available on netlib.
13. G. Globisch, *PARMESH a parallel mesh generator*, Tech. Report SPC 93-3, DFG-Forschergruppe "Scientific Parallel Computing", TU Chemnitz-Zwickau, Fakultät Mathematik, June 1993.
14. B. Hendrickson and R. Leland, *A multilevel algorithm for partitioning graphs*, Tech. Report SAND 93-1301, Sandia Natl. Lab., Albuquerque, NM, June 1993.
15. D. C. Hodgson and P. K. Jimack, *Parallel generation of partitioned, unstructured meshes*, Tech. Report 94/19, University of Leeds, School of Computer Studies, June 1994.

16. Z. Johan, K.K. Mathur, and S.L. Johnsson, *An efficient communication strategy for finite element methods on the connection machine CM-5 system*, Tech. Report TR 256, Thinking Machines, 245 First Street, Cambridge MA 02142, 1993.

17. G. Karypis and V. Kumar, *A fast and high quality multilevel scheme for partitioning irregular graphs*, Tech. Report 95-035, Dept. Computer Science, University of Minnesota, Minneapolis, MN, 1995.

18. B. Kernighan and S. Lin, *An efficient heuristic procedure for partitioning graphs.*, Bell System Technical Journal **29** (1970), 291–307.

19. A.I. Khan and B.H.V. Topping, *Parallel adaptive mesh generation*, Computing systems in Engineering **2** (1991), no. 1, 75–101.

20. R. Leland and B. Hendrickson, *An empirical study of static load balancing algorithms*, Scalable High-Performance Computing Conf. 1994, IEEE Computer Society Press, May 1994, pp. 682–685.

21. T. Lengauer, *Combinatorical algorithms for integrated circuit layout*, Teubner-Verlag, Stuttgart, 1990.

22. R. Loehner, J. Camberos, and M. Merriam, *Parallel unstructured grid generation*, Computer Methods in Applied Mechanics and Engineering **95** (1992), 343–357.

23. S. Moitra and A. Moitra, *Parallel grid generation algorithm for distributed memory computers*, Tech. Report 3429, NASA, NASA, Langley research Centre, Hampton VA, 23681-0001, February 1994.

24. J. Olden, *Finite-Element-Analyse von Plattentragwerken durch adaptive Software-Techniken*, Ph.D. Thesis, Darmstadt, Techn. Hochsch., 1996.

25. A. Pothen, H.D. Simon, and K.-P. Liou, *Partitioning sparse matrice with Eigenvectors of graphs*, SIAM J. Matrix Appl. **11** (1990), no. 3, 430–452.

26. H.D. Simon, *Partitioning of unstructured problems for parallel processing.*, Computing Systems in Engineering **2** (1991), no. 2/3, 135–148.

27. E.A. Varvarigos and D.P. Bertsekas, *Communication algorithms for isotropic tasks in hypercubes and wraparound meshes*, Parallel Computing **18** (1992), 1233–1257.

28. C. Walshaw, M. Cross, and M.G. Everett, *A localized algorithm for optimising unstructured mesh partitions*, Int. J. Supercomputer Appl. **9** (1996), no. 4, 280–295.

29. D.R. Williams, *Performance of dynamic load balancing algorithms for unstructured mesh calculations.*, Concurrency: Practice and Experience **3** (1991), no. 5, 457–491.

30. _____, *Adaptive parallel meshes with complex geometry*, Tech. report, Concurrent Supercomputing Facilities, California Institute of Technology, Pasadena CA, 1992.

31. P. Wu and E.N. Houstis, *Parallel adaptive mesh generation and decomposition*, Engineering with Computers **12** (1996), 155–167.

Efficient Massively Parallel Quicksort

Peter Sanders and Thomas Hansch

University of Karlsruhe, Department of Computer Science, 76128 Karlsruhe, Germany

Abstract. Parallel quicksort is known as a very scalable parallel sorting algorithm. However, most actual implementations have been limited to basic versions which suffer from irregular communication patterns and load imbalance. We have implemented a high performance variant of parallel quicksort which incorporates the following optimizations: Stop the recursion at the right time, sort locally first, use accurate yet efficient pivot selection strategies, streamline communication patterns, use locality preserving processor indexing schemes and work with multiple pivots at once. It turns out to be among the best practical sorting methods. It is about three times faster than the basic algorithm and achieves a speedup of 810 on a 1024 processor Parsytec GCel for the NAS parallel sorting benchmark of size 2^{24}. The optimized algorithm can also be shown to be asymptotically optimal on meshes.

Keywords: Implementation of parallel quicksort, sorting with multiple pivots, locality preserving indexing schemes, load balancing, irregular communication patterns.

1 Introduction

Sorting has always been an important area of research both in theoretical and practical parallel computing. A disappointing observation is that some of the best practical algorithms like sample-sort are not very good from a theoretical point of view. These algorithms only work well for large amounts of data, i.e., they do not have a good scalability (except on powerful models like CREW-PRAMS). This is a particular problem for applications which use sorting of small or medium amounts of data as a frequently needed subroutine. On the other hand, there are algorithms which are theoretically very scalable. For example, parallel quicksort [15].[1] However, implementations of parallel quicksort (e.g. [6,5]) are so far more a proof of concept than a competitive practical sorting algorithm. They have an irregular communication pattern, require higher communication bandwidth than other algorithms and suffer from problems associated with load balancing.

In this paper, we describe our experiences with implementing a high performance parallel quicksort. Our work has been influenced by a number of interesting comparisons between parallel sorting algorithms regarding their practical

---

[1] Due to their poor scalability, we do *not* consider the abundant versions of parallel quicksort which only parallelize the recursive calls of sequential quicksort.

usefulness ([2,7,4]). However, rather than screening a number of quite different algorithms, we intensively study one basic strategy – namely parallel quicksort.

In Section 2 we introduce some notation and describe our experimental setup. The basic algorithm for parallel quicksort is described in Section 3. Section 4 discusses the simple yet important measure to switch from quicksort to some specialized algorithm when few PEs are involved. Section 5 explains why it is advantageous to sort the locally present data before starting the main parallel algorithm. Particularly on mesh-connected machines, the locality preserving PE indexing schemes discussed in Section 6 significantly reduce communication expense. Strategies for selecting good pivots as described in Section 7 turn out to be a prerequisite for simplifications in the data exchange patterns introduced in Section 8. Finally, using multiple pivots at once, opens up a new world of algorithmic variants in Section 9. The bottom line performance of the implementation is compared to previous work in Section 10. After a short discussion of algorithmic variants for more general networks in Section 11, Section 12 summarizes the results and outlines additional optimization opportunities. For readers interested in more details we sometimes make reference to 72 data sheets which are available electronically under `http://liinwww.ira.uka.de/~sanders/sorting/`.

2 Basic Terminology

We consider a distributed memory MIMD-machine with P PEs numbered 0 through $P - 1$. Often, we additionally assume the interconnection network to be an r-dimensional mesh and that $P^{1/r}$ is a power of 2. (Generalizations are outlined in Section 11.) Initially, each PE holds n data items. After sorting, the PEs should collectively hold a permutation of the data such that the items are sorted locally and items on PE i must not be larger than items on PE j for $i < j$.

All measurements have been done on a Parsytec GCel with 16 × 16 respectively 32 × 32 PEs.[2] The programs are written in C and use a portable communication layer [13] so that they run on many abstract machines (e.g. MPI, PVM, Parix). On the GCel, the operating system COSY [3] is used. Although the *performance* of this configuration is outdated by todays standards, it is still a good tool for parallel computing research. There are no MIMD computers in Europe with more PEs, and the ratio between communication performance and computation performance is comparable to modern machines ($\approx$10 MIPS/PE versus $\approx$1000MIPS/PE, bandwidth/PE $\approx$1MB/s versus $\approx$100 MB/s and latency for short globally communicated messages $\approx$1ms versus $\approx$10μs.[3]) The data items are generated according to the rules of the NAS Parallel Benchmark [1]. We use the ANSI-C library function `qsort` for sequential sorting.

---

[2] We would like to thank the Paderborn Center for Parallel Computing for making this machine available.

[3] Shared memory systems like the Cray T3E and some experimental message passing systems are even faster while MPI or PVM communication is regretably slower. There is also a discrepancy for short messages between neigboring PEs where transputers are relatively fast. But these do not play a big role in this paper.

3 Parallel Quicksort

We now describe a portable algorithm for parallel quicksort which is similar to the algorithm described in [8] for 2D-Meshes and in [6] (where it is augmented by a load balancer). The sorting is performed by a recursive procedure which is called on a segment of P' PEs numbered P_0 through $P_0 + P' - 1$. Each PE holds n' local items d_i for $0 \le i < n'$. Initially, $P_0 = 0$, $P' = P$ and $n' = n$. However, in general n' may take different values on each PE:

- If $P' = 1$ then sort locally and return.
- Else,
  - Collectively select one pivot p for all P' PEs in the segment.
  - Partition the d_i into n'_s *small* items s_j $(0 \le j < n'_s)$ and n'_l *large* items l_k $(0 \le k < n'_l)$ such that $s_j \le p$ and $l_k \ge p$.
  - Let N'_s denote the total number of small items and N'_l denote the total number of large items (taken over the entire segment). Split the PEs in the segment into $P_{\text{split}} := \text{round}\left(P' \frac{N'_s}{N'_s + N'_l}\right)$ PEs for the small items and $P' - P_{\text{split}}$ PEs for the large items.
  - Redistribute the items such that "small" PEs receive only small items and "large" PEs receive only large items. Furthermore, each "small" ("large") PE ends up with $\left\lfloor \frac{N_s}{P_{\text{split}}} \right\rfloor$ or $\left\lceil \frac{N_s}{P_{\text{split}}} \right\rceil$ $\left(\left\lfloor \frac{N_l}{P' - P_{\text{split}}} \right\rfloor$ or $\left\lceil \frac{N_l}{P' - P_{\text{split}}} \right\rceil\right)$ items.
  - Call quicksort recursively (in parallel) for the segments of "small" and "large" PEs.

The necessary coordination between PEs can be performed using the collective operations broadcast, reduce and prefix sum on segments of PEs. It can be shown that even for a very simple pivot selection strategy (i.e. random pivots), the expected parallel execution time is bounded by

$$\mathbf{ET}_{\text{par}} \in O(n \log n + \log P(T_{\text{routing}}(n) + T_{\text{coll}}))$$

where $T_{\text{routing}}(n)$ is the time required for exchanging n elments on each PE and T_{coll} is the time required for collective operations. The $n \log n$ term is due to sequential sorting within each processor when $P' = 1$ and the $\log P$ factor comes from the expeced recursion depth. For example, on a butterfly network, this reduces to $\mathbf{ET}_{\text{par}} \in O\big(n(\log n + (\log P)^2)\big)$.

4 Breaking the Recursion

When the PEs segments are split, there will in general be a roundoff error with the effect that one segment will get about $n/2$ more data items than the other segment. This error happens for every split and finally the PE which ends up with the largest number of items, dominates the time required for local sorting. We have observed a maximum final load between $1.77n$ and $3.16n$ depending on

n and the quality of pivot selection. (For example, refer to data sheets 1 and 12.) This problem can be alleviated by stopping the recursion for segments of small size and switching to a different sorting algorithm. We use merge-splitting sort (e.g. [16]). Depending on other optimizations, it turns out to be best to stop the recursion for segment sizes between two and four. (The main issue for this decision is load balancing so that it not very machine dependent.) The final algorithm has almost no data imbalance for large n.

5 Sort Locally First

Imbalance of the final data distribution has several disadvantages, but for large n its main impact on execution time is the increased time for local sorting which grows superlinearly with the imbalance. This problem can be solved by doing the local sorting initially before any communication takes place. To maintain the invariant that all items are sorted locally, the sorted sequences of items received during data redistributions are merged. The additional expense for merging is offset by the time saved for partitioning data. Partitioning is now possible in logarithmic time using binary search. Furthermore, more accurate pivot selection strategies are now feasible because the median of the locally present items can be computed in constant time. For large n (2^{15}), the overall improvement due to this measure is about 20 %. (For example refer to data sheets 37 and 55).

6 Locality Preserving Indexing Schemes

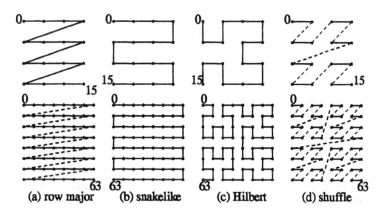

Fig. 1. PE numbering schemes for $P = 16$ and $P = 64$.

In this section we assume that the interconnection network is a square mesh and $\log \sqrt{P} \in \mathbb{N}$. The usual way to number the PEs of a mesh is the row-major ordering shown in Fig. 1-(a), i.e., the PE at row i and column j is numbered

$i\sqrt{P}+j$. This has the disadvantage that the diameter of the subnetworks involved in data exchange decreases only very slowly, although the number of PEs involved decreases exponentially with the recursion depth. Even when there are only a constant number of PEs left, they may be at a distance of $\sqrt{P}$.

A slight improvement is the snake-like ordering depicted in Fig. 1-(b) where a segment of k consecutive PEs never has a diameter of more than $k-1$. But nevertheless there is only a constant factor decrease in the diameter of segments until the segment sizes fall below $\sqrt{P}$.

What we would like to have is a scheme where any segment of PEs numbered $i,\ldots,j$ has a diameter in $O(\sqrt{j-i})$. In [10] it is shown that the *Hilbert-indexing* depicted in Fig. 1-(c) has a worst case segment diameter of exactly $3\sqrt{j-i}-2$. This result can also be generalized for three-dimensional Hilbert-indexings.

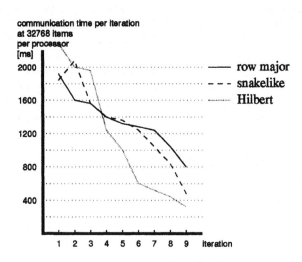

Fig. 2. Communication expense for each level of recursion for row-major, snakelike and Hilbert indexing schemes (refer to data sheet 1–3 for details).

The consequence of using these three different indexing schemes is shown in Fig. 2. For row-major and snakelike ordering, the communication expense for redistribution decreases only slowly with decreasing segments sizes. For the Hilbert-indexing, we have an exponential decrease of communication expense with the recursion depth. This property also has theoretical appeal. For example, it can be shown that using random pivot selection, the execution time is in $O\left(n\log n + n\sqrt{P}\right)$ with high probability for every legal input. This is asymptotically optimal. Analogous results can be obtained for higher dimensions. However, we observe the practical problem that in the first partitioning step, the Hilbert indexing is slowest. This is due to a more irregular communication pattern and offsets the advantage of the Hilbert indexing for small P.

In conjunction with the optimizations described in the next two sections, the shuffle indexing scheme depicted in Fig. 1-(d) (also refer to [8]) is also a good choice although it is not locality preserving in the strict sense used above.

7 Pivot Selection

We have implemented a number of increasingly accurate pivot selection strategies. In the simplest case we take the *3-median* of d_0, $d_{n'/2}$ and $d_{n'}$ on PE P_0. On the other end of the spectrum, we use the median of the local medians of all PEs. For small n (e.g. $n = 128$) local strategies are faster. Nevertheless, it always pays off to invest more effort in median selection than using the 3-median. For large n, the more accurate global strategies are always preferable. Very accurate pivot selection strategies are also a prerequisite for the algorithmic simplifications described next.

8 Simplifying Communication Patterns

In [11,8] a simple form of parallel quicksort for hypercubes is described. But it can also be used on other machines. In particular, when P is a power of two: Simply set $P_{\text{split}} := P'/2$. A "small" PE with number i sends its large items to PE $i + P_{\text{split}}$. A "large" PE with number i sends its small items to PE $i - P_{\text{split}}$. This has the following advantages over the more complicated algorithm:

- Ideally, only half the data is moved for each level of recursion.
- The prefix sums and other collective communications are saved.
- Every PE sends and receives exactly one packet.
- The communication pattern is very regular. This effect significantly increases the effective network bandwidth.
- On a 2-D mesh, the overall distance traveled by a data item through all $\log P$ levels of recursion is in $O(\sqrt{P})$ even for row-major indexing. For row-major and shuffle indexing it is at most $2\sqrt{P} - 2$, i.e., the network diameter – thus matching a trivial lower bound. Since the latter two indexings additionally imply more regular communication patterns than the Hilbert indexing they are the best choice now.

The drawback of this simple approach is that the load imbalance can grow exponentially with the recursion depth. But for randomly generated data, the median-of-medians pivot selection strategy described above is so accurate that the resulting imbalance is quite tolerable. For randomly generated data, the imbalance even *decreases*. Apparently, the property of random data distribution is destroyed by the prefix sum based algorithm while it is maintained by the simplified algorithm. Fig. 3 compares the final load imbalance and the communication expense per recursion level with and without simplified communication. (256 PEs, row-major indexing, median of medians pivot selection. For details refer to data sheets 41 and 50.)

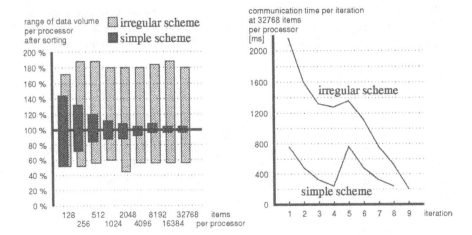

Fig. 3. Final load imbalance and communication expense per recursion level with and without simplified communication for 256 PEs, row-major indexing and median of medians pivot selection. For details refer to data sheets 41 and 50.)

When the input is not randomly generated, the simplified algorithm can still be applied by performing a random permutation of the data before starting the sorting algorithm. For large n, the additional cost for this permutation is offset by saving more than 50 % of the communication costs in the main algorithm.

9 Multiple Pivots

Suppose we not only have an estimate for the median of the data items, but estimates for the $\frac{i}{k}$-quantiles[4] of the data for $0 < i < k$. Then we can split the data into k partitions using a single redistribution. For $k = P$ this technique is known as sample sort. Some of the (asymptotically) best sorting algorithms also use this approach with $k \ll P$ [12]. By increasing k, the recursion depth decreases but the data has to travel on longer paths and in smaller packets. Furthermore, it becomes more and more difficult to find accurate pivots.

We have implemented a multi-pivot quicksort using the medians of the local $\frac{i}{k}$-quantiles as pivots. Choosing $k = 4$ turned out to be the best choice in all cases. This value is particularly "magical" for the shuffle indexing scheme. Together with the simplified communication scheme of the previous section it turns out that the communication patterns of all iterations are identical except that the distances are halved in each iteration. While the shuffle indexing did not yield an improvement for the previously considered algorithmic variants, it now turns out to be the overall "winner". (Although the improvement is not large in our measurements.) It implies a more regular communication pattern than the

---

[4] An α-quantile p_α of m items has the property that αm items are not larger than p_α and $(1 - \alpha)m$ items are not smaller than p_α.

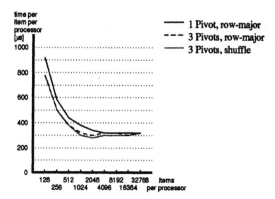

Fig. 4. Execution time for different input sizes for $k = 2$ and $k = 4$. (For details refer to data sheets 50, 55, 66 and 69.)

Hilbert indexing and it is superior to row-major indexing because it can exploit both horizontal and vertical network interconnections in every iteration. Fig. 4 shows the execution times per item per PE on 16×16 PEs for the simplified communication pattern using $k = 2$ respectively $k = 4$. Multiple pivots yield improvements for small and medium input sizes. In the multi-pivot case, the shuffle indexing slightly outperforms row-major indexing. (For a single pivot the timings are identical for both indexing schemes. Refer to data sheets 50, 55, 66 and 69 for details.)

10   Overall Performance

Figure 5-(a) compares the performance of the basic parallel quicksort (no specialized routine for small numbers of PEs, median-of-3 pivot of a single PE, local sorting at the end) with the best variant we have implemented (merge-splitting sort for two PEs, initial local sorting, 3 pivots using medians of quantiles, simplified communication, shuffle indexing) on 16×16 PEs. The final algorithm is about three times faster than our basic algorithm. (Also compare data sheets 1 and 69). For large n it also achieves a high efficiency. Figure 5-(b) and (c) show the speedup for 256 respectively 1024 PEs.[5]

Figure 5-(d) compares our final algorithm on 32×32 PEs (data sheet 72) with the timings measured in [4] for five other sorting algorithms on the same machine[6] and the same measurement approach. The basic algorithm would just

---

[5] Due to memory restrictions the dotted parts of the line had to be estimated by fitting a curve of the form $an + bn \log n$ to the execution time of sequential **qsort**.

[6] However, the Paderborn group uses Parix rather than CoSY as a communication system. Parix has lower latencies so that our results look worse for small n. CoSY allows for a higher communication bandwidth. But even artificially halving our communication bandwidth did not change the relative performance of the algorithms.

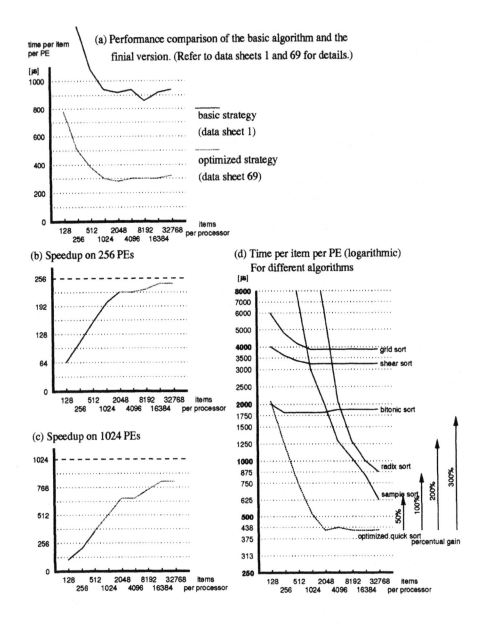

Fig. 5. Overall performance of the optimized algorithms.

barely be able to compete for medium n. The optimized algorithm is the best algorithm for the entire range of n measured.[7] For medium sized inputs it is three times faster than the best of the other algorithms. This demonstrates that quicksort has to be counted among the best practical sorting algorithms – in particular for meshes – but also that careful optimizations can be as important for comparing sorting algorithms as the basic algorithmic approach.

There are also good algorithmic reasons for the merits of parallel quicksort. It moves the data less frequently than grid sort, shear sort and bitonic sort. Sample sort (and to a lesser degree radix sort) can move the data even less frequently but over the same total distance and at the price of either sending many messages ($O(P^2)$ in a naive implementation) or also sending the data in several intermediate steps. Quicksort only needs $O(P \log P)$ messages. Since message startup overheads are quite high on many machines, this is an important point. For large P, sample sort has the additional disadvantage that sorting samples and distributing P splitters gets considerably more expensive than the $O(\log P)$ subsequent pivot selections used in quicksort.

11 Coping with More General Networks

Adapting the algorithmic measures described above for meshes with higher dimensions is straightforward. The same holds for non-square meshes as long as P is a power of two. For other cases and small n, straightforward measures like concentrating (locally sorted) data in a square submesh might be adequate. But at least for large n the available computation and communication capacity should be fully exploited. Fortunately, for large numbers of randomly distributed locally sorted items, we can adapt the pivot selection strategy. Consider an $a \times b$ mesh. Split the data between four submeshes of size $\lfloor a/2 \rfloor \times \lfloor b/2 \rfloor$, $\lfloor a/2 \rfloor \times \lceil b/2 \rceil$, $\lceil a/2 \rceil \times \lfloor b/2 \rfloor$, $\lceil a/2 \rceil \times \lceil b/2 \rceil$ respectively. Choose the medians of the local $\frac{\lfloor a/2 \rfloor \lfloor b/2 \rfloor}{ab}$-quantiles, $\frac{\lfloor a/2 \rfloor \lfloor b/2 \rfloor + \lfloor a/2 \rfloor \lceil b/2 \rceil}{ab}$-quantiles and $\frac{\lfloor a/2 \rfloor \lfloor b/2 \rfloor + \lfloor a/2 \rfloor \lceil b/2 \rceil + \lceil a/2 \rceil \lfloor b/2 \rfloor}{ab}$-quantiles. The single-pivot variant of this approach somewhat resembles the QSP-2-algorithm described in [15]. However, the new algorithm avoids the large load imbalance observed for QSP-2 because of better pivots and because the recursion is stopped for small meshes. Furthermore, the QSP-2 is not a sorting algorithm in the strict sense since it does not use a fixed PE-indexing.

Even networks which are not a mesh can profit from specialized subdivision strategies. For example, a cluster computer often consists of multiple small tightly coupled parallel machines interconnected by a relatively low bandwidth network with high startup overheads. In this case, it makes sense to partition the data in such a way that all subsequent communication is within the tightly coupled subnetworks.

---

[7] No data for larger n are available due to memory limitations of the GCel (in particular for Parix). Our measurements already show close to perfect speedup for $n = 2^{15}$ so that measurements with larger data sets would also not be very interesting.

12 Conclusions and Future Work

Our results considerably strenghten the view that parallel quicksort has to be counted among the best parallel sorting algorithms known. From a theoretical point of view, it is interesting to oberserve how self similar indexing schemes like the Hilbert numbering are able to preserve locality of communication despite of the irregularity of the communication patterns – they make quicksort one of the simplest known algorithms for meshes which is among those which are asymptotically optimal. Unlike other asymptotically optimal algorithms, the data items have to be moved only $O(\log P)$ times.

This also makes the algorithm appealing from a practical point of view because message startup overheads are high on todays machines. However, certain optimizations like breaking the recursion and choosing good pivots are required for achieving high performances in terms of constant factors. Sorting locally first reduces the impact of load imbalance and simplifies pivot selection. For randomly generated inputs, a large improvement can be achieved by *simplifying* the algorithm at the right place. These improvements are so large, that it seems worthwhile to add an explicit random permutation of the data in this case. Interestingly, although sorting requires a very irregular data movement on the microscopic level, for large inputs, the problem can be solved efficiently using a very regular communication pattern. The only remaining irregularity observable on the top level is a slight load imbalance. There are many additional optimizations which might be interesting for future research:

- If we are willing to accept that some PEs are responsible for two data partitions, quicksort can be implemented without any imbalance at all. This possibility has been considered in [15] but the required additional communication is overestimated there for large n because a PE with a partition boundary exchanges smaller messages.
- We could use a better algorithm like sample sort for the base case.
- Currently, the median selection strategies are implemented by gathering the required data on a single PE. For large P, parallelizing pivot selection could be an improvement. This is also a prerequisite for reconciling the theoretical analysis with the practically useful optimizations. For example, the parallel median selection algorithm described in [14] might be useful.
- Try pivot selection strategies based on random samples which do not require a random data distribution.
- Tune message exchange and merging steps such that copying is minimized.
- We currently use a relatively simple, high level implementation for collective communication. Exploiting tuned implementations (which are available in high quality MPI implementations for example) should yield significant improvements for small n.

Ultimately, the goal could be a toolbox of reusable components which can be configured to yield a very efficient sorting algorithm on many different architectures and for different input specifications. We expect that for very large n or small P, sample sort or one of its deterministic relatives (e.g. [9]) will be the method of choice. For smaller n or for sorting samples, quicksort will be better – in particular on meshes. Models for the algorithmic components and the machines could be calibrated using profiling data in order to make it possible to automatically plan an optimal combination of methods for every situation.

References

1. D. Bailey, E. Barszcz, J. Barton, D. Browning, and R. Carter. The NAS parallel benchmarks. Technical Report RNR-94-007, RNR, 1994.
2. G. E. Blelloch, C. E. Leiserson, B. M. Maggs, C. G. Plaxton, S. J. Smith, and M. Zagha. A comparison of sorting algorithms for the connection machine CM-2. In *ACM Symposium on Parallel Architectures and Algorithms*, pages 3–16, 1991.
3. R. Butenuth, W. Burke, and H.-U. Heiß. COSY – an operating system for highly parallel computers. *ACM Operating Systems Review*, 30(2):81–91, 1996.
4. R. Diekmann, J. Gehring, R. Lüling, B. Monien, M. Nübel, and R. Wanka. Sorting large data sets on a massively parallel system. In *6th IEEE Symposium on Parallel and Distributed Processing*, pages 2–9, 1994.
5. S. Goil, S. Aluru, and S. Ranka. Concatenated parallelism: A technique for efficient parallel divide and conquer. In *Proceedings of the 8th IEEE Symposium on Parallel and Distributed Processing*, pages 488–495, 1996.
6. J. Hardwick. An efficient implementation of nested data parallelism for irregurlar divide-and-conquer algorithms. In *Workshop on High-Level Programming Models and Supportive Environments*, Honolulu, Hawaii, 1996.
7. W. L. Hightower, J. F. Prins, and J. H. Reif. Implementations of randomized sorting on large parallel machines. In *ACM Symposium on Parallel Architectures and Algorithms*, pages 158–167, 1992.
8. V. Kumar, A. Grama, A. Gupta, and G. Karypis. *Introduction to Parallel Computing. Design and Analysis of Algorithms*. Benjamin/Cummings, 1994.
9. H. Li and K. C. Sevcik. Parallel sorting by overpartitioning. In *ACM Symposium on Parallel Architectures and Algorithms*, pages 46–56, Cape May, New Jersey, 1994.
10. R. Niedermeier and P. Sanders. On the Manhattan-distance between points on space-filling mesh-indexings. Technical Report IB 18/96, Universität Karlsruhe, Fakultät für Informatik, 1996.
11. M. J. Quinn. Analysis and benchmarking oaf two parallel sorting algorithms: hyperquicksort and quickmerge. *BIT*, 25:239–250, 1989.
12. S. Rajasekaran and S. Sen. Random sampling techniques and parallel algorithm design. In H. Reif, editor, *Synthesis of Parallel Algorithms*, chapter 9, pages 411–451. Morgan Kaufmann, 1993.
13. P. Sanders. A scalable parallel tree search library. In S. Ranka, editor, *2nd Workshop on Solving Irregular Problems on Distributed Memory Machines*, Honolulu, Hawaii, 1996.
14. P. Sanders. Randomized priority queues for fast parallel access. Technical Report IB 7/97, Universität Karlsruhe, Fakultät für Informatik, 1997.
15. V. Singh, V. Kumar, G. Agha, and C. Tomlinson. Efficient algorithms for parallel sorting on mesh multicomputers. *International Journal of Parallel Programming*, 20(2):95–131, 1991.
16. T. Umland. Parallel sorting revisited. *Parallel Computing*, 20(1):115–124, 1994.

Practical Parallel List Ranking

Jop F. Sibeyn[*]   Frank Guillaume[*]   Tillmann Seidel[*]

Abstract

Parallel list ranking is a hard problem due to its extreme degree of irregularity. Also because of its linear sequential complexity, it requires considerable effort to just reach speed-up one (break even). In this paper, we address the question of how to solve the list-ranking problem for lists of length up to $2 \cdot 10^8$ in practice: we consider implementations on the Intel Paragon, whose PUs are laid-out as a grid.

It turns out that pointer jumping, independent-set removal and sparse ruling sets, all have practical importance for current systems. For the sparse-ruling-set algorithm the speed-up strongly increases with the number k of nodes per PU, to finally reach 27 with 100 PUs, for $k = 2 \cdot 10^6$.

1 Introduction

Lists and List Ranking. A *linked list*, hereafter just *list*, is a basic data structure: it consists of nodes which are linked together, such that every node has precisely one predecessor and one successor, except for the *initial node*, which has no predecessor, and the *final node*, which has no successor. The *list ranking problem* consists of determining the rank for all nodes. The *rank* of a node is its distance to either the first or last node of its list (these notions are easily convertible). We work under the following assumptions:

- There may be many lists. The sum of the lengths of all lists, the *total length*, is N.

- A node i only knows its successor, $suc(i)$. The final nodes have a special end-of-list marker in their successor field.

- Every PU holds the data related to $k \geq 1$ nodes. The data of node i are initially stored in the PU with index $\lfloor i/k \rfloor$.

Importance. Lists play a central role, both in sequential and in parallel algorithms. Sequentially they constitute the basic dynamic structure. In parallel computation they have several additional applications. For example, they are used in the Euler-tour technique (see [6]), which is of outstanding importance in the theory of parallel computation: it is applied as a subroutine in many parallel algorithms for problems on trees and graphs.

The basic operations on lists (except for insertions and deletions), are the determination of the first (or last) elements, and the list ranking problem. The first problem we call *list rooting*. It arises in parallel algorithms for tree rooting. If there may be more than one list, list-rooting is only slightly easier than list-ranking In most applications both problems have to be solved at the same time: knowing that a node is at distance r from the first or last node on its list is not very useful if we do not know to which list it belongs. Thus, henceforth, we consider the following generalized problem, which we will continue to call "list ranking":

---

[*]Max-Planck-Institut für Informatik, Im Stadtwald, 66123 Saarbrücken, Germany. E-mail: jopsi, guillaum, seidel@mpi-sb.mpg.de. URL: http://www.mpi-sb.mpg.de/~jopsi/

Definition 1 *For a set of linked lists,* list ranking *consists of (1) selecting for every list a unique node which is known by all its nodes; and (2) for all nodes their* rank, *the distance from the selected node.*

Once the nodes have been ranked, the lists can be transformed into arrays, on which many parallel operations can be performed more efficiently. We summarize our main arguments:

- List ranking is of fundamental importance in all sorts of applications.

- List ranking is an excellent representative from the class of irregular problems.

Limiting our attention to distributed-memory machines (and even more specifically to grids), is motivated by the fact that efficient shared-memory machines appear not to be constructable far beyond their actual size (see [8] for a treaty on machine balance').

Algorithms. On synchronous parallel computers equipped with a shared-memory, *PRAMs*, the basic approach is 'pointer jumping'. This technique can be used in a list-ranking algorithm which runs in $\mathcal{O}(\log N)$ time with $\mathcal{O}(N \cdot \log N)$ work on an 'EREW' PRAM. Using 'accelerated cascading', the work of this algorithm is reduced to the optimal $\mathcal{O}(N)$, while maintaining running time $\mathcal{O}(\log N)$ [3, 1]. These improved algorithms start by repeatedly selecting an 'independent set'. The selected nodes are excluded: the links of their predecessors are set to their successors. In the meantime it is recorded over how many other nodes a link jumps. In this way the size of the graph is reduced by a constant factor in every phase. Then, if it has been reduced to $N/\log N$, pointer jumping is applied. See [6, 10] for detailed descriptions of this algorithm and definitions of the technical terms.

More realistic than PRAMs are parallel computers consisting of N processing units, *PUs*, that communicate through an interconnection network. By the extremely non-local nature of list ranking, it is very hard to achieve good results on such networks. Several algorithms have been developed [11, 2, 4], but they all suffer from too large leading constants, even though they establish the optimal time order. A more useful algorithm is the "*sparse-ruling-set*" algorithm from [12]. On a two-dimensional mesh, for $k = \omega(1)$, near-optimal performance is achieved: $(1/2 + o(1)) \cdot k \cdot n$ steps.

This Paper. We have carefully designed programs for pointer-jumping, independent-set removal and the sparse-ruling-set algorithm, which are theoretically analyzed in [12].

In the following sections, the approaches are described one by one, together with a description of numerous modifications and experimental results. We will see that for each approach there is a range of P and k values, in which it is better than the others. As a benchmark, we take the maximal speed-up[1] that can be achieved for $P = 100$. With pointer-jumping, for $k = 32,000$, the speed-up is 5; with independent-set removal and $k = 512,000$ it is 14. For large problems, by far the best performance is achieved with the the sparse-ruling-set algorithm: speed-up 27 for $k = 2048,000$.

Before all this, we present routing schemes, that are essential for the good performances.

2 Routing Random Distributions

We present algorithms for routing k packets from each PU on an $n \times n$ grid to randomly chosen distributions. The algorithms are tuned to the features of the Intel Paragon, but these are not exceptional in interconnection networks. So, we may assume that they have validity beyond the scope of this architecture.

Basic Properties of the Paragon. The PUs of the Paragon are interconnected by a two-dimensional grid, and the routing is performed according to a trivial XY-scheme: packets are

---

[1]All speed-up results are given with respect to an optimized implementation of the basic sequential algorithm, running on a single node of the Paragon.

first routed along the rows to their destination columns, and then along the columns to their destinations.

The time for a routing pattern increases linearly with the maximum number of bytes going over a connection and the maximum number of packets sent and received by the PUs:

$$T_{\text{send}}(sr, a, m) = sr \cdot (t_s + t_f \cdot m) + a \cdot t_{\text{th}} \cdot m. \tag{1}$$

Here m is the size of the packets; sr is the maximum numbers of packets sent and received by a PU; and a is the maximum number traversing a connection. $t_s \simeq 1.4 \cdot 10^{-4}$ is called the *start-up time*; $t_f \simeq 1.2 \cdot 10^{-8}$ the *feeding time*; and $t_{\text{th}} \simeq 8.3 \cdot 10^{-9}$ the *throughput time*.

(1) shows that apparently the network is relatively powerful: only for large partitions the finite capacity of the connections becomes noticeable. Much more important is the effect of the start-ups: only for packets of more than 10KB, the feeding time exceeds the start-up time.

Random Packet Distributions. If each PU holds k packets with random destinations, then the expected number of packets between a pair of PUs equals k/P. To estimate the maximum arising number, we can use the Chernoff bounds, as given in [5], to derive

Lemma 1 *On a processor network consisting of P PUs, we are performing a routings. In each routing every PU sends k packets to randomly selected destinations. We may assume that the number of packets send between any pair of PUs is bounded by*

$$k/P + (3 \cdot k/P \cdot \ln(10^7 \cdot a \cdot P^2))^{1/2}.$$

As in all our applications $a \cdot P^2 < 10^7$, we may assume that the amount sent from one PU to another does not exceed $k/P + 10 \cdot \sqrt{k/P}$. Only for $k/P < 10{,}000$ does the second term play a role, but for such small packets the effect of the start-up time is even larger.

One-Phase Router. For routing random k-distributions, the powerful network does not leave much room for clever routing strategies: simply routing all packets to their destinations performs quite well. We refer to this approach as ONE_PHASE_ROUTE. Because of the random distribution, we may even assume that all packets are the same size.

The most important technical detail is that a 'receive' instruction has to be issued, before the corresponding packet starts to arrive at a PU. Furthermore, it turns out to be the best to work in rounds: in Round t, $1 \leq t \leq P-1$, PU i sends a packet to PU $(i+t) \mod P$. And, of course, all packets traveling from one PU to another have to be bundled into one 'superpacket'.

We have implemented ONE_PHASE_ROUTE on the Paragon. For square $n \times n$ partitions, the routing time is accurately (deviations of less than 2%) predicted by

$$T_{\text{one}}(k, n) = 2.1 \cdot 10^{-4} \cdot (n^2 - 1) + 8.3 \cdot 10^{-8} \cdot k \cdot (1 - 1/n^2) + 0.9 \cdot 10^{-8} \cdot k \cdot bf \cdot (1 - 1/n^2).$$

Here k gives the number of integers that each PU has to send. bf is the 'bisection factor': $\max\{0, n-3\}$, which gives a refinement on (1). The expression nicely shows the three contributions: start-ups, time to handle the data, and time for the data to cross the bisection. A start-up requires more than t_s because every send from a PU P to a PU P' is preceded by a 'hand-shaking' operation, in which P' sends a null-message to P.

Log-Phase Router. The large number of start-ups in ONE_PHASE_ROUTE leads to a poor performance when in (1) the contribution from the start-up time becomes considerable.

On grids consisting of P PUs $P = 2^p$, the routing can be performed with only p, instead of $P-1$, start-ups. The idea is to simulate the algorithm for routing on a binary hypercube: a packet traveling from PU i to PU j, with $i = \sum_{l=0}^{p-1} i_l \cdot 2^l$ and $j = \sum_{l=0}^{p-1} j_l \cdot 2^l$, is sent in Phase l, $0 \leq l \leq p-1$, from PU $(i_{p-1}, \ldots, i_{l+1}, i_l, j_{l-1}, \ldots, j_0)$ to PU $(i_{p-1}, \ldots, i_{l+1}, j_l, j_{l-1}, \ldots, j_0)$. Call this algorithm LOG_PHASE_ROUTE.

A drawback is the large number of phases: at the end of every phase, each PU has to store the received packets, and to rearrange them to create new packets. For this reason LOG_PHASE_ROUTE should be used only for small packets.

Two-Phase Algorithm. For a large intermediate range of packet sizes, we need an algorithm that finds a compromise between the number of start-ups and the number of phases. The most practical such algorithm, is one with two phases: first the packets are routed within the rows, then within the columns. In this way, the number of start-ups on an $n \times n$ partition equals $2 \cdot n - 2$, which is generally much smaller than n^2.

An accurate description of the time consumption on the Paragon is given by

$$T_{\text{two}}(k, n) = 2.1 \cdot 10^{-4} \cdot (2 \cdot n - 2) + 28 \cdot 10^{-8} \cdot k \cdot (1 - 1/n) + 1.7 \cdot 10^{-8} \cdot k \cdot bf \cdot (1 - 1/n).$$

Comparison. On a network with P PUs, the log-phase router is the best for $k \leq 50 \cdot P$, the two-phase router for $50 \cdot P < k \leq 1000 \cdot P$, and the one-phase router for $k > 1000 \cdot P$, Here k gives the total number of (4 bytes) integers that have to be routed. For each algorithm, there are k values, such that it is more than twice as fast as the other two.

3 Pointer Jumping

3.1 Technique

We have a set of lists of total length N. $mst(p)$ denotes the master of a node p. Initially, all non-final nodes are active and have $mst(p) = suc(p)$. The final nodes are passive and have $mst(p) = p$. The following process of repeatedly doubling is called *pointer jumping*.

repeat $\lceil \log N \rceil$ **times**
    for all active p **do**
        if $mst(p)$ passive **then** p becomes passive **fi** ;
        $mst(p) := mst(mst(p))$ **od** .

Hereafter, for all p, $mst(p)$ gives the final node on the list of p (see, for example, [6, 10]). Keeping track of some additional information, the algorithm can easily be modified to compute the rank, the distance to the final node.

As the algorithm is given, it consists of $\lceil \log N \rceil$ exclusive read steps. On an interconnection network, each read means routing a request from the position where the value is needed to the position where it is stored, and back again. As long as we may assume that the requested data is more or less randomly selected from the complete set of stored data, a trivial approach performs well. In the case of list ranking, the communication pattern depends on the distribution of the nodes over the PUs. For lists with a special structure, it may be preferable to first randomly redistribute the nodes, but in our case this is not necessary.

3.2 Improvements

Hopping Twice. In the above pointer-jumping algorithm, we have phases of the following type:

    1. Update the data in the PUs and send new requests: node p sends a request to node $mst(p)$,

    2. Create answers to the received requests: node $mst(p)$ sends a packet with the value of $mst(mst(p))$ to p.

We can also perform an extended version, by proceeding after Step 1 with

    2. For every request from p, $mst(p)$ sends a super-request to $mst(mst(p))$.

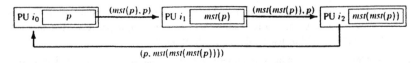

Figure 1: The sending pattern for the improved pointer-jumping algorithm: hopping twice, before returning the answer. In the indicated packets, the first field indicates the destination node, the second field the transferred value.

3. Create answers to the received super-requests: node $mst(mst(p))$ sends a packet with the value of $mst(mst(mst(p)))$ to p.

The sending pattern is illustrated in Figure 1. With the modification, after i rounds, a node can look 3^i positions ahead, instead of 2^i. So, the extended algorithm requires only $\lceil \log_3 N \rceil$ instead of $\lceil \log_2 N \rceil$ rounds. The total number of routing operations is now $3 \cdot \log_3 N = 3 \cdot \log_3 2 \cdot \log_2 N \simeq 1.89 \cdot \log_2 N$. So, depending on the circumstances, this may implie an improvement. Making more than three hops will not be profitable.

Minimizing the Volume of Transferred Data. In all implemented algorithms it turns out that with a proper organization many data are *known* beforehand, and do not have to be transferred.

In Figure 1, we indicated packets consisting of two numbers: the indices of the source and the target nodes. If the packets are routed back by simply reversing all operations, then the index of the source node can be omitted, saving a factor two. Unfortunately this approach appears to be incompatible with the two-hop algorithm.

In a list-ranking algorithm we do not only want to know the index of the last node on the list, but also the distance to it. At a first glance, this requires that additional information is transferred. But, also here most information is already known. After round i, an active node p, knows that the distance to $p' = mst(p)$ is exactly 2^{i+1}. Only when p' is passive, meaning that $p'' = mst(p')$ is a final node, must p know the distance from p' to p''. In the packet that is sent from p' to p, the first bit is reserved for the status (active or passive) of p'. The distance from p' to p'' is then asked for and transferred in the next sending round. This gives:

Theorem 1 *Consider a list whose N nodes are distributed randomly. Applying pointer jumping, the list-ranking problem can be performed in $\lceil \log N \rceil + 1$ rounds in which two packets with at most $(4 + o(1)) \cdot N/P^2$ bytes are routed between any pair of PUs.*

3.3 Experimental Results

We have implemented the pointer-jumping algorithm with minimized data transfer. For the routing we apply the best of the presented algorithms, depending on the size of the packets. The memory requirement of our implementation is just over $20 \cdot N/P$ bytes per PU. This includes all the routing buffers. This is quite good, considering that the sequential algorithm requires $12 \cdot N$ bytes. An implementation with $16 \cdot N/P$ bytes per PU is straightforward, but then we would have to do more work to trace information back to the requesting nodes.

Good fits (deviations less than 10%) through the resulting ranking times when applying the one- and two-phase router, respectively, are obtained by

$$T_{POJ1} = -0.02 + 2.6 \cdot 10^{-6} \cdot k \cdot \log N + 4.9 \cdot 10^{-4} \cdot \log N \cdot P.$$
$$T_{POJ2} = -0.03 + 3.0 \cdot 10^{-6} \cdot k \cdot \log N + 13.4 \cdot 10^{-4} \cdot \log N \cdot (\sqrt{P} - 1).$$

These formulas reflect the essential features of the algorithms: the logarithmic increase that becomes noticeable for $N > 10 \cdot 10^6$, and the effect of start-ups. With ONE_PHASE_ROUTE, the start-ups become dominating for about $k < 1000 \cdot P$. For sufficiently large N, the speed-up is about $1.7/\log N \cdot P$. Figure 2 gives a plot. It clearly illustrates the effect of the start-ups.

This effect rapidly increases with increasing P. One can also see the effect of the logarithmic increase of the ranking time for very large N.

Reliability of Experimental Results. Due to space limitations, we cannot give the results of our measurements. They will be made available in a separate technical report. At this point we want to state that all these results are characterized by a high degree of predictability. Repeating any of our experiments for the same N and P, gives almost identical results (deviations less than 1%). Systematic deviations from the formulas, are far more important. The most important effect is caching: all algorithms become considerably faster if the total size of all data in a PU is less than 256K.

3.4 Double Pointer Jumping

In practice, pointer jumping will be applied only to the small subproblems that remain after reduction with one of the algorithms considered in the following sections. Therefore, it may be worthwhile to consider an alternative that minimizes the number of start-ups.

The two-hop approach gives a 5% reduction, but we can do much better. In the above algorithm, each PU performs $2 \cdot \lceil \log N \rceil + 2$ all-to-all sending operations: an alternation of requests and answers. If a node p would know the node p' that is interested in knowing $mst(p)$ beforehand, then the requests could be omitted. We describe how to achieve this.

Algorithm and Analysis. Initially, every non-final node p sends a packet to $p' = suc(p)$, informing p' that p is its predecessor. Every node p has a master $mst(p)$, which is initialized as before, and a subject $sbj(p)$. Every non-initial node p (recognizable by the fact that it has received a packet) sets $sbj(p)$ equal to its predecessor. All nodes have two status bits indicating whether their subjects are initial nodes, and whether their masters are final nodes or not. The initial nodes still must inform their successors about this.

After these initializations the algorithm performs $\lceil \log N \rceil$ rounds, in which a node informs its master about its subject and vice-versa. The only restriction is that initial and final nodes do not send and are not sent to. A few extra bits suffice to inform a node that its next subject is an initial node, or that its next master is a final node. The distances can be handled as before.

An invariant of this 'double' pointer jumping algorithm, is that for every non-initial node p, with $p' = mst(p)$ not a final node, $sbj(p') = p$. This implies that a PU can exactly determine for which nodes it will receive packets from any other PU. Thus, as before, the index of the sending nodes does not need to be sent along. This gives the following analogue of Theorem 1:

Theorem 2 *Consider a list whose N nodes are distributed randomly. Applying double pointer jumping, the list-ranking problem can be performed in $\lceil \log N \rceil + 3$ rounds in which one packet with at most $(8 + o(1)) \cdot N/P^2$ bytes are routed between any pair of PUs.*

So, the total volume of communication is about the same as before. The additional cost of handling the subjects is comparable to the cost of answering before. However, in order to trace a packet from $mst(p)$ back to p, we now have to sort the packets. This requires some extra time. Furthermore, double pointer jumping requires four bytes extra for every node. This brings the memory requirement for every PU to $24 \cdot N/P$ bytes. This is not serious, because double-pointer jumping is applied only for small k.

Experimental Results. We have implemented the double-pointer-jumping algorithm. For most values of N and P, the ranking time is somewhat slower than for pointer jumping. However, double pointer jumping is almost twice as fast for small inputs. The smallest N for which double pointer jumping is faster than the sequential algorithm is $N = 50,000$.

4 Independent Set Removal

Pointer jumping is a nice and simple technique, but it is very communication and computation intensive. Its weak point is that over the rounds there is no substantial reduction of the amount of work that has to be done. Optimal PRAM algorithms [3, 1] are obtained by first reducing the number of active nodes by a factor $\log N$, and applying pointer jumping only afterwards. Such algorithms have the following structure:

Algorithm RULING_SET_REDUCTION

1. $rnd := 0$;
   while more than N_{end} nodes remain **do**
   $\quad rnd := rnd + 1$;
   $\quad$ select ruling set;
   $\quad$ exclude non-selected nodes.

2. rank the remaining nodes.

3. **for** $i := rnd$ **downto** 1 **do**
   reinsert the nodes that were excluded in Round i.

Definition 2 *For a graph $G = (V, E)$, a subset $\mathcal{I} \subset V$ is an independent set, if no two elements of $\mathcal{I}$ are connected by an edge $e \in E$. A subset $\mathcal{R} \subset V$ is a ruling set if each element of V is adjacent to an element of $\mathcal{R}$.*

Let f be the fraction of the nodes that are selected in the ruling set, and let T_{sl}, T_{ex} and T_{re} give the times for selection, exclusion and reinsertion, respectively. If we assume that these times are linear in N, and that the costs of the final ranking are small, then

$$T_{rank}(N) \simeq \frac{1}{1-f} \cdot (T_{sl}(N) + T_{ex}(N) + T_{re}(N)). \tag{2}$$

4.1 Selection and Exclusion

Various ruling-set algorithms, deterministic and randomized, are considered in [12]. The randomized algorithms are clearly more efficient. The complement of an independent set is a ruling set, thus one can also exclude an independent set.

The simplest independent-set algorithm chooses with probability $1/2$ a color white or black for every node. Then a node is selected in the independent set if it is black itself, while its predecessor is white. This selection is not very effective: the size of the independent set lies close to $N/4$. Hence, the resulting ruling set has size $3/4 \cdot N$: in (2) we get $1/(1 - f) = 4$. Notwithstanding its inefficiency, this approach outperforms all alternatives, due to the extremely low costs of the individual operations. It is of great importance, that with this simple independent-set selection, the list does not need to have forward and backward links. This does not only reduce the memory requirements, but particularly also reduces the amount of data that have to be sent during the exclusion.

The basic idea is as follows: first every active node is given a color. Then in a PU P, all white nodes with successor in PU P' are determined. The indices of these successors are written in a packet and sent to P'. P' looks the referenced nodes up. If such a node p is black, then P' writes $suc(p)$ in the place where formerly the index of p was written, and sets the first bit to one as a flag. p will now be excluded from the list. Otherwise P' writes zero in this place. After going through the whole packet, it is returned to P. P can now easily update the successors of its nodes, because the order has not been disturbed.

On this already efficient approach, we can give a further improvement: it is a waste to reserve a whole word just for sending back the information that an element was not selected in the independent set. However, if we would just leave it away, then we could not lead the returned information back to the nodes it is destined for. The solution is to send along an array consisting of all the first bits. With a word-length of 32 bits, this gives a considerable saving.

In addition, for every selected node p, we must also send back the length of the link to $suc(p)$. Now, we see that if all PUs hold k active nodes, the total amount of data sent by a PU during these steps is bounded by approximately $k/2 + k/4 + k/4 + k/(2 \cdot word\text{-}length) \simeq k$ words. For the exclusion we do not need any further communication.

4.2 Reinsertion

The nodes are reinserted round by round: the slaves that 'wake up' inform their rulers to put their links back to them. Their masters inform them about the index of the final node of their lists. Also computing the distances is straight-forward, but requires a considerable amount of additional communication. There is an interesting alternative: first perform the reduction and the final algorithm without keeping track of distance information. Consider the N_{end} remaining nodes. A node p has rank $rnk(p)$. Then we perform the following steps

Algorithm REINSERT
$shift := 2^{rnd}$.
For each node p, $rnk(p) := rnk(p) \cdot shift$.
for $i := rnd$ **downto** 1 **do**
    $shift := shift/2$.
    Reinsert the nodes excluded in Round i:
        a node p that is reinserted after p' sets $rnk(p) := rnk(p') + shift$.

Lemma 2 *After* REINSERT, *for any two nodes p and p' belonging to the same list, $rnk(p') < rnk(p)$ iff p' stands earlier in the list than p'.*

Proof: During the reinsertion, the difference of the ranks of any two consecutive elements of the list is at least two. Hence, the rank of a reinserted element does not overlap with the rank of its successor. □

The created total order can be turned into a ranking by sorting the nodes on their rnk values. Because the numbers are more or less regularly distributed between 0 and $2^{rnd} \cdot N_{end}$, this can efficiently be achieved by a radix sort. It depends on the details of the implementation and the parallel computer whether REINSERT (together with the sorting) is faster. The only condition for the application of REINSERT is that $2^{rnd} \cdot N_{end}$ does not exceed the maximum integer value. For $N_{end} = 2^{16}$, and 32 useful bits, this gives the condition $rnd \leq 16$. Thus, this technique works for N upto $N_{max} \simeq 2^{16}/0.75^{16} > 6.5 \cdot 10^6$. As RULING_SET_REDUCTION has practical importance only for moderate values of N, this condition on N is not too serious.

4.3 Experimental Results

Whether sophisticated choices of the ruling set, or a reduction of the volume of transferred data at the expense of a sorting operation at the end, give an improvement in practice, depends on the parallel computer under consideration. On the Paragon, with its extremely fast network, the simplest strategy is slightly better than all alternatives.

Practically it is also important how the exclusion is performed: either the active nodes can be rearranged such that they stand in a compact block of the memory at all times, or we can construct a field pointing to the active nodes. The first solution requires considerable data movement, the second solution leads to many additional cache misses. In practice both solutions are about equally fast.

For our implementation we approximately (deviations less than 10%) have

$$T_{ISR} \simeq 0.13 + 28 \cdot 10^{-6} \cdot N/P + 42 \cdot 10^{-4} \cdot P.$$

A plot is given in Figure 2. Comparison of the two pictures shows that independent-set removal is faster than pointer jumping except for small problems on large networks. For sufficiently large problems, the speed-up is about $0.13 \cdot P$.

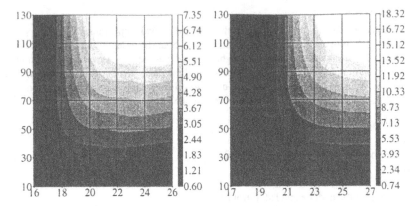

Figure 2: **Practical results for the pointer-jumping (on the left) and independent-set removal (on the right): the x-axis gives $\log N$, the y-axis P, and the gray-tones the speed-up.**

5 Sparse Ruling Set Approach

The strong point of independent set removal is that its time consumption is absolutely linear. Its weak point is that the size of the active set is reduced only by a small factor in every round. In this section we describe a simplified version of the algorithm in [12], which we call the *sparse ruling-set approach*. In spirit it is similar to the algorithms in [1, 9].

5.1 Algorithm

We have removed some features from the algorithm of [12], that are of theoretical interest, but in practice complicate the algorithm without making it faster. What remains is simple:

Algorithm SPARSE_RULING_SET

1. Determine the initial elements.

2. Randomly choose R nodes as rulers.

3. Each PU writes in its out-buffer, for every ruler or initial node r it holds, a pair $(suc(r), r)$.

4. $t := 0$. While there is still a non-empty out-buffer, set $t := t + 1$, and perform the following steps:

   4.a. Send the packets to their destination PUs.

   4.b. Upon reception of a packet holding (p, r), set $mst(p) := r$, $del := t$. If p is not a ruler or a final element, then write $(suc(p), r)$ in the out-buffer.

5. Perform a suitable terminal algorithm on the sublist of the rulers. In this sublist the pointers are directed backwards, and may have any length ≥ 1. Afterwards, for every ruler r, the first element on its list is stored in $mst(r)$, and the distance thereof in $del(r)$.

6. All non-rulers p that were not reached from an initial node. send a packet to $r = mst(p)$.

7. Upon reception of a packet holding r, a PU returns $(mst(r). del(r))$.

8. Upon reception of the answer $(mst(r), del(r))$. a non-ruler p with $mst(p) = r$, sets $del(p) := del(p) + del(r)$ and $mst(p) := mst(r)$.

SPARSE_RULING_SET determines the distance and the index of the node initial of every list. We assume that the initial nodes are evenly distributed.

5.2 Refinements and Analysis

In Step 4, for every node only one packet is send, but it may happen that the distance between a pair of rulers is large. This would lead to many passes through the loop, and thus to many start-ups. It is better to bound the number of passes, and to treat the unreached nodes specially.

Lemma 3 *After* $\ln(N/R) \cdot N/R$ *passes through the loop in Step 4, the expected number of unreached nodes is approximately R or less.*

Proof: Consider a node p that is not within distance $\ln(N/R) \cdot N/R$ of an initial node. The probability that there is no ruler in the d positions preceding p equals $(1 - R/N)^d \simeq e^{-d \cdot R/N}$. Thus, for $d = \ln(N/R) \cdot N/R$, this probability is just R/N. □

The unreached nodes belong to many sublists, starting in the last nodes that have been reached, and ending in the following ruler. On these nodes we apply the same algorithm as in Step 5.

Lemma 4 *Let $T_{srs}(N, P, R)$ denote the time for running SPARSE_RULING_SET on a network with P PUs, and $T_{rank}(N, P)$ the minimal time for a ranking problem of size N. Then, applying in Step 4 a routing algorithm that requires st start-ups, for a small constant c,*

$$T_{srs}(N, P, R) \simeq c \cdot 10^{-6} \cdot N/P + 2 \cdot 10^{-4} \cdot \ln(N/R) \cdot N/R \cdot st + 2 \cdot T_{rank}(R, P). \quad (3)$$

Proof: Every PU sends and receives about N/P packets, and performs $\mathcal{O}(N/P)$ assignments and additions. During the routing the packets may have to be rearranged a certain number of times. There are $\ln(N/R) \cdot N/R$ sending operations, and finally two problems of expected size R must be solved. □

SPARSE_RULING_SET consists of a single reduction followed by a terminal algorithm. Of course, we can also apply several reductions. The first reduction is simpler than the following ones: all links have length one, and this information does not need to be sent along. In subsequent reductions, the links have lengths larger than one. This requires that the distance from a ruler is sent along with a packet, and that *del* is set equal to this distance.

The disadvantage of having many relatively small packets, also has a positive side: the size of the routing buffers is much smaller. With the sparse ruling-set approach, a PU has to send at most $2 \cdot R$ integers in a round. This means that we need hardly more memory space than for the sequential algorithm: just over $12 + 8 \cdot R$ bytes per node. This more efficient use of the memory allows us to penetrate into a range of N values where the algorithm is truly powerful.

5.3 Applying Terminal Algorithms

It is important to consider the 'direction' in which a terminal algorithm works: pointer jumping computes the index of the final nodes, and the distance thereof. Thus, for ranking the rulers, it can be applied immediately, but for ranking the unreached nodes, the list must be 'reversed'. The sequential approach determines the initial nodes and the distances thereof: oppositely.

Terminating with the sequential algorithm requires that all remaining nodes are concentrated in one of the PUs, ranked, and sent back. Alternatively, they can be 'gossiped' around such that they become available in *all* PUs. Gossiping has been analyzed intensively (see [7] for gossiping on the Paragon and references). For a rectangular $n_1 \times n_2$ partition, a reasonable compromise is to gossip first in $n_1 - 1$ steps in the columns, and then in $n_2 - 1$ steps in the rows. For ranking the rulers, it has the great advantage that, after ranking them, the packets can look up their ranks locally, saving the massive routings in Steps 6 and 7 of the algorithm.

Unfortunately, the concentrated nodes cannot easily be stored such that the position of a node can be determined from its index. Renumbering or performing some (binary) search is quite expensive. For the rulers (not for the non-reached nodes) we have the following solution: for a (more or less) random input, there is no reason to choose the rulers randomly: we can take the nodes with index $0, 1, \ldots, R - 1$ in every PU. Doing this, the ruler with index (i, j), $0 \leq i < P, 0 \leq j < R/P$, comes in position $i \cdot R/P + j$ of the array holding the rulers.

If the first elements are selected as rulers, then gossiping the ranks of the rulers may even be profitable when pointer jumping is applied for the final ranking. When gossiping, every PU sends and receives about R packets, performing Step 6 and 7, about $2 \cdot N/P$. So, we may expect gossiping to be faster if $N/R > P/2$.

5.4 Experimental Results

SPARSE_RULING_SET is the only parallel algorithm that promises satisfying speed-ups: for $P = 2$ and $k = 2 \cdot 10^6$, the complete algorithm takes $8.4 \cdot k$ μs. As the sequential algorithm takes $3.9 \cdot N$ μs, this means that we achieve a speed-up of about $0.46 \cdot P$.

Looking back at (3), we can understand why such a good performance can be achieved only for very large k. As an example, we estimate the conditions for obtaining speed-up at least $0.30 \cdot P$. This can only be achieved if the second and third term in (3) each contribute at most 20% of the first term. For the third term, this means that we must have $f = N/R \geq 20$. After $2 \cdot f$ iterations of Step 4, the packet sizes have been reduced by a fraction $1 - 1/e^2$ of the packets has been routed, but during these important first iterations, the one-phase router must be applied. Combining these facts, we get the condition

$$2 \cdot 10^{-4} \cdot P \cdot 2 \cdot f \leq 8.4 \cdot 10^{-6} \cdot k/5.$$

Substituting $f = 20$, gives $k > 8000 \cdot P$ (actually, k must be even larger). This simple argument clearly shows the limitations of the sparse-ruling-set approach. In practice, the ranking times are well described by

$$T_{SRS1} = 5.3 \cdot 10^{-6} \cdot k + 1.8 \cdot 10^{-6} \cdot k \cdot \sqrt{P} + 42 \cdot 10^{-4} \cdot P,$$
$$T_{SRS2} = 8.2 \cdot 10^{-6} \cdot k + 1.0 \cdot 10^{-6} \cdot k \cdot \log P + 41 \cdot 10^{-4} \cdot P.$$

Here SRS1 and SRS2, refer to the algorithm with one and two rounds of reduction, respectively. In the formulas we find a linear main term, a 'loss' term, and a start-up term. It is hard to model the real cost of these algorithms because the routers are changed dynamically. The adaptations were obtained by fitting several possible functions with a format similar to (3) through sets of 50 results in the range $16 \cdot 1000 \leq k \leq 2048 \cdot 1000$ and $2 \leq N \leq 128$. All deviations are less than 12%, and most of them less than 5%.

Figure 3 gives a plot based on the minimum of T_{SRS1} and T_{SRS2}. It shows that in contrast to the previous algorithms, the effect of start-ups is not localized: up to the highest values of N, the performance continues to improve. Actually, for $P = 100$, with $k = 2 \cdot 10^6$, SPARSE_RULING_SET requires about 29 seconds. That is 27 times faster than the sequential algorithm. The irregularity of the problem, and the high start-up cost of the Paragon, make it hard to do better.

6   Conclusion

We have given a detailed analysis resulting in a complete view of the current world of list ranking. Finally, we can answer the guiding question of this paper: "Which algorithm solves a list-ranking problem of size N, on a parallel computer with P PUs fastest?"

Figure 3 shows that each algorithm has a region in which it out-performs the other two. When many PUs are available, then pointer jumping is the best for small problems. Sparse ruling sets is the best whenever k is sufficiently large. Independent set removal is the best in an intermediate range. The sequential algorithm dominates when $P < 4$ or $N < 32,000$.

These numbers have no universal value, but the observed tendencies hold generally. The relative importance of the algorithms depends on the ratio t_s/t_f. For smaller ratios, all boundary lines in Figure 3 are shifted to the left: the importance of the sparse-ruling-set algorithm increases, and the regions of the other algorithms may be squeezed.

36

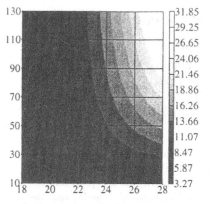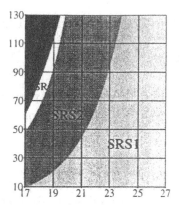

Figure 3: Practical results for the sparse-ruling-set algorithm (on the left): the x-axis gives $\log N$, the y-axis P, and the gray-tones the speed-up. On the right we indicate where each of the algorithms is best.

References

[1] Anderson, R.J., G.L. Miller, 'Deterministic Par. List Ranking,' *Algorithmica*, 6, pp. 859–868, 1991.

[2] Atallah, M.J., S.E. Hambrusch, 'Solving Tree Problems on a Mesh-Connected Processor Array,' *Information and Control*, 69, pp. 168–187, 1986.

[3] Cole, R., U. Vishkin, 'Deterministic Coin Tossing and Accelerated Cascades: Micro and Macro Techniques for Designing Parallel Algorithms,' *Proc. 18th Symp. on Theory of Computing*, pp. 206–219, ACM, 1986.

[4] Gibbons, A.M., Y. N. Srikant, 'A Class of Problems Efficiently Solvable on Mesh-Connected Computers Including Dynamic Expression Evaluation,' *IPL*, 32, pp. 305–311, 1989.

[5] Hagerup, T., C. Rüb, 'A Guided Tour of Chernoff Bounds,' *IPL*, 33, 305–308, 1990.

[6] JáJá, J., *An Introduction to Parallel Algorithms*, Addison-Wesley Publishing Company. Inc., 1992.

[7] Juurlink, B., P.S. Rao, J.F. Sibeyn, 'Gossiping on Meshes and Tori,' *Proc. 2nd Euro-Par Conference*, LNCS 1123, pp. 361–369, Springer-Verlag, 1996.

[8] McCalpin, J.D., 'Memory Bandwidth and Machine Balance in Current High Performance Computers,' *IEEE Technical Committee on Computer Architecture Newsletter*, pp. 19–25, 12-1995.

[9] Reid-Miller, M., 'List Ranking and List Scan on the Cray C-90,' *Proc. 6th Symposium on Parallel Algorithms and Architectures*, pp. 104–113, ACM, 1994.

[10] Reid-Miller, M., G.L. Miller, F. Modugno, 'List-Ranking and Parallel Tree Contraction,' in *Synthesis of Parallel Algorithms*, J. Reif (ed), pp. 115–194, Morgan Kaufmann, 1993.

[11] Ryu, K.W., J. JáJá, 'Efficient Algorithms for List Ranking and for Solving Graph Problems on the Hypercube,' *IEEE Trans. on Parallel and Distributed Systems*, Vol. 1, No. 1, pp. 83–90, 1990.

[12] Sibeyn, J.F., 'List Ranking on Interconnection Networks,' *Proc. 2nd Euro-Par Conference*, LNCS 1123, pp. 799–808, Springer-Verlag, 1996. Full version in *Technical Report 11/1995, SFB 124-D6*, Universität Saarbrücken, Saarbrücken, Germany, 1995.

On Computing All Maximal Cliques Distributedly[*]

Fábio Protti
Felipe M. G. França
Jayme Luiz Szwarcfiter
Universidade Federal do Rio de Janeiro, Brazil
email: {fabiop, felipe, jayme}@cos.ufrj.br

Abstract. A distributed algorithm is presented for generating all maximal cliques in a network graph, based on the sequential version of Tsukiyama et al. [TIAS77]. The time complexity of the proposed approach is restricted to the induced neighborhood of a node, and the communication complexity is $O(md)$ where m is the number of connections, and d is the maximum degree in the graph. Messages are $O(\log n)$ bits long, where n is the number of nodes (processors) in the system. As an application, a distributed algorithm for constructing the *clique graph* $K(G)$ from a given network graph G is developed within the scope of dynamic transformations of topologies.

1 Introduction

The generation of all possible configurations with a given property is a problem that occurs frequently in many distinct situations. In terms of distributed systems, the recognition of all subgraphs possessing special features within a network graph may simplify distributed algorithms developed for the network [CG90]. Among the different types of subgraphs, maximal cliques (completely connected subgraphs) play a major role. For instance, the analysis of maximal cliques provides a better understanding of the behavior of scheduling mechanisms [BG89] and the efficiency of mappings [FF95] for neighborhood-constrained systems.

In this paper, we propose a distributed algorithm for generating all maximal cliques of an arbitrary network graph. A previous approach to parallel distributed computation of all maximal cliques of a given graph was presented in [JM92]. The latter is based on the parallel version proposed in [DK88] and in order to compute all maximal cliques, divide-and-conquer strategy is used by dividing the node set into two disjoint subsets V_1, V_2, finding the maximal cliques in V_1, V_2 recursively, and joining these cliques via maximal bipartite complete subgraphs of the graph $G'(V_1 \cup V_2, E')$, where E' is the subset of edges with one endpoint in V_1 and other endpoint in V_2.

The present method is based on the concurrent execution of a local sequential algorithm [TIAS77] for all induced neighborhoods. The time complexity is restricted to the induced neighborhood of a node, and the communication complexity is $O(md)$ where m is the number of connections, and d is the maximum

---

[*] This work has been partially supported by the Conselho Nacional de Desenvolvimento Científico e Tecnológico (CNPq) Brazil.

degree in the graph. Messages are $O(\log n)$ bits long, where n is the number of nodes (processors) in the system.

This work is organized as follows. Section 2 describes the distributed processing environment. Section 3 reviews the sequential algorithm of Tsukiyama et al. [TIAS77]. Section 4 describes the present algorithm and analyzes its complexity. Section 5 contains the following application: a distributed algorithm for constructing the *clique graph* $K(G)$ from a given network graph G. The graph $K(G)$ has the set of maximal cliques of G as node-set, and there exists an edge (C, C') in $K(G)$ iff $C \cap C' \neq \emptyset$. Some remarks form the last section.

2 The Distributed Processing Model

Consider an arbitrary connected graph $G = (V, E)$, $V = \{1, 2, \ldots, n\}$, where each node in V corresponds to a complete local processing environment containing a processor and sufficient local memory to perform computations. Notice that we assume implicitly a set of distinct identifications of nodes having a *total ordering*.

Two nodes i and j are *neighbors* if $(i, j) \in E$. The adjacency set $N(i) = \{j \in V | (i, j) \in E\}$ is the *neighborhood* of node i. The cardinality of $N(i)$ is the *degree* d_i of node i. If $X \subseteq V$, we define $G[X]$ as the *subgraph of G induced by X*: its node-set is X, and a connection $e = (i, j)$ in E belongs to the edge-set of $G[X]$ if and only if $i, j \in X$. For each node i, $G[N(i)]$ is the induced neighborhood of i.

A *clique* $C \subseteq V$ is a completely connected set of nodes. That is, if $C = \{i_1, i_2, \ldots, i_k\}$, then $(i_r, i_s) \in E$ for all $r, s \in \{1, 2, \ldots, k\}$, $r \neq s$. A clique C is *maximal* if, whenever we have $C \subseteq C'$ for a clique C' in V, then $C = C'$.

Messages can flow independently in both directions between neighboring nodes through the connection linking them. Each node knows its neighborhood, and maintains internally an input buffer for arriving messages.

We will assume that each node of G is able to execute the communication instruction $BROADN(msg)$ [BF88, BDH94]. Used in some massively parallel applications, $BROADN(msg)$ works simply as follows: when executing it, node i sends a message msg to all of its neighbors.

We will also assume that in the distributed model all nodes execute the same algorithm. Message transmission time is not taken into account, since the distributed algorithm is totally message-driven, not only inside each of its local steps, but also between the steps.

3 The Sequential Version

First, we review the sequential algorithm for generating all maximal independent sets developed in [TIAS77] (see also [L76], [PU59]), in its maximal clique version. The proofs of lemmas in this section can be found in [L76].

Let H be a connected graph with node-set $V(H) = \{1, 2, \ldots, p\}$. Let C_j denote the collection of all maximal cliques in $H[\{1, 2, \ldots, j\}]$. The idea is to compute C_{j+1} from C_j, eventually computing C_p.

Let $C \in C_j$. If $C \cap N(j+1) = C$, then $C \cup \{j+1\} \in C_{j+1}$. If $C \cap N(j+1) \neq C$, then $C \in C_{j+1}$ and $(C \cap N(j+1)) \cup \{j+1\}$ is a clique (not necessarily maximal). Let $C'_{j+1} = \{C' | C' = (C \cap N(j+1)) \cup \{j+1\}, C \in C_j\}$.

The observations above lead to the following lemma:

Lemma 1. $C_{j+1} \subseteq C_j \cup C'_{j+1}$.

Lemma 1 provides a way of generating algorithmically C_{j+1} from C_j. One point to be considered is how to avoid inclusion of nonmaximal sets in C_{j+1}, since $C_j \cup C'_{j+1}$ may contain sets of that sort. This is the subject of Lemma 2.

Lemma 2. Let $C \in C_j$. Then, $C' = (C \cap N(j+1)) \cup \{j+1\} \in C_{j+1}$ if and only if for all $k < j, k \notin C$, we have $N(k) \cap C' \neq C'$.

Another point is how to avoid inclusion of a maximal set which has already been included. To deal with this second question, we may regard C'_{j+1} as a multiset containing *duplicates*. That is, a given set $C' \in C'_{j+1}$ may be obtained from distinct sets $C^1, C^2 \in C_j$ such that $C' = (C^1 \cap N(j+1)) \cup \{j+1\} = (C^2 \cap N(j+1)) \cup \{j+1\}$. In order to avoid duplicates, we use the following rule: given a set $C \in C_j$, we include $C' = (C \cap N(j+1)) \cup \{j+1\}$ into C'_{j+1} *only if $C - N(j+1)$ is lexicographically smallest among all sets $C \in C_j$ for which $C \cap N(j+1)$ is the same* (all sets are implicitly organized as lists in increasing order). Lemma 3 tells us how to check this property for a given set.

Lemma 3. A set $C \in C_j$ satisfies the lexicographic condition if and only if there is no node $k < j+1, k \notin C$, such that:
a) k is adjacent to $X = C \cap N(j+1)$, that is, $N(k) \cap X = X$;
b) k is adjacent to all lower-numbered nodes in $C - N(j+1)$, that is, $N(k) \cap Y = Y$, where $Y = \{1, 2, \ldots, k-1\} \cap (C - N(j+1))$.

The algorithm below follows from Lemmas 1-3:

proc *Generate*(C_j, C_{j+1})

    for $C \in C_j$ **do**

        if $C \cup \{j+1\}$ is a clique
            then include $C \cup \{j+1\}$ in C_{j+1}
            else
                include C in C_{j+1};
                $C' := (C \cap N(j+1)) \cup \{j+1\}$;
                if C' and C satisfy Lemmas 2 and 3 **then** include C' in C_{j+1};

end

The following routine computes C_p:

proc *Generate_maximal_cliques*(H, C_p)
% Input: connected graph H with node-set $\{1, \ldots, p\}$
% Output: C_p, the set of all maximal cliques of H

    $C_1 := \{\{1\}\}$;
    for $j := 1$ **to** $p - 1$ **do call** *Generate*(C_j, C_{j+1});
end

The tree of Figure 1 illustrates the computation. Nodes at level j correspond to the sets in C_j. The root of the tree is $\{1\}$, the only set in C_1. Each node has at least one son because for each $C \in C_j$, either C or $C \cup \{j + 1\}$ is in C_{j+1}. The left son of a node $C \in C_j$, if it exists, corresponds to C. It exists only if $C \cap N(j + 1) \neq C$, otherwise C would not be maximal in C_{j+1}. The right son exists only if the maximality and the lexicographic conditions hold. In this case, it corresponds to $(C \cap N(j + 1)) \cup \{j + 1\}$.

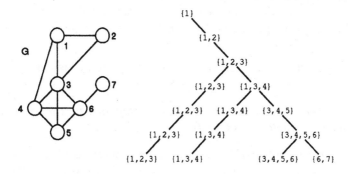

Fig. 1. Tree representing the generation of all maximal cliques for the graph G.

Let q be the number of edges and M be the number of maximal cliques in H. Both tests of Lemmas 2 and 3 can be performed for C' and C in $O(p + q)$ time. Therefore, we can compute C_{j+1} from C_j in $O((p + q)M)$ time. That is, C_p can be obtained in $O((p + q)p)M = O(qpM)$ time. The computation can be done by a depth-first search of the tree. If the leaves are to be output, space requirements are reduced to $O(p)$ cells, each one containing at most p nodes.

4 The Distributed Version

The distributed algorithm consists of two asynchronous steps. At the end, each node will have computed *all the maximal cliques it belongs to*.

4.1 Step 1: Constructing induced neighborhoods

The objective of this step is that all of the system's nodes may know their induced neighborhoods. Therefore, every node i starts by notifying all of its neighbors about its own adjacency relations by executing the following subroutine:

proc *Broadcast_neighborhood(i)*

    for $j \in N(i)$ **do** *BROADN(adj(i, j))*;
    BROADN(broad_end(i));
end

An $adj(i, j)$ message means that i and j are neighbors, and a $broad\_end(i)$ message warns the neighborhood of i to consider the broadcasting of the adjacency relations concluded.

    Next, i is ready to process information that has just been received from its neighbors. Upon receiving a message $broad\_end(j)$ from every node $j \in N(i)$, the construction of $G[N(i)]$ is concluded.

proc *Construct_induced_neighborhood(i)*

% G_i is the induced neighborhood $G[N(i)]$ to be constructed

    for $j \in N(i)$ **do** include edge (i, j) in G_i;
    for an $adj(j, k)$ message arrived before a $broad\_end(j)$ message **do**
        if $j, k \neq i$ **and** $j, k \in N(i)$ **then** include edge (j, k) in G_i;
end

Now, let m be the number of connections and d the maximum degree in the system.

Lemma 4. *The first step takes $O(d_i^2)$ local computation time for node i, and its communication complexity is $O(md)$.*

Proof. Executing *BROADN* instructions and collecting $adj(j, k)$ messages to construct $G[N(i)]$ takes $O(d_i^2)$ time, therefore the first part follows. After all nodes have completed the first phase, the number of messages is $\sum_{(i,j) \in E} d_i + d_j + 2 = O(md)$. Notice that each message can be formed with $O(\log n)$ bits. □

4.2 Step 2: Computing all maximal cliques

Having constructed $G[N(i)]$, node i is able to compute internally and sequentially the maximal cliques it belongs to. The essence of the second step is the concurrent execution of the sequential algorithm we described in Section 3. Let C be a maximal clique containing i. Of course, $G[C - \{i\}] \subseteq G[N(i)]$. This suggests that node i must execute the sequential algorithm applied to its induced neighborhood.

proc *All_maximal_cliques(i)*

    $H \leftarrow \emptyset$; $p \leftarrow d_i$;
    map nodes of $G[N(i)]$ into $\{1, \ldots, d_i\}$ so that $f(j) < f(k)$ iff $j < k$;
    for $(j, k) \in G[N(i)]$ **do** include edge $(f(j), f(k))$ in H;
    call *Generate_maximal_cliques(H, C_p)*; % local sequential algorithm
    for $B \in C_p$ **do**
        let C be the set of nodes in $\{f^{-1}(j) | j \in B\} \cup \{i\}$;

order C increasingly as a list;
store (or output) C;
end

In order to generate each clique as a list in increasing order of nodes, nodes of $G[N(i)]$ are mapped into $V(H) = \{1, 2, \ldots, d_i\}$ in such a way that relative positions between nodes be preserved. After generating a clique, i must itself be included in it.

Lemma 5. *Let m_i be the number of connections in $G[N(i)]$ and M_i the number of maximal cliques containing i. Then, the second step takes $O(m_i d_i M_i)$ local computation time for i, and no messages are sent.*

Proof. Consider the complexity of the algorithm in [TIAS77] when applied to $G[N(i)]$. □

4.3   Analysis of the algorithm

Asynchronicity is an interesting feature of the proposed algorithm, as each node can start independently the first step and enter the second step even though other nodes in the system may have not completed the first step yet. This facilitates considerably the development of actual distributed and parallel applications. Lemmas 4 and 5 lead to the following theorem, which synthetizes the properties of the algorithm.

Theorem 6. *The time complexity of the algorithm is $\max_{i \in V}\{O(m_i d_i M_i)\}$. The communication complexity is $O(md)$, with $O(\log n)$-bit long messages.*

The communication complexity of the algorithm in [JM92] is $O(M^2 n^2 \log n)$, where M is the number of maximal cliques and there is a condition on the messages, *assumed* to be $O(\log n)$ bits long. In [JM92], since there are transmission of messages containing *identifications of cliques*, message length becomes a function of M. Thus, it has been assumed that M is polinomial on n in order to ensure that message length is not higher than $O(\log n)$. In the present algorithm, there is no such restrictions: the communication mechanism does not depend on M, and message length is indeed limited by $O(\log n)$.

Notice that the time complexity of the proposed algorithm is limited to one of the induced neighborhoods of the system. On the other hand, the $O(Mn \log n)$ time complexity of [JM92] refers to message transmission time.

Our algorithm is time efficient, of course, for classes of systems in which the number of maximal cliques is polynomially bounded. The time performance is also good for topologies with constrained connectivity, i.e., when the maximum degree d is limited (a quite natural restriction for many actual systems).

The distributivity of the computation among the processors is weaker in systems with universal nodes (nodes u for which $N(u) = V - u$), since every node of this kind computes all maximal cliques. On the other hand, as the number of universal nodes increases, the number of maximal cliques tends to decrease (in the extreme case of a complete graph, there is one maximal clique only).

5 An application: clique graph construction

The subject of this section is within the scope of dynamic transformations of topologies according to a pre-established rule or *operation*. An operation $op : U \to U$ is a function over the set U of all graphs. Now, let N be a network graph on which a certain operation op will be applied. This means that we will run on N a distributed algorithm A, which by creating and removing nodes and connections, will modify the topology of N transforming it into a new system N' such that $N' = op(N)$. By running A on N', we generate N'', which by its turn corresponds to $op(N')$. That is, repeated executions of A correspond to iterated applications of op.

The operation we consider is the construction of the *clique graph* $K(G)$ from a given graph G. The node-set of $K(G)$ is the set of all maximal cliques in G, and there is an edge between two nodes of $K(G)$ whenever the corresponding cliques in G share a common node. Detailed information on clique graphs can be found in [P95].

The construction of $K(G)$ can be briefly described as follows. The nodes of the connected network graph $G = (V, E)$, $V = \{1, 2, \ldots, n\}$, start running independently copies of the same algorithm. At the end, the system will be identified with $K(G)$, in the sense that new nodes correspond to maximal cliques of G, and new connections to node-sharing between cliques.

Before describing the algorithm, it is necessary to detail an additional set of instructions for communication between nodes (see [BF88, BDH94]). Assume each of the following instructrions is being executed by node i.

- $SEND(j, msg)$: node i sends a message msg destined to node j. If i and j are neighbors the communication is direct. Otherwise, the instruction needs to be re-transmitted by other nodes. This means that routing information indicating the next node to send arriving messages is required in general. This mechanism is independent of internal computations in the nodes, i.e., a node only processes messages sent to it. If n is the number of nodes in the system, an instruction $SEND$ may require $n-1$ transmissions, and this fact must be taken into account when calculating the communication complexity of a distributed algorithm. If j does not exist, msg is discarded after visiting all of the nodes.

- $CREATE(id)$: this is a high-level instruction where node i creates a new neighbor and assigns to it the identification id.

- $LINK_1(k, j)$: this is another high-level instruction where node i creates a connection between nodes k and j belonging to its neighborhood.

- $LINK_2(j)$: node$i$ creates a connection between itself and j. If i and j are already neighbors, this instruction has no effect.

- $UNLINK(j)$: node i removes the connection between itself and j.

The construction of the clique graph has three steps.

5.1 Step 1: local computing of maximal cliques

In this step each node i simply executes the algorithm of the previous section for generating all maximal cliques it belongs to. The only difference is that i

discards every clique C whose first node in the list is different from i (recall that the cliques are generated as lists in increasing order of nodes). This means that the set of all maximal cliques of G becomes *partitioned* among the nodes. Notice that some nodes will store no cliques.

Lemma 7. *Step 1 takes $O(m_i d_i M_i)$ time for node i. The communication complexity of this step is $O(md)$, with $O(\log n)$-bit long messages.*

Proof. See Theorem 6. □

5.2 Step 2: generating clusters

Once i has completed Step 1, it starts immediately Step 2 by generating a *cluster*, a clique formed by new nodes where each one represents a clique C stored in i at the end of Step 1 and receives C as its identification. See Figure 2.

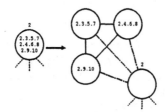

Fig. 2. Generation of a generic cluster formed by 3 nodes.

Observe that each cluster must be completely connected, since it consists of nodes identifying pairwise non-disjoint cliques of G. Observe also that new identifications are formed for ordered lists of nodes in G. This means that the set of identifications of $K(G)$ will also have a total (lexicographic) ordering. The processing of Step 2 can be described in the following way:

proc *Step_2(i)*

    for a clique C stored in i **do** $CREATE(C)$;
    for distinct cliques C, C' stored in i **do** $LINK_1(C, C')$;
end

For the sake of simplicity, from now on we will refer indistinctly to the *node C* in $K(G)$ and the *clique C* in G. Figure 3a shows the system G in Figure 1 after all nodes have completed Step 2.

Lemma 8. *Step 2 takes $O(M_i^2)$ time for node i. No messages are sent in this step.*

Proof. The second instruction **for** consists of M_i^2 iterations. □

5.3 Step 3: making links between clusters

Let us call *leaders* the *original nodes* of the system G. Notice that only leaders run the algorithm.

The processing in Step 3 is the following:

proc $Step\_3(i)$

> % notifying other leaders - - - - -
> **for** a clique C belonging to the cluster of i **do**
> > let j be the highest node in C;
> > **for** $k = i + 1$ **to** j **do** $SEND(k, clique(C))$;
>
> % synchronization - - - - -
> **if** $i = 1$
> > **then** $SEND(i + 1, end\_notify(i))$
> > **else**
> > collect all $clique(\ )$ messages arriving before a $end\_notify(i - 1)$ message;
> > $SEND(i + 1, end\_notify(i))$;
>
> % making links - - - - -
> **for** a $clique(C)$ message collected **do**
> % messages referring only to $SEND(i, clique(C))$ instructions
> > **for** a clique C' belonging to the cluster of i **do**
> > > **if** $C' \cap C \neq \emptyset$ **then**
> > > > $LINK_2(C)$;
> > > > $LINK_1(C', C)$;
> > > > $UNLINK(C)$;
>
> % removing - - - - -
> **for** a leader $j \in N(i)$ **do** $UNLINK(j)$;
> **for** a clique $C \in N(i)$ **do** $UNLINK(C)$;
> **end**

Node i starts notifying other leaders about the nodes belonging to the cluster it leads. A message $clique(C)$ means that C is a clique. For each node C belonging to the cluster of i, all leaders whose clusters may contain some node C' such that $C \cap C' \neq \emptyset$ are notified. Let max be the highest node in the union of the cliques i leads. Notice that there is no use notifying leaders with higher identifications than max, since they cannot lead a clique C' such that $C \cap C' \neq \emptyset$. Also, no leader k with identification smaller than i is notified, since k will certainly notify i in case when there exists a clique C in the cluster of k and a clique C' in the cluster of i such that $C' \cap C \neq \emptyset$.

Next, a synchronization point is needed: each leader, in order to perform the next stages ('making links' and 'removing'), must be sure that no more $clique(\ )$ messages will be received, that is, there are no more $clique(\ )$ messages to be processed.

Now let us deal with the matter of linking the nodes of $K(G)$. Every $clique(\ )$ message is (re-)transmitted between leaders only. Each leader i verifies for each message $clique(C)$ received whether there are cliques in its cluster intercepting C. For each clique C' with this property, an instruction $LINK_1(C', C)$ is executed. Figure 3b shows the system after all the leaders have executed this linking process.

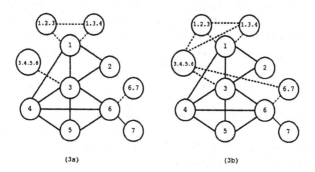

(3a)   (3b)

Fig. 3. The system after generating clusters (3a) and after making links between nodes in distinct clusters (3b).

Finally, notice that by removing the leaders we obtain a system corresponding exactly to the clique graph $K(G)$. Since $K(G)$ is a connected graph whenever G is also connected, the resulting system after removing the leaders is connected.

Lemma 9. *Step_3 takes* $O(d_i M_i \sum_{j<i} M_j) = O(d_i M_i M)$ *time for node* i. *The communication complexity of this step is* $O(ndM)$, *with* $O(d \log n)$-*bit long messages.*

Proof. Notifying other leaders requires $O(nM_i)$ time, and synchronization requires $O(\sum_{j<i} M_j)$ time. With respect to making the links, recall that i does not receive any message $clique(\ )$ from leaders with higher identifications than i. Since the internal **for** requires $O(d_i M_i)$ time, links are made in $O(d_i M_i \sum_{j<i} M_j)$ time. Removing requires simply $O(d_i)$ time, therefore the time complexity of the third phase follows. The number of messages corresponds to the number of $SEND$ instructions executed overall. Each $clique(\ )$ message is retransmitted at most n times, thus the communication complexity of this phase is $O(\sum_{i=1}^{n} nd_i M_i) = O(ndM)$. Since each clique has $O(d)$ nodes, message length is $O(d \log n)$. □

By Lemmas 7,8 and 9, we obtain the complexity for constructing $K(G)$. Compare it with the sequential-time complexity, which is $O(mnM + dM^2)$.

Theorem 10. *The time complexity of the clique graph construction algorithm is* $\max_{i \in V} \{O(m_i d_i M_i + d_i M_i M)\}$, *and its communication complexity is* $O(ndM)$, *with* $O(d \log n)$-*bit long messages.*

It remains to prove the correctness of the construction and that no redundant linking instructions are executed.

Theorem 11. *Let C and C' be two nodes of $K(G)$ created in Step 2.*
a) if a linking instruction between C and C' is executed, then $C \cap C' \neq \emptyset$;
b) if $C \cap C' \neq \emptyset$, then exactly one linking instruction between C and C' is executed.

Proof. The proof of *a)* is straightforward, since any $LINK$ instruction between C and C' is only executed either when constructing a cluster in Step 2 or after an intersection test in Step 3. In both cases, $C \cap C' \neq \emptyset$. For the proof of *b)*, let C and C' be two cliques of $K(G)$ such that $C \cap C' \neq \emptyset$ and C is lexicographically smaller than C'. Let k be the smallest node in $C \cap C'$. There are two cases:

i) k is the smallest node in C. In this case, k is also the smallest node in C'. Therefore, C and C' belong to the same cluster generated in Step 2. By construction of the clusters, exactly one linking instruction between C and C' is executed.

ii) k is not the smallest node in C. Let l and m be the smallest nodes in C and C', respectively. Of course, $l < m \leq k$. Observe that leader l executes a $SEND(m, clique(C))$ instruction, and leader m executes a $LINK(C, C')$ instruction, since C' is a clique in the cluster of m and $C \cap C' \neq \emptyset$. Moreover, every instruction of the form $LINK(*, C')$ is executed only inside node m. Since $SEND(m, clique(C))$ is executed by l exactly once, $LINK(C, C')$ is executed exactly once. □

6 Conclusions

A simple distributed algorithm for generating all maximal cliques of any arbitrary network graph has been presented. As an application, a method for constructing clique graphs distributedly has also been described.

References

[BDH94] V. C. BARBOSA, L. M. de A. DRUMMOND, and A. L. H. HELLMUT, From distributed algorithms to Occam programs by successive refinements, The J. of Systems and Software 26 (1994), pp. 257-272.

[BF88] V. C. BARBOSA and F. M. G. FRANÇA, Specification of a communication virtual processor for parallel processing systems, in Proc. of Euromicro-88 (1988), pp. 511-518.

[BG89] V. C. BARBOSA and E. GAFNI, Concurrency in heavily loaded neighborhood-constrained systems, ACM Transactions on Programming Languages and Systems 11 (1989), pp. 562-584.

[BS96] V. C. BARBOSA and J. L. SZWARCFITER, Generating all acyclic orientations of an undirected graph, Technical Report ES-405/96, COPPE/Federal University of Rio de Janeiro.

[CG90] I. CIDON and I. S. GOPAL, Dynamic Detection of Subgraphs in Computer Networks, Algorithmica 5 (1990), pp. 277-294.

[DK88] E. DAHLHAUS and M. KARPINSKI, A fast parallel algorithm for computing all maximal cliques in a graph and related problems, Proc. of the first Scandinavian Workshop on Algorithm Theory (1988),pp. 139-144.

[FF95] F. M. G. FRANÇA and L. FARIA, Optimal mapping of neighborhood-constrained systems, in A. Ferreira and J. Rolim eds., Lecture Notes in Computer Science 980, pp. 165-170.

[JM92] E. JENNINGS and L. MOTYCKOVA, A distributed algorithm for finding all maximal cliques in a network graph, in I. Simon, ed., Lecture Notes in Computer Science 583, pp. 281-293, 1992.

[JYP88] D. S. JOHNSON, M. YANNAKAKIS, and C. H. PAPADIMITRIOU, On generating all maximal independent sets, Information Processing Letters 27 (1988), pp. 119-123

[L76] E. L. LAWLER, Graphical algorithms and their complexity, Mathematical Centre Tracts 81 (1976), Foundations of Computer Science II Part I, Mathematisch Centrum, Amsterdam, pp. 3-32.

[PU59] M. C. PAUL and S. H. UNGER, Minimizing the number of states in incompletely specified sequential functions, IRE Trans. Electr. Computers EC-8 (1959), pp. 356-357.

[P95] E. PRISNER, Graph Dynamics, Pitman Research Notes in Mathematics Series 338 (1995), Longman.

[TIAS77] S. TSUKIYAMA, M. IDE, H. ARUJOSHI and H. OZAKI, A new algorithm for generating all the maximal independent sets, SIAM J. Computing 6 (1977), pp. 505-517.

A Probabilistic Model for Best-First Search B&B Algorithms [*]

F. Argüello[1], N. Guil[2], J. López[2], M. Amor[1] and E. L. Zapata[2]

[1] Dept. Electrónica y Computación, Universidad de Santiago
15706 Santiago. Spain. elusive,elamor@usc.es
[2] Dept. Arquitectura de Computadores, Universidad de Málaga
P.O. Box 4114, 29080 Málaga. Spain. nico,juan,ezapata@ac.uma.es

Abstract. In this paper we present a probabilistic model for analysing trees generated during problem solving using best-first search Branch and Bound algorithm. We consider the weight associated with the edges as well as the number of children in a node to be random. The expressions obtained for the sequential algorithm are applied to several parallel strategies using a multiprocessor with distributed memory.

1 Introduction

Branch and Bound (BB) algorithms are the most frequently used methods for the solution of combinatorial optimization problems. There are four fundamental ingredients for a BB method: *branching rule, bounding procedure, selection rule* and *termination test*. The *branching rule* takes a given problem G and either solves it directly or derives from it a set of subproblems G_1, G_2, ..., G_a (expansion of G). The *bounding procedure* computes a lower bound on the cost of an optimal solution to a subproblem and eliminates the subproblems that can not generate an optimal solution. The *selection rule* selects one subproblem for the next expansion. The *termination test* ends the process if there are no more subproblems to expand.

A BB computation can be seen as a search through a tree in which the original problem occurs at the root and the children nodes of a given subproblem are those obtained from it by branching. The *bounding procedure* establishes a bound for each new child, inserts these nodes in a list of *eus* (evaluated but unexpanded subproblems) and eliminates those nodes in the *eus* list with a bound smaller than the current upper bound of the optimal solution cost (initially infinity). The leaves of the tree correspond to subproblems that can be solved directly and represent all the feasible solutions (*feas*). When a *feas* is reached, its bound is used as the new upper bound for the *bounding procedure*. The object of the search through the BB tree is to find a leaf of minimum bound.

[*] This work was supported in part by the MES (CICYT) of Spain under contract TIC96-1125-C03, Xunta de Galicia XUGA20605B96 and EC project BRPR-CT96-01070. The authors would like to thank Fujitsu Parallel Computing Research Centre for making an account on the AP1000 multicomputer available to us.

The *selection rule* determines the order in which the nodes of the tree are explored. The usual *selection rule* is called best-first. This algorithm selects the node with lower bound in the *eus* list. Breadth-first (*eus* list is maintained in a FIFO order), depth-first (*eus* list is maintained in a LIFO order) and random search are other proposed algorithms for selection rules. Best-first algorithm minimizes the number of nodes evaluated and depth-first minimizes the size of the *eus* list and therefore, the memory requirements.

Since the BB search trees are irregular, some problems arise when parallel methods are used. Load balance must be achieved without increasing the number of interprocessor communications. On the other hand, when a processor (PE) reaches a feasible solution all the PEs should know their cost in order to ensure that they do not waste time exploring useless nodes. An overall of previous efforts devoted to solve these difficulties during the eighty decade can be found in [7]. More recently synchronous and asynchronous parallel implementations of BB algorithms have been presented [5, 6]. In the synchronous implementation a *macro-fringe* of nodes is distributed among neighbor PEs.

In this work we present a probabilistic model (Section 2) to predict the estimate solution for generic BB algorithms and the number of evaluated nodes for best-first selection rule. A similar study has been carried out for the depth-first strategy but it has not been included to avoid extending this work. Our model is similar to the one proposed by Yang and Das [1], and Smith [2], but is more general. Then, our model is applied to estimate the computational complexity of several parallel algorithms in a multiprocessor with distributed memory (Section 3). For the local best-first broadcast algorithm we get an expression for the expected unbalance. Futhermore, a dynamic load balance algorithm with distributed control is proposed which keep low the number of communications and balance the amount of computations among the PEs. Finally, in Section 4 we establish our conclusions.

2 The Branch and Bound algorithm model

A *branch and bound* (BB) process generates a tree structure. In this section we define a probabilistic model for this kind of tree and analyze some parameters which are independent from the search algorithm used.

We assume the following tree structure model:

1. The tree has n levels.
2. The number of children for each node is not fixed, but can take any positive integer value (including 0) according to a random experiment. P_i will denote the probability of a node having i children.
3. Each edge has an associated weight which is taken from a set of numerable values (positive integers) by a second random experiment. $\{x_i, i = 1, \ldots, \infty\}$ will denote the set of possible weights and Q_i the probability of a given edge having weight x_i.
4. Each node has an associated bound obtained from the sum of the parent node bound and the weight of its edge.

2.1 The estimated value of the solution

The *probability generating function* of a random process is a well known topic in Probability Theory which allows us to easily calculate the mean (or expected value) and the moments of a given random variable.

Definition 1. The probability generating function for the *number of children* in a node is defined by

$$p(s) = \sum_{i=0}^{\infty} P_i s^i, \quad |s| \le 1, \tag{1}$$

where s is a complex variable.

The *iterates* of the generating function $p(s)$ will be defined by

$$
\begin{aligned}
p_0(s) &= s, \\
p_1(s) &= p(s), \\
p_{k+1}(s) &= p[p_k(s)], \quad k = 0, 1, 2, \ldots.
\end{aligned}
\tag{2}
$$

Definition 2. The probability generating function of the *weight of the edges* from a node is defined by

$$q(s) = \sum_{i=1}^{\infty} Q_i s^{x_i}, \quad |s| \le 1, \tag{3}$$

where s is a complex variable. We assume $x_1 < x_2 < x_3 < \cdots$.

The *iterates* of the generating function $q(s)$ will be defined by

$$q_k(s) = (q(s))^k, \quad k = 0, 1, 2, \ldots. \tag{4}$$

It can be easily verified that each of the *iterates* is a probability generating function. The kth iterates $p_k(s)$ and $q_k(s)$ are the generating functions of the *number* and *bounds of the nodes* for level k, respectively.

The mean value of the *number of children* can be obtained by $p'(1)$ where p' denotes the derivative of p in respect to s. Analogously, the expected value for the number of nodes of level (generation) k is $p'_k(1) = (p'(1))^k$, as follows from expression (2). Similarly, the mean value of the *weight of the edges* is $q'(1)$ and the mean value of *the bounds of the nodes* at level k is $q'_k(1) = kq'(1)$.

Example 1. Let us consider a binary tree with edges taking their weights from the set $\{0, 1\}$, with probabilities $1/2$ and $1/2$, respectively. The generating functions for the *number of children* and the *weight of the edges* in this kind of tree are

$$p(s) = s^2 \quad \text{(binary tree)}, \tag{5}$$

$$q(s) = \frac{1}{2} + \frac{1}{2}s. \tag{6}$$

In order to obtain the solution's expected value for a BB algorithm, we define a new probability generating function for the *edges' minimum weights* in the branch of a node:

Definition 3. The probability generating function of *the edges' minimum weight* of a node $r(s)$, is defined by the operation $\overset{\min}{\oplus}$

$$r(s) = p(s) \overset{\min}{\oplus} q(s), \quad |s| \leq 1, \tag{7}$$

The operation $\overset{\min}{\oplus}$ acts upon two generating functions. Without loss of generality, for the generating functions $p(s)$ and $q(s)$ we define

$$\left\{ \sum_{i=0}^{\infty} P_i s^i \right\} \overset{\min}{\oplus} (A_1 s^{a_1} + A_2 s^{a_2} + \cdots) = \sum_{i=0}^{\infty} P_i (A_1 s^{a_1} + A_2 s^{a_2} + \cdots)^{(i)\min}, \tag{8}$$

with

$$(A_1 s^{a_1} + A_2 s^{a_2} + A_3 s^{a_3} \cdots)^{(i)\min} = \left(\sum \begin{array}{l} \text{products of } i \text{ coefficients} \\ \text{containing } A_1 \end{array} \right) s^{a_1}$$
$$+ \left(\sum \begin{array}{l} \text{products of } i \text{ coefficients} \\ \text{containing } A_2 \text{ but not } A_1 \end{array} \right) s^{a_2}$$
$$+ \left(\sum \begin{array}{l} \text{products of } i \text{ coefficients} \\ \text{containing } A_3 \text{ but not } A_1, A_2 \end{array} \right) s^{a_3}$$
$$+ \cdots. \tag{9}$$

In equation (9) we calculate all the variations with repetition of i coefficients. When two coefficients are multiplied (product terms), the result is associated to the lower exponent value. Thus, if a product term contains the coefficient A_1, then the exponent s^{a_1} is assigned to it; the remaining terms containing the coefficient A_2 are assigned the exponent s^{a_2}; the remaining terms containing the coefficient A_3 are assigned the exponent s^{a_3}; and so on.

The *iterates* of the generating function $r(s)$ will be defined by

$$r_0(s) = 1,$$
$$r_{k+1}(s) = p(s) \overset{\min}{\oplus} (q(s) r_k(s)), \quad k = 0, 1, 2, \ldots. \tag{10}$$

Example 2. Let us consider the generating functions $p(s)$ and $q(s)$ from Example 1. Then, the generating function for the minimum weight of the edges will be

$$r_1(s) = p(s) \overset{\min}{\oplus} q(s) = \left(\frac{1}{2} + \frac{1}{2} s \right)^{(2)\min} = \frac{3}{4} + \frac{1}{4} s. \tag{11}$$

The iterates for a two-level tree will be:

$$q(s)r_1(s) = \left(\frac{1}{2} + \frac{1}{2}s\right)\left(\frac{3}{4} + \frac{1}{4}s\right) = \frac{3}{8} + \frac{4}{8}s + \frac{1}{8}s^2, \tag{12}$$

$$r_2(s) = p(s) \overset{min}{\oplus} (q(s)r_1(s)) = \left(\frac{3}{8} + \frac{4}{8}s + \frac{1}{8}s^2\right)^{(2)min} = \frac{39}{64} + \frac{24}{64}s + \frac{1}{64}s^2.$$

For a three-level tree,

$$q(s)r_2(s) = \left(\frac{1}{2} + \frac{1}{2}s\right)\left(\frac{39}{64} + \frac{24}{64}s + \frac{1}{64}s^2\right) = \frac{39}{128} + \frac{63}{128}s + \frac{25}{128}s^2 + \frac{1}{128}s^3,$$

$$r_3(s) = p(s) \overset{min}{\oplus} (q(s)r_2(s)) = \left(\frac{39}{128} + \frac{63}{128}s + \frac{25}{128}s^2 + \frac{1}{128}s^3\right)^{(2)min}$$

$$= \frac{8463}{16384} + \frac{7245}{16384}s + \frac{675}{16384}s^2 + \frac{1}{16384}s^3. \tag{13}$$

The following theorem shows how the iterates $r_k(s)$ are the generating functions for the minimum bound of the leaves of a k-level tree. In other words, the coefficient of s^{x_1} gives us the probability of a k-level tree having at least one leaf with an x_1 bound; the coefficient for s^{x_2} gives us the probability of having at least one leaf with an x_2 bound and no other leaf with bound x_1 and so on, successively.

Theorem 4. *The probability generating function of the minimum bound of the nodes in an n-level BB tree is $r_n(s) = p(s) \overset{min}{\oplus} (q(s)r_{n-1}(s))$. Thus, the solution's estimated value of an n-level BB algorithm is $r'_n(1)$.*

Proof. We can consider the level of a tree as a recursive process including the two following operations in each iteration

1. insertion of an edge that joins the k-level tree with the root of a $\{k+1\}$-level tree, having $q(s)r_k(s)$ as the generating function, and

2. the branching of the root of the $\{k+1\}$-level tree (operator $p(s) \overset{min}{\oplus}$).

The probability of having i edges with $a_{j_1}, a_{j_2}, \ldots, a_{j_i}$ weights is $A_{j_1} A_{j_2} \cdots A_{j_i}$, while the probability of having at least one with a minimum weight of a_1 among the i edges is the sum of all the product terms containing A_1. On the other hand, in a set of i edges the probability of having no edges with the minimum weight, but with the next smaller weight (a_2), is the sum of the terms which do not contain A_1 but which have A_2. Continuing in this fashion we obtain equation (9). Then, we obtain $r_{k+1}(s) = p(s) \overset{min}{\oplus} (q(s)r_k(s))$ (expression (10) and figure 1).

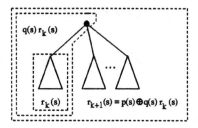

Fig. 1. Obtaining the function of the minimum bound of the BB tree nodes ($r_k(s)$).

Example 3. The expression (11) from the previous example gives a $\frac{3}{4}$ probability of having a tree whose edges have a minimum weight of 0 (trees whose edges are 00, 01 and 10) and a $\frac{1}{4}$ probability of having a tree with a minimum edge of 1 (trees of edges 11). Expression (13) shows the probability of obtaining trees whose nodes at level 2 have a minimum bound of 0 ($\frac{39}{64}$), 1 ($\frac{24}{64}$) and 2 ($\frac{1}{64}$). As a consequence, the solution's expected value for a two-level BB is $r_2'(1) = \frac{13}{32}$. Similarly, for a three-level tree is $r_3'(1) = 0.5247$ (see equation (13)).

Before ending this subsection we will define another probability generating function, similar to those shown in Definition 3, expression (7), for the maximum bound of the leaves of a BB tree. These generating functions will be used in section 3 to calculate the maximum expected value in the leaves of a BB tree.

Definition 5. The probability generating function of the *maximum weight of the edges* of a node, $t(s)$, is given by

$$t(s) = p(s) \overset{max}{\odot} q(s), \quad |s| \leq 1 \tag{14}$$

We define operation $\overset{max}{\odot}$ as follows

$$\left\{ \sum_{i=0}^{\infty} P_i s^i \right\} \overset{max}{\odot} (A_1 s^{a_1} + A_2 s^{a_2} + \cdots) = \sum_{i=0}^{\infty} P_i (A_1 s^{a_1} + A_2 s^{a_2} + \cdots)^{[i]max}, \tag{15}$$

where

$$(A_1 s^{a_1} + A_2 s^{a_2} + A_3 s^{a_3} \cdots)^{[i]max} = \left(\sum \begin{array}{c} \text{products of } i \text{ coefficients} \\ \text{containing only } A_1 \end{array} \right) s^{a_1} +$$
$$+ \left(\sum \begin{array}{c} \text{products of } i \text{ coefficients} \\ \text{containing only } A_1 \text{ and } A_2 \end{array} \right) s^{a_2} + \left(\sum \begin{array}{c} \text{products of } i \text{ coefficients} \\ \text{containing only } A_1, A_2 \text{ and } A_3 \end{array} \right) s^{a_3}$$
$$+ \cdots. \tag{16}$$

The difference from $\overset{min}{\oplus}$ is that in $\overset{max}{\odot}$ the product terms are associated to the highest exponent during the multiplication operation.

The iterates of the generating function $t(s)$ are given by

$$t_0(s) = 1$$
$$t_{k+1}(s) = p(s) \overset{\max}{\odot} (q(s)t_k(s)), \quad k = 0, 1, 2, \cdots \tag{17}$$

2.2   The number of evaluated nodes

Any BB algorithm explores feasible solutions (*feas*) before finding the final solution. Tree nodes whose bounds are smaller than the value of the current *feas* must be expanded and their children evaluated. If best-first is the search strategy used, the number of *feas* is usually small so the final solution is obtained at an early stage of the search. Therefore, we can consider that the nodes and their children to be evaluated by the best-first search BB algorithm will be those whose bounds are smaller than the estimated value of the solution. From now on $a = p'(1)$ will denote the mean value of the number of children and $b_n = r'_n(1)$ will denote the solution for the BB algorithm on an n level tree.

First, we will determine the distribution of the node bounds of the whole tree (from the root to level m.) This will be achieved by adding up the bound generating functions of the nodes at each level $(q_k(s))$ and multiply them by the mean number of nodes at that given level $(p'_k(1) = a^k)$. The following function is obtained

$$N_m(s) = \sum_{k=0}^{m} a^k q_k(s) = \frac{(a \cdot q(s))^{m+1} - 1}{a \cdot q(s) - 1}. \tag{18}$$

By expanding (18) into powers of s, we obtain

$$N_m(s) = \sum_i N_i s^{n_i}, \tag{19}$$

where N_i will provide the estimated number of tree nodes with bound n_i. $N_n(1)$ gives us the estimated number of nodes in the entire tree.

In the best-first search strategy any node with a bound smaller than the estimated solution is expanded. Therefore, the number of such nodes is obtained by truncating the expression (19), and only keeping the terms corresponding to those nodes. If we denote this new function as $\lfloor N_m(s) \rfloor_{b_n}$, we obtain

$$\lfloor N_m(s) \rfloor_{b_n} = \sum (\text{terms of } N_m(s) \text{ with exponents} \le b_n) = \sum_{i \mid n_i \le b_n} N_i s^{n_i}. \tag{20}$$

The number of nodes evaluated by the best-first search BB algorithm can be obtained by adding the children number that have a bound smaller or equal than b_n and, furthermore, are between the levels 1 and $n - 1$. Having into account the root node, the estimated number of nodes evaluated by the best-first search BB algorithm is

$$N_n^{bf} = 1 + a \lfloor N_{n-1}(1) \rfloor_{b_n}. \tag{21}$$

Table 1 shows the comparison between theoretical and experimental values.

n	Theor.	Exper.	(min,max)
10	226	225	(43,541)
12	467	480	(101,1149)
14	954	1000	(159,2457)
16	1938	2067	(231,4903)
18	3926	4226	(729,10853)
20	7943	8625	(1239,22329)

(a)

n	Theor.	Exper.	(min,max)
1	6.5	6.5	(2,11)
2	18	19	(3,60)
4	85	96	(5,318)
6	355	372	(31,1429)
8	1368	1272	(74,5601)
10	5152	4313	(255,20991)

(b)

Table 1. Comparison between the number of nodes evaluated with best-first search on a tree in the theoretical (model) and experimental cases. The weights of the edges range from 1 to 10 with a 0.1 probability for each of them. 4096 experiments were carried out for each instance. The table shows the minimum, mean and maximum number of nodes needed to obtain all the solutions. (a) Binary tree. (b) From 1 to 10 children, with a 0.1 probability for each of them.

3 Parallel computation

In this section we show two different parallel implementations of the Branch and Bound algorithm on a parallel computer with M PEs. Computational complexity is calculated by using the expressions obtained from the sequential model. We have studied the following parallel algorithms:

Local best-first search broadcast feasible solutions algorithm $(LB - BF)$. In this instance each PE carries out the subtree computation derived from the first node assigned, but the PEs exchange feasible solutions as soon as they are found.

Dynamic and distributed best-first search algorithm $(DD - BF)$. In this algorithm we use a dynamic re-balancing of the load so as to distribute the computation between the PEs in a more suitable way, at the same time keeping a low value as regards the number of communications.

3.1 Local best-first search broadcasting feasible solutions

Best-first search is local and each PE works with its own nodes, but once a feasible solution is reached this is broadcasted to all the other PEs. In this way we guarantee all PEs will use the expected value of the solution, b_n, to carry out the cutoff of the nodes. We also assume that $m = \log_a M$ steps are initially needed to expand M nodes. Each node is assigned to a PE. However, we cannot guarantee that all the PEs will evaluate the same number of nodes, since when we initially distributed the first M nodes, some PEs were favored and received nodes with upper bounds. The number of processing cycles will be given by the PE initially receiving the smallest bound node. The value of this node is b_m and thus, the estimated number of nodes in this PE is (see expression 21):

$$N_{n,m}^{LB-BF} = m + 1 + a\lfloor N_{n-m-1}(1)\rfloor_{\{b_n-b_m\}}. \tag{22}$$

Load unbalancing in the $LB - BF$ algorithm If the only information exchange allowed among the PEs is the *feas* as they are calculated, a clear imbalance is going to take place in the PEs' computational load since the number of nodes to be computed by each PE is different. This imbalance will be given by the difference between the most and the least loaded PE, that is

$$\delta_{n,m}^{LB-BF} = N_{n,m}^{bf\ \max} - N_{n,m}^{bf\ \min}, \tag{23}$$

where $N_{n,m}^{bf\ \max}$ and $N_{n,m}^{bf\ \min}$ indicate the number of nodes evaluated by the most and least loaded PE, respectively. The $N_{n,m}^{bf\ \max}$ mean value is given by the PE receiving the node with the lowest bound ($b_m = r'_m(1)$, equation (7)) and $N_{n,m}^{bf\ \min}$ is given by the PE receiving the node with the highest bound ($c_m = t'_m(1)$, equation (14)):

$$N_{n,m}^{bf\ \max} = 1 + a\lfloor N_{n-m-1}(1)\rfloor_{\{b_n-b_m\}}$$
$$N_{n,m}^{bf\ \min} = 1 + a\lfloor N_{n-m-1}(1)\rfloor_{\{b_n-c_m\}}. \tag{24}$$

n	12					16					20				
PEs	Theo.	Exp.	δ_T	δ_E	T(s.)	Theo.	Exp.	δ_T	δ_E	T(s.)	Theo.	Exp.	δ_T	δ_E	T(s.)
1	467	413	-	-	3.00	1938	1862	-	-	15.0	7944	7413	-	-	75.6
2	267	299	90	88	1.82	1106	1123	378	375	7.30	4528	4535	1555	1572	36.0
4	176	182	116	85	1.10	726	816	488	390	5.13	2970	2990	2003	1776	22.4
8	102	117	81	70	0.70	477	526	397	349	3.25	1949	2027	1586	1301	13.9
16	68	73	59	42	0.44	314	336	286	179	2.05	1279	1497	1159	890	9.88
32	47	52	38	33	0.32	208	227	195	167	1.37	841	940	805	709	5.96
64	37	35	30	27	0.22	138	162	132	124	0.98	554	671	538	511	4.18

Table 2. Theoretical and experimental number of nodes and imbalance δ_T, δ_E, and execution times for different binary trees to obtain the first solution of the $LB - BF$ algorithm. In each node 2500 floating point multiplications are carried out (Fujitsu AP-1000 parallel computer).

In Table 2 we show the theoretical and experimental results regarding imbalance for a binary tree with edge weights ranging from 1 to 10. As can be seen the values predicted by probability theory are very similar to those obtained experimentally.

3.2 Dynamic load balancing algorithm with distributed control

Several parallel algorithms can be implemented which carry out dynamic load balancing by using centralized and distributed control strategies. In this work we have developed a dynamic algorithm with distributed control to execute BB problems, which we call the Dynamic Distributed BB algorithm, $DD-BF$. This algorithm involves low computational overheads.

We now explain how the algorithm works. Initially, the root node is divided into a large enough number of children. A child node is then sent to each PE. In the distributed control parallel algorithm for the dynamic distribution of the load, the re-balancing process stages are as follows [4]

Determination of the benefit in the load distribution. Taking into account the difficulties to carry out the load measurements we have implemented a simple strategy. Thus, when a processor becomes idle, it ask the rest of the processors for new load. Thus, rebalancing is carried out whenever a processor demands it. This policy is useful in this case because the information to be transferred is not very large and it simplifies the rebalancing mechanism used.

Task migration strategy. In our case, the PE that starts the re-balancing is the destination PE. In this way, when a PE becomes idle, it sends a request to the rest of PEs in the parallel computer in order to ask for new nodes to compute. Load requests can be limited to a subset of PEs, especially if the number of PEs is very high. In this way less communications are generated. The subset can be variable to prevent sending requests to the same region of the parallel computer.

Task selection strategy. The source PEs that receive the request for transferring one node must select the most appropiate available node. The criteria we are going to follow is to choose the first node in the *eus* list.

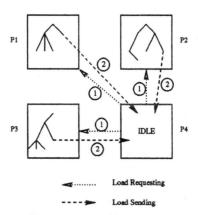

Fig. 2. Distributed management of the load distribution.

In Figure 2 we display an outline of the load distribution process. PE P4 is momentarily left with no load and therefore sends in instant 1 a request for load to the remaining PEs. These PEs are evaluating their particular tree: the computation is being carried out in the first *eus* list node. The next first nodes in the *eus* list are transferred to PE P4 in instant 2.

n	12		16		20	
PEs	δ	T (s.)	δ	T(s.)	δ	T(s.)
1	0	3.00	0	15.03	0	75.60
2	18	1.62	30	7.33	315	37.66
4	19	0.91	74	4.14	253	19.99
8	18	0.53	76	2.31	309	10.60
16	20	0.39	64	1.41	268	6.50
32	22	0.32	57	0.98	183	3.71
64	23	0.29	59	0.80	160	2.61

Table 3. Mean load balancing in nodes and execution time to obtain the first solution of the $DD - BF$ algorithm (binary tree) on an Fujitsu AP-1000. 1000 iterations have been carried out for each item of data obtained. In each node 2500 floating point multiplications are carried out.

The $DD - BF$ algorithm has been executed on an AP-1000 multiprocessor. Table 3 shows the results of this execution for a different number of PEs and tree levels. Each result has been obtained after 1000 simulations. Moreover, 2500 floating point multiplications were carried in each node. The algorithm's speed-up is indicated in Figure 3.a. Note that as the number of computations increases (more levels in the tree) the better values are obtained regarding benefits.

Table 3 also shows the unbalancing of the $DD - BF$ algorithm. In figure 3.b we can observe how this algorithm's values are noticeably better than those obtained by the LB algorithm. The reason behind the persistence of imbalance in the dynamic algorithm is that communication consumes a certain amount of time, which cannot be dedicated to computation. Moreover, the PEs that are idle must wait for the load requests termination.

4 Conclusions

A new model for best-first search Branch and Bound algorithm has been proposed. The model is based on the generating function concept and allows to study the generated trees using both a random number of children and a random weight of edges. From the experimental results we can conclude that it permits to predict the behavior of the sequential Branch and Bound problems with a high accuracy. Furthermore, it has been successfully applied to parallel

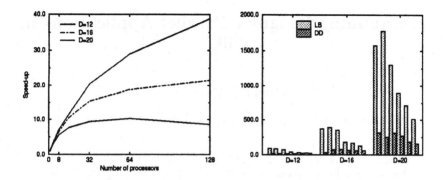

Fig. 3. (a) Speed-up of the $DD - BF$ algorithm on the AP-1000 and (b)imbalance comparison between $LB - BF$ and $DD - BF$ algorithm.

computation experiments for distributed memory machines using two different parallel algorithms.

References

1. M. K. Yang and C. R. Das, "Evaluation of a Parallel Branch-and-Bound Algorithm on a Class of Multiprocessors". *IEEE Trans. on Parallel and Distributed Systems.* vol. 5, No. 1, pp. 74-86. January 1994.
2. D. R. Smith, "Random Trees and the Analysis of Branch and Bound Procedures", *J. of the ACM*, vol. 31, No. 1, pp. 163-168, January 1984.
3. T. E. Harris, *The Theory of Branching Processes*, Dover Publications, Inc, New York, 1989.
4. M.H. Willebeek-LeMair and A.P. Reeves. "Strategies for Dynamic Load Balancing on Highly Parallel Computers". IEEE Transactions on parallel and distributed processing, vol. 4, No. 9, pp. 979-993, 1993.
5. R. Correa and A. Ferreira, "A Distributed Implementation of Asyncronous Parallel Branch and Bound". In Ferreira and J.D.P. Rolim (eds.), *Parallel Algorithms for Irregular Problems* (pp. 157-176), Kluwer Academic, 1995.
6. C.G. Diderich and M. Gengler, "Experiments with a parallel synchronized Branch and Bound algorithm". In Ferreira and J.D.P. Rolim (eds.), *Parallel Algorithms for Irregular Problems* (pp. 177-193), Kluwer Academic, 1995.
7. Y. Zhang, "Parallel Algorithms for Combinatorial Search Problems", PhD. Thesis, University of California at Berkley, 1989.

Programming Irregular Parallel Applications in Cilk

Charles E. Leiserson

MIT Laboratory for Computer Science, Cambridge, MA 02139, USA

Abstract. Cilk (pronounced "silk") is a C-based language for multi-threaded parallel programming. Cilk makes it easy to program irregular parallel applications, especially as compared with data-parallel or message-passing programming systems. A Cilk programmer need not worry about protocols and load balancing, which are handled by Cilk's provably efficient runtime system. Many regular and irregular Cilk applications run nearly as fast on one processor as comparable C programs, but the Cilk programs scale well to many processors.

1 Background and goals

Cilk is an algorithmic multithreaded language. The philosophy behind Cilk is that a programmer should concentrate on structuring his program to expose parallelism and exploit locality, leaving the runtime system with the responsibility of scheduling the computation to run efficiently on a given platform. Thus, the Cilk runtime system takes care of details like load balancing and communication protocols. Unlike other multithreaded languages, however, Cilk is algorithmic in that the runtime system's scheduler guarantees provably efficient and predictable performance.

Cilk grew out of theoretical work [1, 5, 6] on the scheduling of multithreaded computations. The basis of Cilk is a provably good scheduling algorithm that has been the cornerstone of Cilk system development. Cilk's provably good scheduler engendered a performance model that accurately predicts the efficiency of a Cilk program using two simple parameters: *work* and *critical-path length* [1, 4, 6]. More recent research has included *page faults* as a measure of locality [2, 3, 11].

The first implementation of Cilk was a direct descendent of PCM/Threaded-C [9, 11], a C-based package which provided continuation-passing-style threads on Thinking Machines Corporation's Connection Machine Model CM-5 Supercomputer [12] and which used "work-stealing" as a scheduling policy to improve the load balance and locality of the computation. With the addition of a provably good scheduler and the incorporation of many other parallel programming features, the system was rechristened "Cilk-1." Notable among the platforms that supported Cilk-1 was an adaptively parallel and fault-tolerant network-of-workstations implementation, called Cilk-NOW [1, 7].

The next release, Cilk-2, featured full type-checking, supported all of ANSI C in its C-language subset, and offered call-return semantics for writing multi-threaded procedures [13]. Instead of the C preprocessor used in Cilk-1, the Cilk-2 language was compiled into C by the `cilk2c` type-checking preprocessor, which made Cilk into a real language. The runtime system was made more portable, and the release included support for several architectures other than the CM-5.

Cilk-3 featured an implementation of "dag-consistent" distributed shared memory [3, 2, 11]. With the addition of shared memory, Cilk could be applied solve a much wider class of applications. Dag-consistency is a weak but nonetheless useful consistency model, and its relaxed semantics allows for an efficient, low-overhead, software implementation.

In Cilk-4, we changed our primary development platform from the CM-5 to the Sun Microsystems UltraSPARC Enterprise 5000 symmetric multiprocessor (SMP). The compiler and runtime system were completely reimplemented, eliminating continuation passing as the basis of the scheduler, and instead embedding scheduling decisions directly into the compiled code. The overhead to spawn a parallel thread in Cilk-4 cost less than 3 times that of an ordinary C procedure call, so Cilk programs could "scale down" to run on one processor with nearly the efficiency of analogous C programs.

In the current release, Cilk-5, the runtime system has been rewritten to be more flexible and portable. Included in this release is a novel debugging tool, called the Nondeterminator [8], which helps Cilk programmers to localize determinacy-race bugs in their code.

To date, prototype applications developed in Cilk include graphics rendering, protein folding, backtracking search, n-body simulation, and dense and sparse matrix computations. Our largest engineering effort has been on a series of chess-playing programs. An earlier program, $\star$Socrates [10], took second place in the 1995 ICCA World Computer Chess Championship in Hong Kong running on the 1824-node Intel Paragon at Sandia National Laboratories in New Mexico. Our most recent program, Cilkchess, was the 1996 Open Dutch Computer Chess Champion, undefeated with 10 out of a possible 11 points. For the competition, Cilkchess ran on a 12-processor 167-megahertz UltraSPARC Enterprise 5000 Sun SMP with 1 gigabyte of memory.

The Cilk-5 release includes the Cilk runtime system, the `cilk2c` compiler, a collection of example programs, a manual, and the Nondeterminator debugging tool. The Cilk runtime system included in the release runs on Sun SPARC machines running Solaris, Silicon Graphics machines running Irix, and x86 machines running Linux. Earlier versions of Cilk have been ported to other platforms—including PVM, the Intel Paragon, and networks of workstations—but these ports are not available for this release, which runs only on SMP's. We are currently working on a version for distributed memory multiprocessors, however.

Cilk software, documentation, publications, and up-to-date information are available via the World Wide Web at `http://theory.lcs.mit.edu/cilk`. The two MIT Ph.D. theses [1, 11] contain more detailed descriptions of the foundation and history of early Cilk versions.

2   The Cilk language

The basic Cilk language is extremely simple. It consists of C with the addition of three new keywords to indicate parallelism and synchronization. A Cilk program, when run on one processor, has the same semantics as the C program that results when the Cilk keywords are deleted. In addition, the Cilk system extends serial C semantics in a natural way for parallel execution. For example, C's stack memory is implemented as a "cactus" stack in Cilk. This section overviews how Cilk extends the C language and programming environment.

One of the simplest examples of a Cilk program is a recursive program to compute the Fibonacci numbers. A C program to compute the Fibonacci numbers is shown in Figure 1(a), and Figure 1(b) shows a Cilk program that does the same computation in parallel. Notice how similar the two programs look. In fact, the only differences between them are the inclusion of the library header file and the Cilk keywords `cilk`, `spawn`, and `sync`.

The keyword `cilk` identifies a Cilk *procedure*, which is the parallel analog of a C function. A Cilk procedure may spawn subprocedures in parallel and contain synchronization points. A Cilk procedure definition is identified by the keyword `cilk` and has an argument list and body just like a C function.

Most of the work in a Cilk procedure is executed serially, just like C, but parallelism is created when the invocation of a Cilk procedure is immediately preceded by the keyword `spawn`. A spawn is the parallel analog of a C function call, and like a C function call, when a Cilk procedure is spawned, execution proceeds to the child. Unlike a C function call, however, where the parent is not resumed until after its child returns, in the case of a Cilk spawn, the parent can continue to execute in parallel with the child. Indeed, the parent may continue to spawn off children, producing a high degree of parallelism. Cilk's scheduler takes the responsibility of scheduling the spawned procedures on the processors of the parallel computer.

A Cilk procedure cannot safely use the return values of the children it has spawned until it executes a `sync` statement. If all of its children have not completed when it executes a `sync`, the procedure suspends and does not resume until all of its children have completed. The `sync` statement is a local "barrier," not a global one as, for example, is sometimes used in message-passing programming. In Cilk, a `sync` waits only for the spawned children of the procedure to complete, not for the whole world. When all of its children return, execution of the procedure resumes at the point immediately following the `sync` statement. In the Fibonacci example, a `sync` statement is required before the statement `return (x+y)`, to avoid the anomaly that would occur if `x` and `y` were summed before each had been computed. A Cilk programmer uses `spawn` and `sync` keywords to expose the parallelism in a program, and the Cilk runtime system takes the responsibility of scheduling the procedures efficiently.

Cilk uses a *cactus stack* for stack-allocated storage, such as is needed for procedure-local variables. As is shown shown in Figure 2, from the point of view of a single Cilk procedure, a cactus stack behaves much like an ordinary stack. The procedure can allocate and free memory by pushing and popping the stack.

```
#include <stdlib.h>
#include <stdio.h>

int fib (int n)
{
    if (n<2) return (n);
    else
    {
        int x, y;
        x = fib (n-1);
        y = fib (n-2);
        return (x+y);
    }
}

int main (int argc, char *argv[])
{
    int n, result;
    n = atoi(argv[1]);
    result = fib (n);
    printf ("Result: %d\n", result);
    return 0;
}
```

(a)

```
#include <stdlib.h>
#include <stdio.h>
#include <cilk.h>

cilk int fib (int n)
{
    if (n<2) return n;
    else
    {
        int x, y;
        x = spawn fib (n-1);
        y = spawn fib (n-2);
        sync;
        return (x+y);
    }
}

cilk int main (int argc, char *argv[])
{
    int n, result;
    n = atoi(argv[1]);
    result = spawn fib(n);
    sync;
    printf ("Result: %d\n", result);
    return 0;
}
```

(b)

Fig. 1. Programs to compute the *n*th Fibonacci number. (a) A serial C program. (b) A parallel Cilk program.

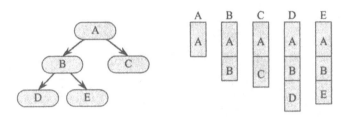

Fig. 2. A cactus stack. Procedure A spawns B and C, and B spawns D and E. The left part of the figure shows the spawn tree, and the right part of the figure shows the view of the stack by the five procedures. (The stack grows downward.)

The procedure views the stack as extending back from its own stack frame to the frame of its parent and continuing to more distant ancestors. The stack becomes a cactus stack when multiple procedures execute in parallel, each with its own view of the stack that corresponds to its call history, as shown in Figure 2.

Cactus stacks in Cilk have essentially the same limitations as ordinary C stacks [14]. For instance, a child procedure cannot return to its parent a pointer to an object that it has allocated, since the object will be deallocated automatically when the child returns. Similarly, sibling procedures cannot reference each other's local variables. Just as with the C stack, pointers to objects allocated on the cactus stack can only be safely passed to procedures below the allocation point in the call tree. Cilk supports heap memory as well as stack memory, however, and a Cilk `malloc()` function is available to programmers.

The Cilk language also supports several advanced parallel programming features. It provides "inlets" as a means of incorporating a returned result of child procedures into a procedure frame in nonstandard ways. Cilk also allows a procedure to abort speculatively spawned work. A procedure can also interact with Cilk's scheduler to test whether it is "synched" without actually executing a `sync`. The Cilk-5 reference manual [15] provides complete documentation of the Cilk language.

3 The Cilk model of multithreaded computation

Cilk supports an algorithmic model of parallel computation. Specifically, it guarantees that programs are scheduled efficiently by its runtime system. To better understand this guarantee, this section surveys the major characteristics of Cilk's algorithmic model.

A Cilk program execution consists of a collection of *procedures*,[1] each of which is broken into a sequence of nonblocking *threads*. In Cilk terminology, a *thread* is a maximal sequence of instructions that ends with a `spawn`, `sync`, or `return` statement. (Arguments to these statements are considered part of the thread preceding the statement.) The first thread that executes when a procedure is

---

[1] Technically, procedure *instances*.

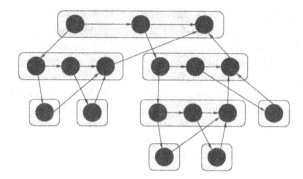

Fig. 3. The Cilk model of multithreaded computation. Each procedure, shown as a rounded rectangle, is broken into sequences of threads, shown as circles. A downward edge indicates the spawning of a subprocedure. A horizontal edge indicates the continuation to a successor thread. An upward edge indicates the returning of a value to a parent procedure. All three types of edges are dependencies which constrain the order in which threads may be scheduled.

called is the procedure's *initial thread*, and the subsequent threads are *successor threads*. At runtime, the binary "spawn" relation causes procedure instances to be structured as a rooted tree, and the dependencies among their threads form a dag embedded in this *spawn tree*, as is illustrated in Figure 3.

A correct execution of a Cilk program must obey all the dependencies in the dag, since a thread cannot be executed until all the threads on which it depends have completed. These dependencies form a partial order, permitting many ways of scheduling the threads in the dag. The order in which the dag unfolds and the mapping of threads onto processors are crucial decisions made by Cilk's scheduler. Every active procedure has associated state that requires storage, and every dependency between threads assigned to different processors requires communication. Thus, different scheduling policies may yield different space and time requirements for the computation.

It can be shown that for general multithreaded dags, no good scheduling policy exists. That is, a dag can be constructed for which any schedule that provides linear speedup also requires vastly more than linear expansion of space [5]. Fortunately, every Cilk program generates a well-structured dag that can be scheduled efficiently [6].

The Cilk runtime system implements a provably efficient scheduling policy based on randomized *work-stealing*. During the execution of a Cilk program, when a processor runs out of work, it asks another processor chosen at random for work to do. Locally, a processor executes procedures in ordinary serial order (just like C), exploring the spawn tree in a depth-first manner. When a child procedure is spawned, the processor saves local variables of the parent on the bottom of a stack and commences work on the child. When the child returns, the bottom of the stack is popped (just like C) and the parent resumes. When another processor requests work, however, work is stolen from the top of the stack, that is, from the end opposite that which is normally used.

Cilk's work-stealing scheduler executes any Cilk computation in nearly optimal time. From a theoretical perspective, there are two fundamental limits to how fast a Cilk program can run. Let us denote by T_P the execution time of a given computation on P processors. The *work* of the computation is the total time needed to execute all threads in the dag. We can denote the work by T_1, since the work is essentially the execution time of the computation on one processor. Notice that with T_1 work and P processors, the lower bound $T_P \geq T_1/P$ must hold.[2] The second limit is based on the program's *critical-path length*, denoted by T_∞, which is the execution time of the computation on an infinite number of processors, or equivalently, the time needed to execute threads along the longest path of dependency. The second lower bound is simply $T_P \geq T_\infty$.

Cilk's work-stealing scheduler executes a Cilk computation on P processors in time $T_P \leq T_1/P + O(T_\infty)$, which is asymptotically optimal. Empirically, the constant factor hidden by the big O is often close to 1 or 2 [4], and the formula

$$T_P \approx T_1/P + T_\infty \tag{1}$$

is a good approximation of runtime. (This model assumes that the parallel computer has adequate bandwidth in its communication network.) This performance model can be interpreted using the notion of *average parallelism*, which is given by the formula $\overline{P} = T_1/T_\infty$. The average parallelism is the average amount of work for every step along the critical path. Whenever $P \ll \overline{P}$, meaning that the actual number of processors is much smaller than the average parallelism of the application, we have equivalently that $T_1/P \gg T_\infty$. Thus, the model predicts that $T_P \approx T_1/P$ and the Cilk program is guaranteed to run with almost perfect linear speedup. The measures of work and critical-path length provide an algorithmic basis for evaluating the performance of Cilk programs over the entire range of possible parallel machine sizes. Cilk provides automatic timing instrumentation that can calculate these two measures during a run of a program, no matter how many processors the program is run on.

Cilk's runtime system also provides a guarantee on the amount of cactus stack space used by a parallel Cilk execution. Denote by S_P the (cactus) stack space required for a P-processor execution. Then, S_1 is the space required for an execution on one processor. Cilk's scheduler guarantees that for a P-processor execution, we have $S_P \leq S_1 P$, which is to say that the average space per processor is bounded above by the serial space. In fact, much less space may be required for many algorithms (see [2]), but the bound $S_P \leq S_1 P$ serves as a reasonable limit. If a computation uses moderate amounts of memory when on one processor, you can be assured that it will use no more space per processor when run in parallel.

The algorithmic complexity measures of work, critical-path length, and space —together with the fact that a programmer can count on them when designing a program—justify Cilk as an *algorithmic* multithreaded language.

---

[2] This abstract model of execution time ignores memory-hierarchy effects, but is nonetheless quite accurate [4].

4 Experiments

The Cilk distribution contains a variety of example programs which explore the difficulty of solving problems in parallel. Some of these programs, such as the program for computing a sparse Cholesky factorization, have irregular inputs. Others, like the backtrack searching algorithm used to solve the n-queens problem, have an irregular structure in the computation. Because of Cilk's flexibility in expressing parallelism, irregular problems pose no undue hardship on execution efficiency. The minimal loss of performance that is sometimes experienced is generally due to parallel algorithms that are intrinsically less efficient than the serial algorithm they replace. This section describes some preliminary performance measurements taken of the example programs.

The Cilk distribution (available from `http://theory.lcs.mit.edu/ cilk`) includes the following programs:

- `blockedmul` — Block multiplication of two dense $n \times n$ matrices, written by Keith Randall.
- `notempmul` — An slightly less parallel, but more efficient, block multiplication of two dense $n \times n$ matrices, written by Keith Randall.
- `queens` — Backtrack search to solve the problem of placing n queens on an $n \times n$ chessboard so that no two queens attack each other, written by Keith Randall.
- `cilksort` — Sort a random permutation of n 32-bit integers, written by Matteo Frigo and Andrew Stark.
- `knapsack` — Solve the 0-1 knapsack problem on n items using branch and bound, written by Matteo Frigo.
- `lu` — LU-decomposition of a dense $n \times n$ matrix, written by Robert D. Blumofe.
- `cholesky` — Cholesky factorization of a sparse symmetric positive-definite matrix represented as a quad-tree, written by Aske Plaat and Keith Randall.
- `heat` — Heat-diffusion calculation on an $m \times n$ mesh, written by Volker Strumpen.
- `fft` — Fast Fourier transformation of a vector of length n, written by Matteo Frigo.
- `Barnes-Hut` — Barnes-Hut n-body calculation, written by Keith Randall.

Table 1 shows preliminary speedup measurements that were taken for some of the example programs, as well as measurements of work, critical path, and average parallelism. The T_P column gives the time in seconds for a P-processor run. The machine used for the test runs was an otherwise idle Sun Enterprise 5000 SMP, with 8 UltraSPARC 167-megahertz processors, 512 megabytes of main memory, 512 kilobytes of L2 cache, 16 kilobytes of instruction L1 cache, and 16 kilobytes of data L1 cache, running Solaris 2.5 and a version of Cilk-5 that used gcc 2.7.2 with optimization level -O3.

Program	Problem	Work	CP	$\overline{P}$	T_1	T_4	T_8	T_1/T_4	T_1/T_8
blockedmul	1024	31.2	0.0046	6730	29.9	7.9	4.29	3.8	7.0
notempmul	1024	30.7	0.0156	1970	29.7	7.1	3.9	4.2	7.6
queens	22	174.3	0.0021	84335	163.1	41.8	14.1	3.9	11.6
cilksort*	4,100,000	5.37	0.0048	1108	5.4	1.5	0.9	3.5	6.0
knapsack	30	148	0.0016	345099	197.4	54.8	23.8	3.6	8.3
lu*	2048	154.2	0.4161	370	155.8	38.9	20.3	4.0	7.7
cholesky*	BCSSTK29	1645	4.1716	394	359	102.6	62.9	3.5	5.7
–	BCSSTK32	9907	12.9	764	2615		390		6.7
heat	4096×512	61.5	0.16	384	62.3	15.9	9.4	3.9	6.6
fft*	2^{20}	5.1	0.0024	2145	4.3	1.2	0.77	3.4	5.6
Barnes-Hut	2^{16}	136.8	9.3206	15	124.0	39.5	24.9	3.14	4.98

Table 1. The performance of Cilk example programs. Times are in seconds. Measurements are for a complete run of the program, except for those programs that are starred, where because of large setup times, only the core algorithm was measured.

Because these programs have been coded only recently, the following caveats should be taken into consideration when interpreting the results in the table. The speedup is the time of the P-processor run of the parallel program is compared to that of the 1-processor run of the same parallel program, not to that of the best serial algorithm in existence. We plan to make this comparison in the near future. The times measured are those of a complete run, except for cilksort, lu, cholesky, and fft, (starred in the figure). The knapsack and queens programs use speculative parallelism, where the amount of work depends on the schedule, creating the possible impression of superlinear speedup. Sometimes the reported figures appear to be inconsistent. For example, the work and the one-processor time should in theory be identical, but in practice they sometimes differ. Most of this effect is probably due to measurement error in the adding up of the execution times of many small thread and from cache effects due to differences in thread scheduling.

As can be seen from the table, the programs give good speedups. Even the complicated Barnes-Hut code provides a speedup of 5 on 8 processors, which is at least as good as any implementation we have found in the literature or on the World Wide Web. Although it is not shown in the figure, the performance of any of our Cilk programs running on one processor is generally indistinguishable from the performance of the comparable C code. The sorting example and sparse Cholesky factorization are worst cases for this set of examples, and even for these programs, the single-processor Cilk performance is within 30 percent of our fastest C code for the problem. The slowdown of sorting is due to the fact that our parallel algorithm cannot be performed in place, and the slowdown of the Cholesky factorization is due the overhead in our quad-tree represenation of sparse matrices compared with the ordinary linked list implementation.

5 Conclusion

To produce high-performance parallel applications, programmers often focus on communication costs and execution time, quantities that are dependent on specific machine configurations. Cilk's philosophy argues that a programmer should think instead about work and critical-path length, abstractions that can be used to characterize the performance of an algorithm independent of the machine configuration. Cilk provides a programming model in which work and critical path are observable quantities, and it delivers guaranteed performance as a function of these quantities. Moreover, Cilk programs "scale down" to run on one processor with nearly the efficiency of analogous C programs.

Because the semantics of Cilk are a simple and natural extension of C semantics, solving regular and irregular problems in Cilk poses no extra runtime overhead compared to solving the problems in C. Programming in parallel can be harder, however, because sometimes to obtain parallelism, one must change the most efficient serial algorithm in a way that sacrifices certain efficiencies. Our initial experiences in writing Cilk programs for regular and irregular problems, however, lead us to believe that this loss of efficiency is frequently small or negligible. Nevertheless, more experience with Cilk will be required to evaluate its effectiveness across a wide range of applications. The Cilk developers invite you to program your favorite application in Cilk. The freely available Cilk software can be downloaded from `http://theory.lcs.mit.edu/cilk`.

Acknowledgments

I consider myself fortunate to have worked on Cilk with many talented people over the last three years. Thanks in particular to Mingdong Feng, Matteo Frigo, Phil Lisiecki, Aske Plaat, Keith Randall, Bin Song, and Volker Strumpen for their contributions to the Cilk-5 release. Thanks also to Bobby Blumofe, Michael Halbherr, Chris Joerg, Bradley Kuszmaul, Rob Miller, and Yuli Zhou for their contributions to earlier releases.

References

1. Robert D. Blumofe. *Executing Multithreaded Programs Efficiently*. PhD thesis, Department of Electrical Engineering and Computer Science, Massachusetts Institute of Technology, September 1995. Available as MIT Laboratory for Computer Science Technical Report MIT/LCS/TR-677.
2. Robert D. Blumofe, Matteo Frigo, Chrisopher F. Joerg, Charles E. Leiserson, and Keith H. Randall. An analysis of dag-consistent distributed shared-memory algorithms. In *Proceedings of the Eighth Annual ACM Symposium on Parallel Algorithms and Architectures (SPAA)*, pages 297–308, Padua, Italy, June 1996.
3. Robert D. Blumofe, Matteo Frigo, Christopher F. Joerg, Charles E. Leiserson, and Keith H. Randall. Dag-consistent distributed shared memory. In *Tenth International Parallel Processing Symposium (IPPS)*, pages 132–141, Honolulu, Hawaii, April 1996.

4. Robert D. Blumofe, Christopher F. Joerg, Bradley C. Kuszmaul, Charles E. Leiserson, Keith H. Randall, and Yuli Zhou. Cilk: An efficient multithreaded runtime system. In *Proceedings of the Fifth ACM SIGPLAN Symposium on Principles and Practice of Parallel Programming (PPoPP)*, pages 207–216, Santa Barbara, California, July 1995.

5. Robert D. Blumofe and Charles E. Leiserson. Space-efficient scheduling of multithreaded computations. In *Proceedings of the Twenty Fifth Annual ACM Symposium on Theory of Computing (STOC)*, pages 362–371, San Diego, California, May 1993.

6. Robert D. Blumofe and Charles E. Leiserson. Scheduling multithreaded computations by work stealing. In *Proceedings of the 35th Annual Symposium on Foundations of Computer Science (FOCS)*, pages 356–368, Santa Fe, New Mexico, November 1994.

7. Robert D. Blumofe and David S. Park. Scheduling large-scale parallel computations on networks of workstations. In *Proceedings of the Third International Symposium on High Performance Distributed Computing (HPDC)*, pages 96–105, San Francisco, California, August 1994.

8. Mingdong Feng and Charles E. Leiserson. Efficient detection of determinacy races in Cilk programs. In *Proceedings of the Ninth Annual ACM Symposium on Parallel Algorithms and Architectures (SPAA)*, Newport, Rhode Island, June 1997. To appear.

9. Michael Halbherr, Yuli Zhou, and Chris F. Joerg. MIMD-style parallel programming with continuation-passing threads. In *Proceedings of the 2nd International Workshop on Massive Parallelism: Hardware, Software, and Applications*, Capri, Italy, September 1994.

10. Chris Joerg and Bradley C. Kuszmaul. Massively parallel chess. In *Proceedings of the Third DIMACS Parallel Implementation Challenge*, Rutgers University, New Jersey, October 1994.

11. Christopher F. Joerg. *The Cilk System for Parallel Multithreaded Computing*. PhD thesis, Department of Electrical Engineering and Computer Science, Massachusetts Institute of Technology, January 1996. Available as MIT Laboratory for Computer Science Technical Report MIT/LCS/TR-701.

12. Charles E. Leiserson, Zahi S. Abuhamdeh, David C. Douglas, Carl R. Feynman, Mahesh N. Ganmukhi, Jeffrey V. Hill, W. Daniel Hillis, Bradley C. Kuszmaul, Margaret A. St. Pierre, David S. Wells, Monica C. Wong, Shaw-Wen Yang, and Robert Zak. The network architecture of the Connection Machine CM-5. In *Proceedings of the Fourth Annual ACM Symposium on Parallel Algorithms and Architectures (SPAA)*, pages 272–285, San Diego, California, June 1992.

13. Robert C. Miller. A type-checking preprocessor for Cilk 2, a multithreaded C language. Master's thesis, Department of Electrical Engineering and Computer Science, Massachusetts Institute of Technology, May 1995.

14. Joel Moses. The function of FUNCTION in LISP or why the FUNARG problem should be called the environment problem. Technical Report memo AI-199, MIT Artificial Intelligence Laboratory, June 1970.

15. Supercomputing Technology Group, Massachusetts Institute of Technology, 545 Technology Square, Cambridge, Massachusetts 02139. *Cilk-5.0 (Beta 1) Reference Manual*, March 1997. Available on the World Wide Web at URL "http://theory.lcs.mit.edu/~cilk".

A Variant of the Biconjugate Gradient Method Suitable for Massively Parallel Computing

H. Martin Bücker* and Manfred Sauren

Zentralinstitut für Angewandte Mathematik
Forschungszentrum Jülich GmbH, 52425 Jülich, Germany

Abstract. Starting from a specific implementation of the Lanczos biorthogonalization algorithm, an iterative process for the solution of systems of linear equations with general non-Hermitian coefficient matrix is derived. Due to the orthogonalization of the underlying Lanczos process the resulting iterative scheme involves inner products leading to global communication and synchronization on parallel processors. For massively parallel computers, these effects cause considerable delays often preventing the scalability of the implementation. In the process proposed, all inner product-like operations of an iteration step are independent such that the implementation consists of only a single global synchronization point per iteration. In exact arithmetic, the process is shown to be mathematically equivalent to the biconjugate gradient method. The efficiency of this new variant is demonstrated by numerical experiments on a PARAGON system using up to 121 processors.

1 Introduction

Storage requirements as well as computation times often set limits to direct methods for the solution of systems of linear equations arising from large scale applications in science and engineering. A popular class of iterative methods particularly attractive for systems with sparse coefficient matrices is the family of Krylov subspace methods [7, 12] involving the coefficient matrix only in the form of matrix-by-vector products. These methods basically consist of the generation of a suitable basis of a vector space called Krylov subspace and the choice of the actual iterate within that space.

In this note, a Krylov subspace method is derived that generates the basis of its primary subspace by means of the Lanczos biorthogonalization algorithm [10]. Due to the orthogonalization of this underlying procedure the resulting process contains inner products that lead to global communication on parallel processors, i.e., communication of all processors at the same time. For large sparse matrices, parallel matrix-by-vector products can be implemented with communication between only nearby processors. This is the reason why scalability of an implementation on massively parallel processors is usually prevented entirely

---

* The work of this author was supported by the Graduiertenkolleg "Informatik und Technik", RWTH Aachen, 52056 Aachen, Germany.

by the computation of inner product-like operations. There are two strategies to remedy the performance degradation which, of course, can be combined. The first is to restructure the code such that most communication is overlapped with useful computation. The second is to eliminate data dependencies such that several inner products are computed simultaneously. This note is concerned with the latter strategy in order to reduce the number of global synchronization points. A *global synchronization point* is defined as the locus of an algorithm at which all local information has to be globally available in order to continue the computation.

This note which is available in an extended form [3] is organized as follows. Section 2 is concerned with the sketch of a specific version of the Lanczos biorthogonalization algorithm suitable for massively parallel computing. In Sect. 3, this version is used to derive a new process for the solution of systems of linear equations that is shown to be mathematically equivalent to the biconjugate gradient method [11, 5]. Section 4 presents numerical experiments.

2 Lanczos biorthogonalization algorithm

The Lanczos biorthogonalization algorithm [10] (Lanczos algorithm hereafter) was originally proposed to reduce a matrix to tridiagonal form. We here focus on a different application, namely the computation of two bases of two Krylov subspaces. Given a general non-Hermitian matrix $\mathbf{A} \in \mathbb{C}^{N \times N}$ and two starting vectors $\mathbf{v}_1, \mathbf{w}_1 \in \mathbb{C}^N$ satisfying $\mathbf{w}_1^T \mathbf{v}_1 = \delta_1 \neq 0$, the Lanczos algorithm generates two finite sequences of vectors $\{\mathbf{v}_n\}_{n=1,2,3,\ldots}$ and $\{\mathbf{w}_n\}_{n=1,2,3,\ldots}$ with the following three properties:

$$\mathbf{v}_n \in \mathcal{K}_n(\mathbf{v}_1, \mathbf{A}) , \tag{1}$$

$$\mathbf{w}_n \in \mathcal{K}_n(\mathbf{w}_1, \mathbf{A}^T) , \tag{2}$$

$$\mathbf{w}_m^T \mathbf{v}_n = \begin{cases} 0 & \text{if } n \neq m , \\ \delta_n \neq 0 & \text{if } n = m , \end{cases} \tag{3}$$

where $\mathcal{K}_n(\mathbf{y}, \mathbf{A}) = \text{span}\{\mathbf{y}, \mathbf{A}\mathbf{y}, \ldots, \mathbf{A}^{n-1}\mathbf{y}\}$ denotes the nth Krylov subspace generated by the matrix $\mathbf{A}$ and the vector $\mathbf{y}$. The vectors $\mathbf{v}_n$ and $\mathbf{w}_n$ are called Lanczos vectors and their relation (3) is commonly referred to as biorthogonality. Implementations of the Lanczos algorithm differ in various aspects, e.g., the scaling of Lanczos vectors, the length of the vector recurrences used to generate them, and the number of global synchronization points per iteration.

We here state the matrix equations that characterize a particular variant of the Lanczos algorithm developed in [2] and give some comments afterwards:

$$\mathbf{W}_n^T \mathbf{V}_n = \mathbf{D}_n , \tag{4a}$$

$$\mathbf{V}_n = \mathbf{P}_n \mathbf{U}_n , \tag{4b}$$

$$\mathbf{A}\mathbf{P}_n = \mathbf{V}_{n+1} \mathbf{L}_n , \tag{4c}$$

$$\tilde{\mathbf{Q}}_n = \mathbf{W}_n \mathbf{D}_n^{-1} \mathbf{U}_n^T + \frac{\mu_{n+1}}{\delta_{n+1}} \mathbf{w}_{n+1} \mathbf{e}_n^T , \tag{4d}$$

$$\mathbf{A}^T \mathbf{W}_n = \tilde{\mathbf{Q}}_{n-1} \mathbf{L}_{n-1}^T \mathbf{D}_n + \tau_n \delta_n \tilde{\mathbf{q}}_n \mathbf{e}_n^T , \tag{4e}$$

where $\mathbf{e}_n = (0, 0, \ldots, 0, 1)^T$ is the last canonical unit vector of $\mathbb{R}^n$. Equation (4a) is a matrix reformulation of the biorthogonality (3) where the Lanczos vectors are put as columns into matrices $\mathbf{V}_n = [\mathbf{v}_1 \, \mathbf{v}_2 \, \cdots \, \mathbf{v}_n]$ and $\mathbf{W}_n = [\mathbf{w}_1 \, \mathbf{w}_2 \, \cdots \, \mathbf{w}_n]$. Moreover, the diagonal matrix $\mathbf{D}_n = \mathrm{diag}(\delta_1, \delta_2, \ldots, \delta_n)$ is used to scale both sequences of Lanczos vectors to unit length

$$\|\mathbf{v}_n\|_2 = \|\mathbf{w}_n\|_2 = 1 \, , \qquad n = 1, 2, 3, \ldots \, , \tag{5}$$

in order to avoid over- or underflow as recommended by [7, 9]. The process is said to be based on coupled two-term recurrences because of the bidiagonal structure of the matrices

$$\mathbf{L}_n = \begin{pmatrix} \tau_1 & & \\ \gamma_2 & \ddots & \\ & \ddots & \tau_n \\ & & \gamma_{n+1} \end{pmatrix} \in \mathbb{C}^{(n+1)\times n} \quad \text{and} \quad \mathbf{U}_n = \begin{pmatrix} 1 & \mu_2 & & \\ & 1 & \ddots & \\ & & \ddots & \mu_n \\ & & & 1 \end{pmatrix} \in \mathbb{C}^{n \times n} \, . \tag{6}$$

Note that, in finite precision arithmetic, numerical experiments [9] suggest to favor the coupled two-term procedure more than a mathematically equivalent three-term process when used as the underlying process of an iterative method for the solution of linear systems. In addition to the Lanczos vectors, two more vector sequences, $\{\mathbf{p}_n\}_{n=1,2,3,\ldots}$ and $\{\tilde{\mathbf{q}}_n\}_{n=1,2,3,\ldots}$, are generated giving rise to matrices $\mathbf{P}_n = [\mathbf{p}_1 \, \mathbf{p}_2 \, \cdots \, \mathbf{p}_n]$ and $\tilde{\mathbf{Q}}_n = [\tilde{\mathbf{q}}_1 \, \tilde{\mathbf{q}}_2 \, \cdots \, \tilde{\mathbf{q}}_n]$. Equations (4b) and (4c) show that the generation of the vector sequences $\mathbf{v}_n$ and $\mathbf{p}_n$ is coupled and so are $\mathbf{w}_n$ and $\tilde{\mathbf{q}}_n$. The complete derivation of this variant of the Lanczos process is given in [2] where it is shown that the resulting algorithm consists of only a single global synchronization point per iteration. We finally stress that there are look-ahead techniques preventing the Lanczos algorithm from breaking down; see [8] and the references therein. Although any implementation will benefit from such techniques they are beyond the scope of this note.

3 A parallel variant of the biconjugate gradient method

The Lanczos algorithm is now used as the underlying process of an iterative method for the solution of systems of linear equations

$$\mathbf{A}\mathbf{x} = \mathbf{b} \, , \tag{7}$$

where $\mathbf{A} \in \mathbb{C}^{N \times N}$ and $\mathbf{x}, \mathbf{b} \in \mathbb{C}^N$. One of the features of the Lanczos algorithm is the fact that the Lanczos vectors $\mathbf{v}_1, \mathbf{v}_2, \ldots, \mathbf{v}_n$ form a basis of the Krylov subspace generated by $\mathbf{A}$ and the starting vector $\mathbf{v}_1$. Since Krylov subspace methods for the solution of (7) are characterized by $\mathbf{x}_n \in \mathbf{x}_0 + \mathcal{K}_n(\mathbf{r}_0, \mathbf{A})$, where $\mathbf{x}_0$ is any initial guess and $\mathbf{r}_0 = \mathbf{b} - \mathbf{A}\mathbf{x}_0$ is the corresponding residual vector, relation (1) raises hopes to derive a Krylov subspace method if the Lanczos algorithm is started with

$$\mathbf{v}_1 = \frac{1}{\gamma_1} \mathbf{r}_0 \, , \tag{8}$$

where $\gamma_1 = \|\mathbf{r}_0\|$ is a scaling factor. Then, the nth iterate is of the form

$$\mathbf{x}_n = \mathbf{x}_0 + \mathbf{V}_n \mathbf{z}_n \ , \tag{9}$$

where as before the columns of $\mathbf{V}_n \in \mathbb{C}^{N \times n}$ are the Lanczos vectors $\mathbf{v}_1, \mathbf{v}_2, \ldots, \mathbf{v}_n$ and $\mathbf{z}_n \in \mathbb{C}^n$ is a parameter vector to be fixed later. The goal of any Krylov subspace method is in some sense to drive the residual vector $\mathbf{r}_n = \mathbf{b} - \mathbf{A}\mathbf{x}_n$ to the zero vector. Using (4b) and introducing

$$\mathbf{y}_n = \mathbf{U}_n \mathbf{z}_n \ , \tag{10}$$

the iterates according to (9) are reformulated in terms of $\mathbf{y}_n$ instead of $\mathbf{z}_n$ giving

$$\mathbf{x}_n = \mathbf{x}_0 + \mathbf{P}_n \mathbf{y}_n \ . \tag{11}$$

On account of (8) and (4c) the corresponding residual vector is

$$\mathbf{r}_n = \mathbf{V}_{n+1} \left(\gamma_1 \mathbf{e}_1^{(n+1)} - \mathbf{L}_n \mathbf{y}_n \right) \ , \tag{12}$$

where $\mathbf{e}_1^{(n+1)} = (1, 0, \ldots, 0)^T \in \mathbb{C}^{n+1}$. Generating the Krylov subspace by means of the Lanczos algorithm and fixing $\mathbf{y}_n$, and so implicitly $\mathbf{z}_n$ by (10), an iterative method can be derived. Let κ_i denote the ith component of $\mathbf{y}_n$, i.e., $\mathbf{y}_n = (\kappa_1, \kappa_2, \ldots, \kappa_n)^T$. The idea is to drive $\mathbf{r}_n$ to the zero vector by

$$\kappa_i = -\frac{\gamma_i}{\tau_i} \kappa_{i-1} \ , \qquad i = 1, 2, \ldots, n \ , \tag{13}$$

with $\kappa_0 = -1$. This choice of $\mathbf{y}_n$ is motivated by the structure of $\mathbf{L}_n$ given in (6) and zeros out the first n components of the vector $\gamma_1 \mathbf{e}_1^{(n+1)} - \mathbf{L}_n \mathbf{y}_n$ in (12). The residual vector is then given by

$$\mathbf{r}_n = \mathbf{V}_{n+1}(0, 0, \ldots, 0, -\gamma_{n+1}\kappa_n)^T \ . \tag{14}$$

It is important to note that the process of fixing $\mathbf{y}_n$ is easily updated in each iteration step because $\mathbf{y}_{n-1}$ coincides with the first $n-1$ components of $\mathbf{y}_n$. The corresponding recursion is

$$\mathbf{y}_n = \begin{pmatrix} \mathbf{y}_{n-1} \\ 0 \end{pmatrix} + \kappa_n \mathbf{e}_n \ , \tag{15}$$

where $\mathbf{e}_n = (0, \ldots, 0, 1)^T \in \mathbb{C}^n$. Inserting (15) into (11) yields

$$\mathbf{x}_n = \mathbf{x}_0 + \mathbf{P}_{n-1}\mathbf{y}_{n-1} + \kappa_n \mathbf{p}_n$$
$$= \mathbf{x}_{n-1} + \kappa_n \mathbf{p}_n \ . \tag{16}$$

Note that $\|\mathbf{r}_n\|_2$ is immediately available from (14) implying the relation

$$\|\mathbf{r}_n\|_2 = \|\mathbf{v}_{n+1}\|_2 \cdot |\gamma_{n+1}\kappa_n| \ . \tag{17}$$

Input $\mathbf{A}, \mathbf{b}$ and $\mathbf{x}_0$

$\mathbf{p}_0 = \mathbf{q}_0 = 0$, $\tilde{\mathbf{v}}_1 = \tilde{\mathbf{w}}_1 = \mathbf{b} - \mathbf{A}\mathbf{x}_0$

$\gamma_0 = \xi_0 = 0$, $\tau_0 = \rho_0 \neq 0$, $\kappa_0 = -1$

$\gamma_1 = \|\tilde{\mathbf{v}}_1\|_2$, $\xi_1 = \|\tilde{\mathbf{w}}_1\|_2$, $\varrho_1 = \tilde{\mathbf{w}}_1^T \tilde{\mathbf{v}}_1$, $\varepsilon_1 = \left(\mathbf{A}^T \tilde{\mathbf{w}}_1\right)^T \tilde{\mathbf{v}}_1$

for $n = 1, 2, 3, \ldots$ **do**

$$\mu_n = \frac{\gamma_{n-1}\xi_{n-1}\varrho_n}{\gamma_n \tau_{n-1}\varrho_{n-1}}, \quad \tau_n = \frac{\varepsilon_n}{\varrho_n} - \gamma_n\mu_n$$

$$\mathbf{p}_n = \frac{1}{\gamma_n}\tilde{\mathbf{v}}_n - \mu_n\mathbf{p}_{n-1}, \quad \mathbf{q}_n = \frac{1}{\xi_n}\mathbf{A}^T\tilde{\mathbf{w}}_n - \frac{\gamma_n\mu_n}{\xi_n}\mathbf{q}_{n-1}$$

$$\tilde{\mathbf{v}}_{n+1} = \mathbf{A}\mathbf{p}_n - \frac{\tau_n}{\gamma_n}\tilde{\mathbf{v}}_n, \quad \tilde{\mathbf{w}}_{n+1} = \mathbf{q}_n - \frac{\tau_n}{\xi_n}\tilde{\mathbf{w}}_n$$

- $\gamma_{n+1} = \|\tilde{\mathbf{v}}_{n+1}\|_2$, $\xi_{n+1} = \|\tilde{\mathbf{w}}_{n+1}\|_2$

- $\varrho_{n+1} = \tilde{\mathbf{w}}_{n+1}^T\tilde{\mathbf{v}}_{n+1}$, $\varepsilon_{n+1} = \left(\mathbf{A}^T\tilde{\mathbf{w}}_{n+1}\right)^T\tilde{\mathbf{v}}_{n+1}$

$$\kappa_n = -\frac{\gamma_n}{\tau_n}\kappa_{n-1}$$

$$\mathbf{x}_n = \mathbf{x}_{n-1} + \kappa_n\mathbf{p}_n$$

if $\left(|\gamma_{n+1}\kappa_n| < \text{tolerance}\right)$ **then STOP**

endfor

Algorithm 1: A parallel variant of the biconjugate gradient method

Properly putting (13) and (16) on top of the Lanczos algorithm characterized by equations (4) results in Alg. 1. We remark that the underlying Lanczos procedure operates with unit scaling of both sequences of Lanczos vectors, see (5), and is explicitly presented in [2, Alg. 3]. In this case, (17) simplifies to

$$\|\mathbf{r}_n\|_2 = |\gamma_{n+1}\kappa_n| \tag{18}$$

that can be used as a simple stopping criterion. In Alg. 1, the operations leading to global communication on parallel processors are marked with a preceding bullet. The absence of any data dependencies of these operations can be exploited to compute them simultaneously, i.e., by reducing their local partial sums in a single global communication operation.

Having derived the above iterative process for the solution of linear systems, we afterwards realized that Alg. 1 is just a new variant of the biconjugate gradient method [11, 5] whose iterates are defined by the Galerkin type condition

$$\mathbf{w}^T\mathbf{r}_n = 0 \quad \text{for all} \quad \mathbf{w} \in \mathcal{K}_n(\mathbf{w}_1, \mathbf{A}^T), \tag{19}$$

where $\mathbf{w}_1$ is arbitrary, provided $\mathbf{w}_1^T\mathbf{v}_1 \neq 0$, but one usually set $\mathbf{w}_1 = \mathbf{r}_0/\|\mathbf{r}_0\|_2$ as Alg. 1 implicitly does [2]. To see why (19) holds, recall that by (2) every vector $\mathbf{w} \in \mathcal{K}_n(\mathbf{w}_1, \mathbf{A}^T)$ is of the form

$$\mathbf{w} = \mathbf{W}_n\mathbf{s} \quad \text{for some} \quad \mathbf{s} \in \mathbb{C}^n,$$

where $\mathbf{W}_n$ is generated by the Lanczos algorithm. Using (14) and the above argumentation of κ_n as the last component of $\mathbf{y}_n$, the condition (19) is equivalent to

$$-\gamma_{n+1}\kappa_n \mathbf{s}^T \mathbf{W}_n^T \mathbf{v}_{n+1} = 0$$

that is satisfied because of the biorthogonality (4a).

Although the history of the biconjugate gradient method dates back to the 1950s, we are unaware of any implementation of this method with the properties of the above procedure: The possibility to scale both sequences of Lanczos vectors and the independence of all inner product-like operations leading to only a single global synchronization point per iteration on parallel processors. The mathematical equivalence of the principal idea given here and the biconjugate gradient method is mentioned in [6, 4] although applied to the Lanczos algorithm based on three-term recurrences. We finally remark that the Lanczos algorithm characterized by equations (4) is also useful to derive a parallel variant of another iterative method [1].

4   Numerical experiments

This section presents numerical experiments by comparing Alg. 1 with a rescaled version of the biconjugate gradient method described in [3, Alg. 2]. In exact arithmetic, both algorithms are mathematically equivalent and offer the possibility to scale the two sequences of Lanczos vectors. The experiments were carried out on Intel's PARAGON XP/S 10 at Forschungszentrum Jülich in double precision FORTRAN.

As an example consider the partial differential equation

$$-\Delta u - 20 \left(x \frac{\partial u}{\partial x} + y \frac{\partial u}{\partial y} \right) = f \qquad \text{on} \quad \Omega = (0,1) \times (0,1)$$

with Dirichlet boundary condition $u = 0$ and the right-hand side f is chosen so that the solution is $u(x,y) = \frac{1}{2}\sin(4\pi x)\sin(6\pi y)$. We discretize the above differential equation using second order centered differences on a 440×440 grid with mesh size $1/441$, leading to a system of linear equations with unsymmetric real coefficient matrix of order $193\,600$ with $966\,240$ nonzero entries. Simple diagonal preconditioning is used by scaling the rows of $\mathbf{A}$ to unit length in the Euclidean norm. For our test runs we choose $\mathbf{x}_0 = 0$ as initial guess to the exact solution and stop as soon as $\|\mathbf{r}_n\|_2 < 10^{-6}$. Smaller tolerances do not yield better approximations to the exact solution u; the absolute difference between the approximations and the exact solution stagnates at $9 \cdot 10^{-5}$.

To partition the data among the processors the parallel implementation subdivides Ω into square subdomains of equal size. Thus, each of the processors holds the data of the corresponding subdomain. Due to the local structure of the discretizing scheme, a processor has to communicate with at most 4 processors to perform a matrix-by-vector product. Note that, in Alg. 1, there are two independent matrix-by-vector products per iteration which are computed simultaneously in this implementation.

Due to the lack of space, we here concentrate on parallel performance aspects rather than on numerical properties of both versions of the biconjugate gradient method. We remark that both versions show a similar convergence history in terms of the true residual norm $\|b - Ax_n\|_2$; see [3] for more detailed numerical results concerning the above example. The parallel performance results are given in Fig. 1. These results are based on time measurements of a fixed number of iterations. The speedup given on the left-hand side of this figure is computed by taking the ratio of the parallel run time and the run time of a serial implementation. While the serial run times of both variants are almost identical there is a considerable difference concerning parallel run times. For all numbers of processors, the new parallel variant is faster than Alg. 2 of [3]. The saving of run time grows with increasing number of processors; see right-hand side of Fig. 1. More precisely, the quantity depicted as a percentage is $1 - T_1(p)/T_2(p)$, where $T_1(p)$ and $T_2(p)$ are the run times on p processors of Alg. 1 and [3, Alg. 2], respectively. Recognize from this figure that, for a fixed number of iterations, the new variant is approximately 24% faster than [3, Alg. 2] on 121 processors. Note further that there is still room for using even more processors before running into saturation.

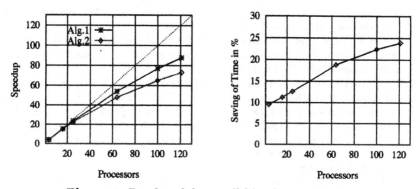

Figure 1. Results of the parallel implementations

5 Concluding Remarks

In its standard form, the biorthogonal Lanczos algorithm makes use of three-term recurrences for the generation of the Lanczos vectors spanning the underlying Krylov spaces. Here, we consider an alternative formulation based on coupled two-term recurrences that, in exact arithmetic, is mathematically equivalent to the three-term form. Taking a specific variant of the coupled two-term process as the starting point, a new iterative scheme for the solution of linear systems with non-Hermitian coefficient matrices is derived that, in exact arithmetic, is mathematically equivalent to the biconjugate gradient method. In this new implementation all inner product-like operations of an iteration step are independent such that the implementation consists of a single global synchronization point per iteration resulting in improved scalability on massively parallel computers. Moreover, the implementation offers the possibility to scale both of the two sequences of Lanczos vectors which is an important aspect concerning numerical stability in practical implementations.

References

1. H. M. Bücker and M. Sauren. A Parallel Version of the Quasi–Minimal Residual Method Based on Coupled Two–Term Recurrences. In J. Waśniewski, J. Dongarra, K. Madsen, and D. Olesen, editors, *Applied Parallel Computing: Industrial Computation and Optimization, Proceedings of the Third International Workshop, PARA '96, Lyngby, Denmark, August 18–21, 1996*, volume 1184 of *Lecture Notes in Computer Science*, pages 157–165, Berlin, 1996. Springer.
2. H. M. Bücker and M. Sauren. A Parallel Version of the Unsymmetric Lanczos Algorithm and its Application to QMR. Internal Report KFA–ZAM–IB–9605, Research Centre Jülich, Jülich, Germany, March 1996.
3. H. M. Bücker and M. Sauren. A Variant of the Biconjugate Gradient Method Suitable for Massively Parallel Computing. Internal Report KFA–ZAM–IB–9702, Research Centre Jülich, Jülich, Germany, 1997.
4. J. K. Cullum and A. Greenbaum. Relations between Galerkin and Norm-Minimizing Iterative Methods for Solving Linear Systems. *SIAM Journal on Matrix Analysis and Applications*, 17(2):223–247, 1996.
5. R. Fletcher. Conjugate Gradient Methods for Indefinite Systems. In G. A. Watson, editor, *Numerical Analysis Dundee 1975*, volume 506 of *Lecture Notes in Mathematics*, pages 73–89, Berlin, 1976. Springer.
6. R. W. Freund. The Look-Ahead Lanczos Process for Large Nonsymmetric Matrices and Related Algorithms. In M. S. Moonen, G. H. Golub, and B. L. R. De Moor, editors, *Linear Algebra for Large Scale and Real-Time Applications*, volume 232 of *NATO ASI Series E: Applied Sciences*, pages 137–163. Kluwer Academic Publishers, Dordrecht, The Netherlands, 1993. Proceedings of the NATO Advanced Study Institute on Linear Algebra for Large Scale and Real-Time Applications, Leuven, Belgium, August 1992.
7. R. W. Freund, G. H. Golub, and N. M. Nachtigal. Iterative Solution of Linear Systems. In *Acta Numerica 1992*, pages 1–44. Cambridge University Press, Cambridge, 1992.
8. R. W. Freund, M. H. Gutknecht, and N. M. Nachtigal. An Implementation of the Look–Ahead Lanczos Algorithm for Non–Hermitian Matrices. *SIAM Journal on Scientific Computing*, 14(1):137–158, 1993.
9. R. W. Freund and N. M. Nachtigal. An Implementation of the QMR Method Based on Coupled Two–Term Recurrences. *SIAM Journal on Scientific Computing*, 15(2):313–337, 1994.
10. C. Lanczos. An Iteration Method for the Solution of the Eigenvalue Problem of Linear Differential and Integral Operators. *Journal of Research of the National Bureau of Standards*, 45(4):255–282, 1950.
11. C. Lanczos. Solutions of Systems of Linear Equations by Minimized Iterations. *Journal of Research of the National Bureau of Standards*, 49(1):33–53, 1952.
12. Y. Saad. *Iterative Methods for Sparse Linear Systems*. PWS Publishing Company, Boston, 1996.

Efficient Implementation of the Improved Quasi-Minimal Residual Method on Massively Distributed Memory Computers

Tianruo Yang

Department of Computer Science, Linköping University
S-581 83, Linköping, Sweden

Hai-Xiang Lin

Department of Technical Mathematics and Computer Science
TU Delft, P.O. Box 5031, 2600 GA Delft, The Netherlands

Abstract. For the solutions of linear systems of equations with unsymmetric coefficient matrices, we has proposed an improved version of the quasi-minimal residual (IQMR) method by using the Lanczos process as a major component combining elements of numerical stability and parallel algorithm design. For Lanczos process, stability is obtained by a couple two-term procedure that generates Lanczos vectors scaled to unit length. The algorithm is derived such that all inner products and matrix-vector multiplications of a single iteration step are independent and communication time required for inner product can be overlapped efficiently with computation time. Therefore, the cost of global communication on parallel distributed memory computers can be significantly reduced. In this paper, we describe an efficient implementation of this method which is particularly well suited to problems with irregular sparsity pattern. The corresponding communication cost is independent of the sparsity pattern with several performance improvement techniques such as overlapping computation and communication, balancing the computational load. The performance is demonstrated by numerical experimental results carried out on massively parallel distributed memory computer Parsytec GC/PowerPlus.

1 Introduction

The quasi-minimal residual (QMR) algorithm [8], which uses the Lanczos process [7] with look-ahead, a technique developed to prevent the process from breaking down in case of numerical instabilities, and in addition imposes a quasi-minimization principle. This combination leads to a quite efficient algorithm for solving linear systems with large and sparse unsymmetric coefficient matrices. These systems arise very frequently in scientific computing, for example from finite difference or finite element approximations to partial differential equations, as intermediate steps in computing the solution of nonlinear problems or as subproblems in linear and nonlinear programming.

On massively parallel computers, the basic time-consuming computational kernels are usually: inner products, vector updates, matrix-vector multiplications. In many situations, especially when matrix operations are well-structured, these operations are suitable for implementation on vector and share memory parallel computers [6]. But for parallel distributed memory machines, the picture is totally different. Vector updates are perfectly parallelizable for large sparse matrices. For matrix-vector multiplication, how the matrices and vectors are distributed over the processors, quite influence the parallel performance of the method although its communication only occur among nearby processors. For inner product operation, even when the matrix operations can be implemented efficiently by parallel operations, we still can not avoid the global communication, i.e. communication of all processors, required for inner product computations. The bottleneck is usually due to inner products enforcing global communication. The detailed discussions on the communication problem on distributed memory systems can be found in [4, 5]. These global communication costs become relatively more and more important when the number of the parallel processors is increased and thus they have the potential to affect the scalability of the algorithm in a negative way [4, 5].

Recently, we have proposed a new improved two-term recurrences Lanczos process without look-ahead technique as the underlying process of QMR. The algorithm is reorganized without changing the numerical stability so that all inner products and matrix-vector multiplications of a single iteration step are independent and communication time required for inner product can be overlapped efficiently with computation time. Therefore, the cost of global communication on parallel distributed memory computers can be significantly reduced. The resulting IQMR algorithm [13] maintains the favorable properties of the Lanczos process while not increasing computational costs.

In [15], a theoretical simple model of computation and communications phases is used to allow us to give a qualitative analysis of the parallel performance with two-dimensional grid topology. The efficiency, speed-up, and runtime are expressed as functions of the number of processors scaled by the number of processors that gives the minimal runtime for the given problem size. The model not only evaluates effectively the improvements in the performance due to the communication reductions by overlapping, but also provides the useful insight in the scalability of IQMR method. But this simple model is quite limited because it does not take into account any communication times during the matrix-vector multiplication. The communication overhead due to this kind of operation is simply ignored. Correspondingly, different data distribution are discussed fully in [14] to exploit the best mapping strategy using the isoefficiency analysis where we put the communication overhead into the isoefficiency concept to model scalability aspects. These techniques are very suitable for well-structured problems usually after some special re-ordering strategies.

In this paper, we mainly consider a more general form of large and sparse matrices, in which the non-zeros are distributed randomly and do not form a regular pattern that can be utilized effectively. Such matrices occur in some applications, notably in linear programming problem. We will describe an efficient implementation of this method which is particularly well suited to this kind of problems with irregular sparsity pattern. The corresponding communication cost is independent of the sparsity pattern. We will also specifically discuss how to overlap communication and computation which thereby reduces the overall run time of the matrix-vector multiplication and balance the computational load. The performance is demonstrated by numerical experimental results carried out on massively parallel distributed memory computer Parsytec GC/PowerPlus.

The paper is organized as follows. In section 2, we will describe fully the improved quasi-minimal residual (IQMR) method. Then the main two computational kernel, namely matrix-vector multiplication and inner product including their communication primitives, basic algorithms, and performance models, are described in detailed in section 3. In section 4, numerical experiments carried out on massively parallel distributed memory computer Parsytec GC/PowerPlus to compare with the theoretical expectations are presented. Finally some concluding remarks and comments are made.

2   Improved Quasi-Minimal Residual Method

We have described the improved Lanczos process by using a couple two-term recurrence procedure in [13, 15] which is used as a major component to a Krylov subspace method for solving a system of linear equations as follows:

$$Ax = b, \quad \text{where} \quad A \in \Re^{n \times n} \quad x, b \in \Re^n. \tag{1}$$

In each step, it produces approximation x_n to the exact solution of the form

$$x_n = x_0 + \mathcal{K}_n(r_0, A), \quad n = 1, 2, \ldots \tag{2}$$

Here x_0 is any initial guess for the solution of linear systems, $r_0 = b - Ax_0$ is the initial residual, and $\mathcal{K}_n(r_0, A) = span\{r_0, Ar_0, \ldots, A^{n-1}r_0\}$, is the n-th Krylov subspace with respect to r_0 and A.

Given any initial guess x_0, the n-th Improved QMR iterate is of the form

$$x_n = x_0 + V_n z_n, \tag{3}$$

where V_n is generated by the improved unsymmetric Lanczos process, and z_n is determined by a quasi-minimal residual property which will be described later.

For improved Lanczos process, the n-th iteration step generates

$$V_{n+1} = [v_1, v_2, \cdots, v_{n+1}] \quad \text{and} \quad P_n = [p_1, p_2, \cdots, p_n],$$

that are connected by

$$P_n = V_n U_n^{-1}, \qquad AP_n = V_{n+1}L_n, \tag{4}$$

where L_n and U_n are the leading principal $(n+1) \times n$ and $n \times n$ submatrices of the bidiagonal matrices L and U generated by the Lanczos algorithm. Note that L_n has full rank since we assume no breakdown occurs. The setting of $y_n = U_n z_n$ can be used to reformulate the improved QMR iterate in term of y_n instead of z_n giving

$$x_n = x_0 + P_n y_n. \tag{5}$$

The corresponding residual vector in term of y_n is obtained by the above scenario, namely

$$r_n = b - A x_n = r_0 - V_{n+1} L_n y_n = V_{n+1}(\gamma_1 e_1^{(n+1)} - L_n y_n), \tag{6}$$

where the improved Lanczos process starts with $v_1 = \frac{1}{\|r_0\|_2} r_0$ and $e_1^{(n+1)} = (1, 0, \ldots, 0)^T$. Rather than minimizing $\|r_n\|_2$ that generally is an expensive task, the quasi-minimal residual property reduces costs by only minimizing the factor of the residual given in parentheses, i.e., y_n is the solution of least squares problem

$$\|\gamma_1 e_1^{(n+1)} - L_n y_n\|_2 = \min_y \|\gamma_1 e_1^{(n+1)} - L_n y\|_2.$$

Since L_n has full rank, the solution y_n is uniquely determined by the coupled iteration derived in [3] by avoiding the standard approach of computing a QR factorization of L_n by means of Givens rotations

$$y_n = \begin{pmatrix} y_{n-1} \\ 0 \end{pmatrix} + g_n, \qquad g_n = \theta_n \begin{pmatrix} g_{n-1} \\ 0 \end{pmatrix} + \kappa_n e_n^{(n)}, \tag{7}$$

where the scalars θ_n and κ_n are supplied from the following expressions

$$\theta_n = \frac{\tau_n^2(1 - \lambda_n)}{\lambda_n \tau_n^2 + \gamma_{n+1}^2}, \qquad \kappa_n = \frac{-\gamma_n \tau_n \kappa_{n-1}}{\lambda_n \tau_n^2 + \gamma_{n+1}^2}, \qquad \lambda_n = \frac{\lambda_{n-1} \tau_{n-1}^2}{\lambda_{n-1} \tau_{n-1}^2 + \gamma_n^2}, \tag{8}$$

with $n \geq 2$, $\lambda_1 = 1$ and $\kappa_0 = 1$.

Inserting the first coupled recurrence relation yields

$$x_n = x_0 + P_{n-1} y_{n-1} + P_n g_n = x_{n-1} + d_n, \tag{9}$$

where $d_n = P_n g_n$ is introduced. Using the second coupled recurrence relation, the vector d_n is updated by

$$d_n = \theta_n P_{n-1} g_{n-1} + \kappa_n P_n e_n^{(n)} = \theta_n d_{n-1} + \kappa_n p_n, \tag{10}$$

where the vector is generated by the improved Lanczos process. Defining $f_n = A d_n$, the residual vector is obtained by

$$r_n = r_{n-1} - f_n, \tag{11}$$

and the corresponding vectors f_n is given

$$f_n = \theta_n f_{n-1} + \kappa_n A p_n = \theta_n f_{n-1} + \kappa_n u_n. \tag{12}$$

The result is an improved QMR method based on coupled two-term recurrences with scaling of both sequence of Lanczos vectors where numerical stability can be maintained so that all inner products and matrix-vector multiplications of a single iteration step are independent and communication time required for inner product can be overlapped efficiently with computation time. The framework of this improved QMR method using Lanczos algorithm based on two-term recurrences as underlying process is depicted in Algorithm 1.

Algorithm 1 Improved Quasi-Minimal Residual Method

1: $\tilde{v}_1 = \tilde{w}_1 = r_0 = b - Ax_0, \quad \lambda_1 = 1, \quad \kappa_0 = 1,$

2: $p_0 = q_0 = u_0 = d_0 = f_0 = 0, \quad \gamma_1 = (\tilde{v}_1, \tilde{v}_1), \quad \xi_1 = (\tilde{w}_1, \tilde{w}_1), s_1 = A^T \tilde{w}_1,$

3: $\rho_1 = (\tilde{w}_1, \tilde{v}_1), \quad \varepsilon_1 = (s_1, \tilde{v}_1), \quad \mu_1 = 0, \quad \tau_1 = \frac{\varepsilon_1}{\rho_1};$

4: **for** n= 1,2,... **do**

5: $\quad q_n = \frac{1}{\xi_n} s_n - \frac{\gamma_n \mu_n}{\xi_n} q_{n-1};$

6: $\quad \tilde{w}_{n+1} = q_n - \frac{\tau_n}{\xi_n} \tilde{w}_n;$

7: $\quad s_{n+1} = A^T \tilde{w}_{n+1};$

8: $\quad t_n = A\tilde{v}_n;$

9: $\quad u_n = \frac{1}{\gamma_n} t_n - \mu_n u_{n-1};$

10: $\quad \tilde{v}_{n+1} = u_n - \frac{\tau_n}{\gamma_n} \tilde{v}_n;$

11: $\quad p_n = \frac{1}{\gamma_n} \tilde{v}_n - \mu_n p_{n-1};$

12: $\quad$ **if** $(r_{n-1}, r_{n-1}) < $ tol **then**

13: $\quad\quad$ quit

14: $\quad$ **else**

15: $\quad\quad \gamma_{n+1} = (\tilde{v}_{n+1}, \tilde{v}_{n+1});$

16: $\quad\quad \xi_{n+1} = (\tilde{w}_{n+1}, \tilde{w}_{n+1});$

17: $\quad\quad \rho_{n+1} = (\tilde{w}_{n+1}, \tilde{v}_{n+1});$

18: $\quad\quad \varepsilon_{n+1} = (s_{n+1}, \tilde{v}_{n+1});$

19: $\quad\quad \mu_{n+1} = \frac{\gamma_n \xi_n \rho_{n+1}}{\gamma_{n+1} \tau_n \rho_n};$

20: $\quad\quad \tau_{n+1} = \frac{\varepsilon_{n+1}}{\rho_{n+1}} - \gamma_{n+1} \mu_{n+1};$

21: $\quad\quad \theta_n = \frac{\tau_n^2 (1 - \lambda_n)}{\lambda_n \tau_n^2 + \gamma_{n+1}^2};$

22: $\quad\quad \kappa_n = \frac{-\gamma_n \tau_n \kappa_{n-1}}{\lambda_n \tau_n^2 + \gamma_{n+1}^2};$

23: $\quad\quad \lambda_n = \frac{\lambda_{n-1} \tau_{n-1}^2}{\lambda_{n-1} \tau_{n-1}^2 + \gamma_n^2};$

24: $\quad\quad d_n = \theta_n d_{n-1} + \kappa_n p_n;$

25: $\quad\quad f_n = \theta_n f_{n-1} + \kappa_n u_n;$

26: $\quad\quad x_n = x_{n-1} + d_n;$

27: $\quad\quad r_n = r_{n-1} - f_n;$

28: $\quad$ **end if**

29: **end for**

Under the assumptions, the improved QMR method using Lanczos algorithm as underlying process can be efficiently parallelized as follows:

- The inner products of a single iteration step (12), (15), (16), (17) and (18) are independent.

- The matrix-vector multiplications of a single iteration step (7) and (8) are independent.

- The communications required for the inner products (12), (15), (16), (17) and (18) can be overlapped with the update for p_n in (11).

Therefore, the cost of communication time on parallel distributed memory computers can be significantly reduced.

3 Data Distribution and Communication Schemes

The IQMR algorithm basically contains three distinct computational tasks per iteration

- Two simultaneous matrix-vector products, $A\tilde{v}_n$ and $A^T\tilde{w}_{n+1}$.

- Five simultaneous inner products, $(\tilde{v}_{n+1}, \tilde{v}_{n+1}), (\tilde{w}_{n+1}, \tilde{w}_{n+1}), (\tilde{w}_{n+1}, \tilde{v}_{n+1})$, $(s_{n+1}, \tilde{v}_{n+1})$ and (r_{n-1}, r_{n-1}).

- Nine vector updates, $q_n, \tilde{w}_{n+1}, u_n, \tilde{v}_{n+1}, p_n, d_n, f_n, x_n$ and r_n.

Here we assume that there is an integer P such that N is evenly divisible by P and that P is an even power of 2 where P denotes the number of processors in the parallel machine and N means the matrix row or column. It is straightforward to relax these restrictions.

3.1 Vector updates operations

Vector is conceptually divided into $\sqrt{P}$ pieces, each with $N/\sqrt{P}$. On massively parallel distributed memory environment, vector updates are computed locally, i.e., the communication and computation times of these operations satisfy

$$T^{v-u}_{comm} = 0, \quad T^{v-u}_{comp} = 2t_{fl}N^2/P. \tag{13}$$

where N^2/P is the local number of unknown of a processor, t_{fl} is the average time for a double precision floating point operation and n_z is the average number of non-zero elements per row of the matrix.

3.2 Inner product operations

Suppose two vectors are distributed over $\sqrt{P}$ processors. The inner product operation of these two vectors are computed by two phases. First, the computation of an inner product is executed by all processors simultaneously without communication or locally. Then, these $\sqrt{P}$ partial results have to be added by a global sum operation involving communication. Conceptually, this communication pattern used in the second phase of the evaluation of an inner product is known as *reduction*.

Due to its global structure, a reduction involves communication times that strongly depend on the way how processors are interconnected. Here we restrict our communication schemes as same as those proposed in [1], namely two dimensional mesh of $\sqrt{P}$ in both dimensions. Furthermore, we assume that a processor can send data on only one of its ports at a time. Receiving data is only permitted on one port at a time as well. However, a processor can send and receive data simultaneously either on the same port or on separate ports.

Also we assume that *cut-through routing* scheme is used in our analysis. In this routing scheme, the time to transfer completely a message of length l between two processors that are d connections away is given by $t_s + lt_w + (d-1)t_h$, where t_s is the start-up time that required to initiate a message transfer at the sending processor, the per-word time t_w is the time each word takes to traverse two directly-connected processors, and the per-hop time t_h is the time needed by the header of a message to travel between two directly-connected processors.

Under the above framework, it is clear that [11] the communication time for reducing messages each containing l words is

$$T_{comm}^{reduction} = (t_s + lt_w)\log_2 P. \tag{14}$$

On most practical parallel computers with *cut-through routing*, the value of t_h is fairly small. We will omit the t_h term from any expression in which the coefficient of t_h is dominated by the coefficients of t_s or t_w. To simplify the further discussion, we use

$$T_{comm}^{inn} \approx (t_s + t_w)\log_2 P, \quad T_{comp}^{inn} = 2t_{fl}N^2/P. \tag{15}$$

as approximations of the communication and computation times of an inner product hereafter.

3.3 Matrix-vector multiplications

Our efficient matrix-vector multiplication algorithm is mainly based on the approach described [10] for the form of $y = Ax$. The operation $y = A^T x$ would be easily generalized according to the approach presented here. Let A be decomposed into square block size $(N/\sqrt{P}) \times (N/\sqrt{P})$, each of which is assigned to one of the P processors. We use the Greek α and *beta* running from 0 to $\sqrt{P} - 1$ to index the row and column ordering of the blocks. The (α, β) block is denoted by $A_{\alpha\beta}$ and owned by the processor $P_{\alpha\beta}$. The input vector x and product vector y are also conceptually divided into $\sqrt{P}$ pieces, each of length $N/\sqrt{P}$, indexed by β and α respectively. With this block decomposition, processor $P_{\alpha,\beta}$ must know x_β in order to compute its contribution to y_α. This contribution is a vector of length $N/\sqrt{P}$ which we denote by $z_{\alpha\beta}$.

There are two important communication primitives used in the matrix-vector multiplication algorithm. The first is an efficient method for summing elements of a vector across multiple processors. In general, if P processors each own a copy of a vector of length N, this primitive will sum the vector copies so that each processor finishes with N/P elements. Each element is the sum of the corresponding elements across all P processors. This operation is called conceptually as *recursive halving* or *fold*.

In the matrix-vector multiplication algorithm, we use this communication operation to sum contributions to y that are computed by the processors that share a given row α. In this case, the *fold* operation occurs between a group of $\sqrt{P}$ processors which is described for processor $P_{\alpha\beta}$ fully in Algorithm 2. This

---

Algorithm 2 The fold operation

Processor $P_{\alpha,\beta}$ knows $z_{\alpha\beta}$.

$z = z_{\alpha\beta}$.

for $i = 0, 1, \ldots, \log_2(\sqrt{P}) - 1$ **do**

Divide the vector z into two equal sized subvectors, z_1 and z_2.

$P = P_{\alpha\beta}$ with i-th bit of β flipped.

if bit of β is 1 **then**

Send z_1 to processor P.

Receive w from processor P.

$z = z_2 + w$.

else

Send z_2 to processor P.

Receive w from processor P.

$z = z_1 + w$.

end if

end for

$y^{\alpha\beta} = z$.

Processor $P_{\alpha\beta}$ now own $y^{\alpha\beta}$.

---

fold operation requires no redundant floating point operations, and the total number of values sent and received by each processor is $N/\sqrt{P} - N/P$.

The second communication primitive is essentially the inverse of the *fold* operation. If each of P processor knows N/P values, the final result of the operation is that all P processors know all N values. This is called conceptually as *recursive doubling* or *expand*. We use this primitive to exchange information among the $\sqrt{P}$ processors sharing each column of the matrix. This operation for processor $P_{\alpha\beta}$ is outlined in Algorithm 3 for communication between processors with the same column index β. Each processor in the column begins with a subvector $y^{\beta\alpha}$ of length N/P. This *expand* operation requires only a logarithmic number of stages and the total number of values sent and received by each processor is $N/\sqrt{P} - N/P$.

The optimal implementation of the *fold* and *expand* operations depends on the machine topology and various hardware considerations. With the communication primitives described above, the matrix-vector multiplication for processor $P_{\alpha\beta}$ is outlined in Algorithm 4.

As we discussed before, the *expand* and *fold* primitives used in this algorithm are most efficient if row and columns of the matrix are mapped to subsets of processors that allows for fast communication, for example, on two-dimensional

---

Algorithm 3 The expand operation

Processor $P_{\alpha,\beta}$ knows $y^{\beta\alpha}$.
$z = y^{\beta\alpha}$.
for $i = \log_2(\sqrt{P}) - 1, \ldots, 0$ **do**
   $P = P_{\alpha\beta}$ with i–th bit of α flipped.
   Send z to processor P.
   bf Receive w from processor P.
   if bit of β is 1 **then**
      w and z are concatenated in the correct order to z.
   else
      z and w are concatenated in the correct order to z.
   end if
end for
$y_\alpha = z$.
Processor $P_{\alpha\beta}$ now own y^α.

---

mesh it is rows and columns. Unfortunately, such a mapping can make the transpose operation inefficient since it requires communication between processors that are architecturally distant. Since we have used the *cut-through routing* for inner product operation, this leads that a single message can be transmitted between non-adjacent processors in nearly the same time as if it were sent between adjacent processors. If multiple messages with *cut-through routing* are simultaneously trying to use the same wire, all but one of them must be delayed. Hence this routing still suffer from serious message congestion. Fortunately, as point out in [10], in this algorithm the message being transposed are shorter than those in the *fold* and *expand* operations by a factor of $\sqrt{P}$. So even if congestion delays the transpose message by a factor of $\sqrt{P}$, the overall communication scaling will not be affected.

The matrix-vector multiplication algorithm has the following drawback that once a processor has sent a message in *fold* or *expand* operations, it is idle until the message from its neighbor arrives. This can be alleviated in the *fold* operation in step (3) of the algorithm by interleaving communication with computation from step (2). Rather than computing all the elements of $z_{\alpha\beta}$ before beginning the *fold* operation, we should compute just those that are about to be sent. Then no matter which value will be sent in the next pass through the fold loop get computed between the send and receive operations in the current pass. In the final pass, the values that the processor will keep are computed. In this way, the total run time is reduced on each pass through the fold loop by the minimum of the message transmission time and the time to compute the next set of elements of $z_{\alpha\beta}$.

Our above discussion is concentrated on the communication requirements of the algorithm, but an efficient algorithm must also ensure that the computational load is well balanced across the processors. For this algorithm, this requires balancing the computations within each local matrix-vector multiplication. If

---

Algorithm 4 The matrix-vector multiplication algorithm

1: **Processor** $P_{\alpha,\beta}$ owns $A_{\alpha\beta}$ and x_{β}.
2: **Compute** $z_{\alpha\beta} = A_{\alpha\beta}x_{\beta}$.
3: **Fold** $z_{\alpha\beta}$ within rows to form $y^{\alpha\beta}$.
4: **Transpose** the $y_{\alpha\beta}$, i.e.
5: a) **Send** $y^{\alpha\beta}$ to $P_{\beta\alpha}$.
6: b) **Receive** $y^{\beta\alpha}$ from $P_{\beta\alpha}$.
7: **Expand** $y^{\beta\alpha}$ within columns to form y_{β}.

---

the region of the matrix owned by a processor has mm nozeros, the number of floating point operations required for the local matrix-vector multiplication is $2mm - N/\sqrt{P}$. There will be balanced if $mm \approx Nn_z/P$ for each processor, where Nn_z is the total number of nonzeros elements in the matrix. For the random or irregular matrices in which $n_z \gg 1$, the load is likely to be balanced. Otherwise, randomly permuting the row and columns gives good balance with high probability suggested by Ogielski and Aiello [12]. A random permutation has the additional advantage that zero values encountered when summing vectors in the *fold* operation are likely distributed randomly among the processors.

The matrix-vector multiplication can be implemented to require the minimal $(2n_z - 1)N$ floating point operation during the calculation of the local multiplication and the rest during the *fold* summations. Here we do not make any assumptions about the data structure used on each processor to compute it local matrix-vector product. This allows for a local optimization for a particular machine. If we assume the computation load is balanced, the time to execute these floating point operations should be nearly $(2n_z - 1)t_{fl}N/P$. Correspondingly the communication time in matrix-vector multiplication is expressed with the computation time as follows:

$$T_{comm}^{mat-vec} = 2(log_2(P) + 1)t_s + \frac{N(\sqrt{P} - 1)}{P}t_w, \quad T_{comp}^{mat-vec} = (2n_z - 1)t_{fl}N/P.$$

4 Numerical Experiments

In this section, the parallel performance of this efficient implementation described above of the improve QMR method is compared with that without so many efficient enhancements, with modified version of QMR proposed by Bucker et al. [2, 3] and the original QMR based on coupled two-term recurrences on a massively massively distributed memory computer Parsytec GC/PowerPlus at National Supercomputing Center, Linköping University, Sweden.

Here we mainly consider the partial differential equation taken from [2]

$$Lu = f, \quad \text{on} \quad \Omega = (0,1) \times (0,1),$$

with Dirichlet boundary condition $u = 0$ where

$$Lu = -\Delta u - 20(x\frac{\partial u}{\partial x} + y\frac{\partial u}{\partial y}),$$

and the right-hand side f is chosen so that the solution is

$$u(x, y) = \frac{1}{2} \sin(4\pi x) \sin(6\pi y).$$

Basically, we discretize the above differential equation using second order centered differences on a 400×400 with mesh size $1/441$, leading to a system of linear equations with unsymmetric coefficient matrix of order 193600 with 966240 nonzero entries. Diagonal preconditioning is used. For our numerical tests, we choose $x_0 = 0$ as initial guess and $tol = 10^{-5}$ as stopping parameter.

The convergence of the improved QMR proposed with this efficient implementation is almost same as the modified QMR suggested by Bucker et al. [2] and original QMR version based on two-term recurrences [9] where $\|r_n\|_2$ is computed recursively. A similar numerical behavior of those variants is recognized. There is hardly any difference to the true residual norm $\|b - Ax_n\|_2$ in those versions. The parallel performance are given in the left-hand side of Fig. 1 where A is the theoretical speedup, E is the speedup of the improved QMR method with efficient implementation, B is the improved QMR method, C is the speedup of the modified QMR method suggested by Bucker et al. [2] and D is the speedup of the original QMR method with two-term recurrences. These results are based on timing measurements of a fixed number of iteration. The speedup in computed by taking the ratio of the parallel execution time and the execution time of a serial computer. From the results, we can see clearly that the modified QMR suggested by Bucker et al. [2] is faster than the original one. Meanwhile the new approach can achieve much better parallel performance with high scalability than the modified one. The new efficient implementation is even better than the improved QMR method. The saving of execution time grows with increasing number of processor in comparison with those two approaches. More precisely, the quantity depicted as a percentage is $1 - T_A(P)/T_B(P)$, where $T_A(P)$ and $T_B(P)$ are the execution times on P processors. In the right-hand of Fig. 1, A shows the percentage of saving of time for improved QMR approach with original one, B shows the percentage of saving of time for modified QMR suggested by Bucker et al. [2] with original one, and C shows the percentage of saving of time for the new proposed efficient implementation described in this paper.

5 Conclusions

For the solutions of linear systems of equations with unsymmetric coefficient matrices, we has proposed an improved version of the quasi-minimal residual (IQMR) method by using the Lanczos process as a major component combining elements of numerical stability and parallel algorithm design. For Lanczos process, stability is obtained by a couple two-term procedure that generates Lanczos vectors scaled to unit length. The algorithm is derived such that all inner products and matrix-vector multiplications of a single iteration step are independent and communication time required for inner product can be overlapped efficiently

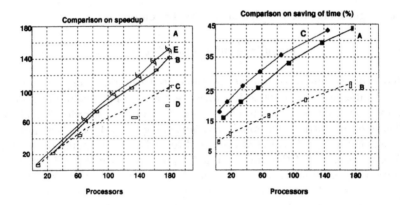

Fig. 1. Experimental results of the parallel implementations

with computation time. Therefore, the cost of global communication on parallel distributed memory computers can be significantly reduced. In this paper, we describe an efficient implementation of this method which is particularly well suited to problems with irregular sparsity pattern with several performance improvement techniques such as overlapping computation and communication, balancing the computational load. The corresponding communication cost is independent of the sparsity pattern. The performance is demonstrated by numerical experimental results carried out on massively parallel distributed memory computer Parsytec GC/PowerPlus.

References

1. H. M. Bucker. Isoefficiciency analysis of parallel QMR-like iterative methods and its implications on parallel algorithm design. Technical Report KFA-ZAM-IB-9604, Central Institute for Applied Mathematics, Research Centre Julich, Germany, January 1996.

2. H. M. Bucker and M. Sauren. A parallel version of the quasi-minimal residual method based on coupled two-term recurrences. In *Proceedings of Workshop on Applied Parallel Computing in Industrial Problems and Optimization (Para96)*. Technical University of Denmark, Lyngby, Denmark, Springer-Verlag, August 1996.

3. H. M. Bucker and M. Sauren. A parallel version of the unsymmetric Lanczos algorithm and its application to QMR. Technical Report KFA-ZAM-IB-9606, Central Institute for Applied Mathematics, Research Centre Julich, Germany, March 1996.

4. E. de Sturler. A parallel variant of the GMRES(m). In *Proceedings of the 13th IMACS World Congress on Computational and Applied Mathematics.* IMACS, Criterion Press, 1991.

5. E. de Sturler and H. A. van der Vorst. Reducing the effect of the global communication in GMRES(m) and CG on parallel distributed memory computers. Technical

Report 832, Mathematical Institute, University of Utrecht, Utrecht, The Netheland, 1994.

6. J. J. Dongarra, I. S. Duff, D. C. Sorensen, and H. A. van der Vorst. *Solving Linear Systems on Vector and Shared Memory Computers*. SIAM, Philadelphia, PA, 1991.

7. R. W. Freund, M. H. Gutknecht, and N. M. Nachtigal. An implementation of the look-ahead Lanczos algorithm for non-Hermitian matrices. *SIAM Journal on Scientific and Statistical Computing*, 14:137–158, 1993.

8. R. W. Freund and N. M. Nachtigal. QMR: a quasi-minimal residual method for non-Hermitian linear systems. *Numerische Mathematik*, 60:315–339, 1991.

9. R. W. Freund and N. M. Nachtigal. An implementation of the QMR method based on coupled two-term recurrences. *SIAM Journal on Scientific and Statistical Computing*, 15(2):313–337, 1994.

10. B. Hendrickson, R. Leland, and S. Plimpton. An efficient parallel algorithm for matrix-vector multiplication. *International Journal of High Speed Computing*, 7(1):73–88, 1995.

11. V. Kumar, A. Grama, A. Gupta, and G. Karypis. *Introduction to Parallel Computing: Design and Analysis of Algorithms*. Benjamin/Cummings, Redwood City, 1994.

12. A. T. Ogielski and W. Aiello. Sparse matrix computations on parallel processor arrays. *SIAM Journal on Scientific and Statistical Computing*, 14:519–530, 1993.

13. T. Yang and H. X. Lin. The improved quasi-minimal residual method on massively distributed memory computers. In Proceedings of *The International Conference on High Performance Computing and Networking (HPCN-97)*, April 1997.

14. T. Yang and H. X. Lin. Isoefficiency analysis of the improved quasi-minimal residual method on massively distributed memory computers. Submitted to *The 2rd International Conference on Parallel Processing and Applied Mathmetics (PPAM-97)*, September 1997.

15. T. Yang and H. X. Lin. Performance evaluation of the improved quasi-minimal residual method on massively distributed memory computers. Submitted to *5th International Conference on Applications of High-Performance Computers in Engineering (AHPCE-97)*, July 1997.

Programming with Shared Data Abstractions

Simon Dobson[1] and Don Goodeve[2]

[1] Well-Founded Systems Unit, CLRC Rutherford Appleton Laboratory, UK
[2] School of Computer Studies, University of Leeds, UK

Abstract. We present a programming system based around a set of highly concurrent, highly distributed data structures. These *shared abstract data types* offer a uniform interface onto a set of possible implementations, each optimised for particular patterns of use and more relaxed coherence conditions in the data. They allow applications written using a shared data model to approach the performance of message passing implementations through a process of pattern analysis and coherence relaxation. We describe the programming system with reference to a sample application solving the travelling salesman problem. Starting from a naïve direct implementation we show significant speed-ups on a network of workstations.

1 Introduction

Parallel computing offers the prospect of high application performance. This can indeed be realised, but at the price of huge development and maintenance costs arising from the complexity of managing inter-processor interaction and application code in the same framework. Ideally an application programmer should only deal with the details of the application problem, reducing complexity to an acceptable and manageable level. Some method for abstracting away from the complexities of inter-processor interaction is therefore required. This has been accomplished most successfully for algorithms having a regular form where static problem decomposition is possible. Producing efficient codes for irregular problems is a significant challenge, as effort must be devoted to the effective management of an evolving computation.

In this paper we present an approach based on *shared abstract data types* or SADTs – an abstract type that can be accessed by many distributed processes concurrently. An SADT insulates an application from many different implementations of the abstraction, which may efficiently support different patterns of use and coherence models. The selection of the representation which best suits the application's exact requirements offers the highest possible performance without compromising the type abstraction boundaries.

Section two introduces an motivating problem. Section three describes the SADT model and its realisation in a prototype programming environment. Section four shows how a simple sequential solver for our illustrative problem may be progressively refined into a high-performance scalable parallel application. Section five compares and contrasts SADTs with related work from the literature. Section six offers some observations and future directions.

2 The Travelling Salesman Problem

The Travelling Salesman Problem (TSP) is a well-known combinatorial optimisation problem. Given a set of cities and the distances between them an algorithm must find the shortest "tour" that visits all cities exactly once. In principle the entire search tree must be explored if an optimal solution is to be found, making the problem NP-complete. Problems of this form are often encountered in the context of organisational scheduling[18]. There are a number of possible approaches to the solution of the problem which broadly fall under the headings of *stochastic* algorithms, such as simulated annealing[17], and *deterministic* algorithms.

Deterministic algorithms have the advantage of guaranteeing that the optimal solution is found, but usually can only deal with modest problem sizes due to time and space constraints. Stochastic algorithms trade improved performance for sub-optimal solutions.

One of the best-known deterministic algorithms is due to Little[14]. This algorithm employs two heuristics to guide the search for a solution by refining a representation of the solution space as a search tree. A node in the tree represents a set of possible tours, with the root representing the set of all possible tours. A step of the algorithm involves selecting a node in the tree and choosing a path segment not contained in this set. The path segment is selected by a heuristic designed so that the set of tours including the selected segment is likely to contain the optimal tour, and conversely that the set of tours that excludes it is unlikely to contain the optimal tour. A step therefore expands a node and replaces it by its two children – one including and one excluding the selected path segment. A second heuristic directs the algorithm to work on the most promising node next. A lower-bound on the length of any tour represented by a node can be computed, and the node selected to be explored next is that with the lowest lower-bound. Once a complete tour has been found, its length can be used to bound future exploration.

Little's algorithm is of particular interest for two reasons:

1. it is a *fine-grained algorithm* – a step of the algorithm involves only a relatively short computation; and
2. it is *irregular* – the evolution of the data cannot be predicted statically, and hence a significant amount of dynamic data management is involved.

These make it a challenging case study for any novel programming environment.

3   Shared Abstract Data Types

We have been exploring the use of *pattern-optimised representations* and *weakened coherence models* to allow a distributed data structure to optimise both the way its data is stored and the degree of coherence between operations on different nodes. These optimisations may occur without changing the type signature of the structure, allowing progressive optimisation. We refer to types with these properties as *shared abstract data types*, or SADTs[6]. The programming model for SADTs is the common one of a collection of shared data objects accessed concurrently by a number of processes independent of relative location or internal representation.

We have implemented a prototype programming environment to test these ideas, using the Modula-3[16] language with network objects[3] to implement bag, accumulator, priority queue and graph SADTs. This has proved to be an excellent and flexible test-bed. (We have also demonstrated more restricted environment, intended for supercomputers, using C.) There were four main design goals:

1. *integration* of SADTs into the underlying programming model and type system;
2. *extensibility* of the environment to derive new SADTs from the existing set, and allow additional base SADTs to be generated where necessary;
3. *incremental development* of applications using progressive optimisation; and
4. *high performance* applications when fully optimised.

We regard these goals as being of equal importance in any serious programming system.

3.1 Architecture

The heart of the SADT architecture is a network of *representatives*, one per processor per SADT. Each representative stores some part of the SADTs contents, and co-operates with the other representatives to maintain the shared data abstraction. Interactions with and between representatives occur through *events*, representing the atomic units of activity supported over the SADT. Events may be created, mapped to actions on values, buffered for later application, and so forth: the way in which representatives handle events defines the behaviour of the SADT, and allows subtly different behaviours to be generated from the same set of events.

The programmer's view of this structure is a local instance of some *front-end* type – usually a container type such as a bag of integers. The front-end provides a set of methods through which the shared data may be accessed: internally each is translated into one or more events which are passed to the representative for the SADT on the local node. It is important to realise that the events of the SADT do not necessarily correspond exactly to the user-level operations.

3.2 Events and Event Processing

Each SADT defines its own set of events, capturing the basic behaviour of the type being represented. For example, a bag SADT might define events to add an element and remove an element. These events and their eventual mapping to actions characterise the SADT.

There is a small core set of events which are applicable across all SADTs. One may regard these core events as providing the essential services to support efficient shared type abstraction. Many are "higher order", in the sense of taking other events as parameters. The set includes:

LocalSync – causes a representative to synchronise its view of the SADT with the other representatives. Typically this involves reading new values into local caches, or flushing pending events.

Span(ev) – performs event ev on all representative concurrently.

Broadcast(ev) – like Span(), but orders broadcasts at each representative (so all representatives will observe the same sequence of broadcasts).

Bulk(ev1, ev2, ...) – submits a set of events in a single operation.

These events may be used to build higher-level operations. For example, an SADT may implement global synchronisation by issuing a Span(LocalSync) event[1], and fully replicated representations will often broadcast update events to all representatives. These events have efficient implementations based on concurrent spanning trees.

The critical importance of events is that they define "first-class" operation requests: representatives may manipulate events in complex ways before resolving them to an action. Event processing is at the core of the SADT model. Using the same set of events with the same mechanisms for rendering events into actions – the same type, in fact – an SADT may deploy several representations which differ in the exact way the events are buffered and combined. The different representations may efficiently support different coherence constraints and patterns in the use of the type.

There are a number of common processes which may be used to manipulate events: they may be *resolved* into an action on some storage, *forwarded* to another representative, *deferred* in a buffer for later handling, *changed* or *combined* or *discarded*. Taken together with event descriptions these processes form the building blocks of a simple specification framework for SADT representations which is sufficiently formal to allow automatic skeleton code generation and a certain amount of rigorous analysis to be applied to each representation[8]. This formal basis is essential for reasoning in the presence of weak coherence.

---

[1] Or sometimes Broadcast(LocalSync), but for many SADTs the extra ordering overhead is unnecessary.

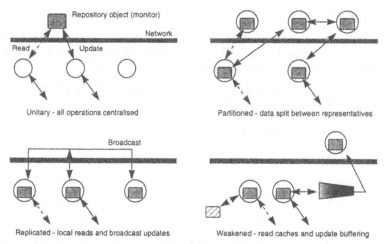

Figure 1. Different representation schemes for an SADT

3.3 Programmability

The creation of an SADT is obviously more complex than ordinary object creation, since a distributed family of objects must be created. We encapsulate this complexity by using an *abstract factory* class for each SADT[11]. The factory combines a language for expressing constraints – such as the degree of weakness acceptable to an algorithm or a pattern of use in the operations – with a decision procedure to determine which of the available representations of the SADT is appropriate to best satisfy these constraints. This decouples the algorithmic constraints from the strategy used to select a representation, allowing both constraint language and representation populations to evolve over time.

Sub-typing is an essential component of large-scale object-oriented applications. Sub-types may derived from SADT front-end objects in the usual way, with the SADT environment creating the appropriate factory object. Sub-types have access to all the representations available to the parent type, and possibly to extra, more specific representations. New representations for SADTs are generated by providing event-handling code and extending the factory decision procedure to allow the new representation to be selected when appropriate constraints are met. Event handlers have access to the core events, which are implemented by the common super-type of representatives. Completely new SADTs may be defined similarly, starting from a description of the type-specific events.

The prototype environment makes extensive use of Modula-3's generic interfaces and powerful configuration control language, together with simple generation and analysis tools.

3.4 Example: the Accumulator SADT

The accumulator[12] is a type which combines a single value with an update function. The value is updated by combining the submitted new value with the current value through the update function. An example is a minimising accumulator, where the update function returns the smaller of the current and submitted values.

The simplest accumulator stores the value in a single representative, performs all updates as they are received, and always returns the current value. Variations[9] allow the semantics to be weakened with various combinations of replication, update deferral and value combination. These variations lead to subtly different accumulators, which may be more efficient than the simple case whilst still providing acceptable guarantees for applications.

4 TSP with SADTs

To illustrate the issues outlined above we present a case study of implementing Little's algorithm on a network of workstations. The starting point for the application code is a sequential version of the program, which uses two key abstract data types:

1. a *priority queue* used to hold the population of nodes. Nodes are dequeued from to be expanded, and the resulting child nodes enqueued. The priority property keeps the "best" node as the head element; and
2. an *accumulator* used to hold the length of the shortest known tour.

The user algorithm functions exactly as outlined in section 2.

4.1 Direct Implementation

A parallel application can quickly be produced by converting the priority queue and accumulator into SADTs and using a number of identical worker processes co-ordinated through these two abstractions.

The simplest implementation of the required SADTs is to implement an object in one address space in the system, and use the mechanisms of the runtime system to provide access to this object across the system. By protecting the object with a monitor, sequential consistency for the SADT is maintained. We term this the *unitary* or *shared object* style.

A problem immediately arises with this simple implementation. Termination in the sequential case is signalled by the priority queue being empty and unable to service a *dequeue* request; in the parallel case this condition may arise in the normal course of the algorithm and correctly detecting termination involves predicting when the structure *is and will remain* empty. This can be achieved by the use of an additional accumulator counting the number of paths expanded.

The application was run on a set of SGI O2 workstations connected by a 10MBit/sec Ethernet running TCP/IP – a platform providing high compute performance but relatively poor communication. Achieving reasonable performance for a fine-grained algorithm such as the TSP solver therefore poses a significant challenge. Indeed, the simple implementation actually exhibits a considerable slow-down as processors are added! Profiling the application exposes the reasons for this. Figure 2 shows a percentage breakdown of the execution time of the application on a 4-processor system by the different SADT operations that are invoked. Shortest is the accumulator for the current best-known solution, Counter is the accumulator used for termination detection, and Store is the priority queue. The final columns on the histogram represents the available CPU time remaining to devote to the application problem once the SADT operations have been taken into account. Clearly, the operations on the SADTs dominate the runtime of the application, leaving a very small proportion of the time for running the actual application-specific code.

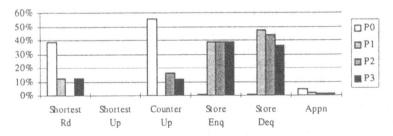

Figure 2. Profiling of naïve implementation.

4.2 Optimised Implementation

To improve the performance of the parallel application, the excessive costs of the SADT operations must be addressed. We may make a number of observations about the TSP algorithm. Firstly, there are significant patterns in the use of the shared data. We may reduce overheads by matching an appropriate implementation of the SADTs to their observed patterns of use. Secondly, several of the "normal" coherence constraints are not actually required for the algorithm to function correctly, and relaxing them again gives scope for performance improvements. Relaxation trades-off a decrease in the efficiency of the heuristics against reduced overheads in maintaining the shared data abstraction. We implement these optimisations by making additional, optimised representations of the SADTs, and the appropriate decision procedure, available to the run-time environment.

4.2.1 Replicated and Weakened Accumulators

From inspection it may be seen that the accumulator holding the shortest known tour is read each time an element is read from the priority queue, to see whether the path should be pruned; it is only updated when a shorter complete tour has been found, which typically happens only a few times during the application. The frequency of reads is much higher than that of updates, so by replicating the accumulator (figure 1) a read may be rapidly serviced locally.

By broadcasting updates the accumulator remain strongly coherent. Weaker approaches apply updates locally and periodically synchronise the accumulator to propagate the best global value, or apply several updates *en bloc* in a single network transaction.

4.2.2 Partitioned Priority Queue

Similar arguments apply to the priority queue: it may be distributed across the system, and relax the requirement that a dequeue operation returns the overall best element to allow the element returned to be locally best but not necessarily globally optimal. Distribution also makes the memory of all nodes available for representing the queue, rather than bounding its size to the memory of a single node. A background task at each representative periodically shuffling elements randomly between representatives. This shuffling improves data balance and, by shuffling high-priority elements, ensures that the algorithm as a whole focuses on the elements with the highest global priority. An important optimisation of shuffling is not to shuffle the *highest* priority elements, but instead to shuffle from one or two elements down. This ensures that a high-priority element is expanded, and is not constantly shuffled.

4.2.3 Synchronising Priority Queue

Termination relies on the counter accumulator, which must be strongly coherent in order to solve the consensus problem[13]. In the single-bus system the unitary implementation appears to be the best that can be done: once the other SADTs have been optimised, however, operations on the counter dominate the performance of the code.

The original reason for introducing the counter accumulator was to solve the distributed termination detection in the face of transient emptiness in the queue. A key observation is that termination can be correctly inferred when the queue *is and will remain* empty, so no new elements can appear "spontaneously" due to other threads. Without additional information about the algorithm, the only condition of the SADT in which we know this to be the case is when *all* processes sharing the queue are attempting to dequeue from it: this is the termination state, as all work has been consumed and no more can arise. On detecting this condition we can release all the processes simultaneously – by returning NIL, a logical extension of the semantics of the sequential dequeue operation to the concurrent case.

The interface to the priority queue is extended so that each thread sharing the SADT is known. A mechanism is then implemented to detect when a representative is empty and all local threads are performing dequeue operations. Combining these local results using a consensus protocol completes the system and removes the need for the counter accumulator.

This insight – the usefulness of *scoped deadlocks* – appears to capture a very general property of termination in parallel algorithms which we are investigating further.

4.2.4 Performance

When more optimal SADTs are used, performance improves dramatically (figure 3). It is clear that the optimisation of the SADTs results in a dramatic increase in the proportion of the total CPU time available for running the (unchanged) application code.

Figure 4 shows the speedup of the application relative to 1 processor when running on 2, 4, 8 and 12 O2 workstations solving a 21-city problem. Real speedups are now achieved. The curve *Optimised SADTs 1* shows the performance obtained when using a random data balancing strategy within the priority queue SADT; *Optimised SADTs 2* shows the performance obtained using simple load information to reduce the chances of starvation..

The *Indy SADTs 2* curve shows the speedups achieved when running the same code on a network of SGI Indy workstations. The Indy has roughly half the raw compute speed of the O2 with the same communication bandwidth, giving a better computation-to-communication balance. (This curve is normalised to a speedup of 2 on 2 processors, as the single processor code exceeds physical memory limits and incurs overheads that are avoided using several processors.)

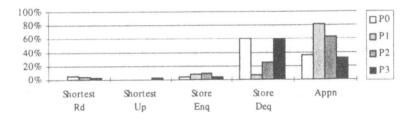

Figure 3. Profiling of the fully optimised implementation.

It is obvious that the improvements in performance trail off with increasing system size. The reason for this is that the priority queue's balancing strategy becomes less effective as size increases, due to both the inadequacy of algorithm and to the limited performance of the underlying network transport. The shuffling on which the current strategy is based results in a network load of about 80KBytes/sec per workstation: a small number of workstations can thus saturate the Ethernet. This situation could be improved by a more informed shuffling strategy, a higher bandwidth transport, or a transport with better scaling properties such as a point-to-point ATM network – all issues demanding further work.

The code implementing the algorithm is almost identical in sequential and parallel systems – the only extra code being algorithmic pattern and coherence constraints. The resulting parallel application achieve high performance for a fine-grained algorithm on an coarse-grained architecture. There is obviously significant scope for improving performance further: however the key principle of optimising performance through constrained representation selection for otherwise fully abstract SADTs has been clearly demonstrated.

5 Related Work

A great many systems have been proposed to allow objects to be manipulated concurrently and location-independently – notable are the CORBA standard, the Orca language[1] and Modula-3 network objects. These systems provide the mechanisms required to build shared data structures, but do not directly address the design of efficient systems. The "single abstraction *via* a community of objects" approach builds on this substrate: it is used by SADTs as well as a number of other systems[2][4][7][10], and is perhaps best exemplified by Concurrent Aggregates[5].

A slightly different approach is taken by languages such as HPF and HPC++, providing a library of effectively sequential types together with a small number of *bulk operations*. The operations hide their efficient parallel implementations from the programmer (both syntactically and semantically), preventing exploitation of more relaxed coherence models.

Berkeley's Multipol[21] provides a library of types for irregular computations. A Multipol object is *multi-ported* in the sense that it exports both a global view and a view of the locally-held data. The SADT approach hides the local view, substituting bulk operations and weakening instead.

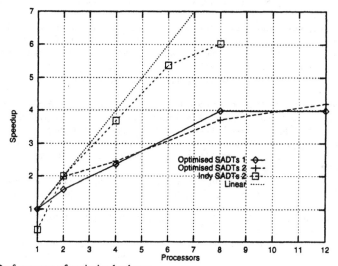

Figure 4. Performance of optimised scheme

The canonical example of weakening is the Bulk Synchronous Parallelism model, in which processes' view of shared state may drift between global synchonisation points. Recent work[19] has defined a set of BSP collection types, which provide bulk operations and high-level interfaces but are still constrained to use the globally-agreed coherence model.

6 Conclusion

We have presented a programming environment based around a collection of types providing a sound basis for parallel programming, whose characteristics are well-matched to the network-of-workstations architecture. Common programming techniques are incorporated in a semantically consistent way, allowing them to be deployed without unduly complicating the application code. Preliminary results indicate that the system can achieve relatively high performance by progressive refinement encapsulated within the run-time system. We are currently performing more extensive evaluations based on the "Cowichan problems" – and it is encouraging to note that the problems identified in these applications are well addressed by the SADT approach[20].

A major future direction is to improve our ability to detect patterns of use of an SADT for which a known efficient representation exists. This would free the programmer of the need to provide hints to the run-time system, at the cost of typing the library closely to the analysis system. We are also seeking means of integrating more closely into the type system the semantic constraints currently supplied to SADT factory objects.

7 References

[1] Henri Bal, Andrew S. Tanenbaum and M. Frans Kaashoek, "*Orca: a language for distributed programming*," ACM SIGPLAN Notices **25**(5) (May 1990) pp.17-24.

[2] J.K. Bennett, J.B. Carter and W. Zwaenpoel, "*Munin: distributed shared memory based on type-specific memory coherence*," ACM SIGPLAN Notices **25**(3) (March 1990) pp.168-176.

[3] Andrew Birrell, Greg Nelson, Susan Owicki and Edward Wobber, "*Network objects*," Research report 115, Digital SRC (1994).

[4] John Chandy, Steven Parkes and Prithviraj Banerjee, "*Distributed object oriented data structures and algorithms for VLSI CAD*,", pp.147-158 in Parallel algorithms for irregularly structured problems, **LNCS 1117**, ed. A. Ferreira, J. Rolim, Y. Saad and T. Yang, Springer Verlag (1996). Proceedings of IRREGULAR'96.

[5] A.A. Chien and W.J. Dally, "*Concurrent Aggregates*," ACM SIGPLAN Notices **25**(3) (March 1990) pp.187-196.

[6] John Davy, Peter Dew, Don Goodeve and Jonathan Nash, "*Concurrent sharing through abstract data types: a case study*,", pp.91-104 in Abstract machine models for parallel and distributed computing, ed. M. Kara, J. Davy, D. Goodeve and J. Nash, IOS Press (1996).

[7] Simon Dobson and Andy Wellings, "*A system for building scalable parallel applications*,", pp.218-230 in Programming environments for parallel computing, ed. N. Topham, R. Ibbett and T. Bemmerl, North Holland Elsevier (1992).

[8] Simon Dobson and Chris Wadsworth, "*Towards a theory of shared data in distributed systems*,", pp.170-182 in Software engineering for parallel and distributed systems, ed. I. Jelly, I. Gorton and P. Croll, Chapman and Hall (1996).

[9] Simon Dobson, "*Characterising the accumulator SADT*," TallShiP/R/22, CLRC Rutherford Appleton Laboratory (1996).

[10] Stephen Fink, Scott Baden and Scott Kohn, "*Flexible communication mechanisms for dynamic structured applications*,", pp.203-213 in Parallel algorithms for irregularly structured problems, **LNCS 1117**, ed. A. Ferreira, J. Rolim, Y. Saad and T. Yang, Springer Verlag (1996). Proceedings of IRREGULAR'96.

[11] Erich Gamma, Richard Helm, Ralph Johnson and John Vlissides, "*Design patterns: elements of reusable object-oriented software*," Addison-Wesley (1995).

[12] Don Goodeve, John Davy and Chris Wadsworth, "*Shared accumulators*," in Proceedings of the World Transputer Congress (1995).

[13] Maurice Herlihy, "*Wait-free synchronisation*," ACM Transactions on Programming Languages and Systems **11**(1) (1991) pp.124-149.

[14] John D.C. Little, Katta G. Murty, Dura W. Sweeney and Caroline Karel, "*An algorithm for the travelling salesman problem*," Operations Research **11** (1993) pp.972-989.

[15] W.F. McColl, "*Bulk synchronous parallel computing*," pp.41-63 in Abstract machine models for highly parallel computers, ed. J.R. Davy and P.M. Dew, Oxford Science Publishers (1993).

[16] Greg Nelson, "*Systems programming with Modula-3*," Prentice Hall (1993).

[17] William H. Press, Brian P. Flannery, Saul A. Teukolsky and William T. Vetterling, "*Numerical recipes in C*," Cambridge University Press (1989).

[18] V.J. Rayward-Smith, S.A. Rush and G.P. McKeown, *"Efficiency considerations in the implementation of parallel branch-and-bound,"* Annals of Operations Research **43** (1993).

[19] K. Ronald Sujithan and Jonathan Hill, *"Collection types for database programming in the BSP model,"* in Proceedings of IEEE Euromicro (1997).

[20] Greg Wilson and Henri Bal, *"Using the Cowichan problems to assess the usability of Orca,"* IEEE Parallel and Distributed Technology **4**(3) (1996) pp.36-44.

[21] Katherine Yelick, Soumen Chakrabarti, Etienne Deprit, Jeff Jones, Arvind Krishnamurthy and Chih-Po Wen, *"Parallel data structured for symbolic computation,"* in Workshop on Parallel Symbolic Languages and Systems (1995).

EXPLORER: Supporting Run-Time Parallelization of DO-ACROSS Loops on General Networks of Workstations   *

Yung-Lin Liu and Chung-Ta King

Department of Computer Science National Tsing Hua University
Hsinchu, Taiwan 30043, R.O.C.

Abstract. Performing runtime parallelization on general networks of workstations (NOWs) without special hardware or system software supports is very difficult, especially for DOACROSS loops. With the high communication overhead on NOWs, there is hardly any performance gain for runtime parallelization, due to the latter's large amount of messages for dependence detection, data accesses, and computation scheduling. In this paper, we introduce the EXPLORER system for runtime parallelization of DOACROSS and DOALL loops on general NOWs. EXPLORER hides the communication overhead on NOWs through multithreading — a facility supported in almost all workstations. A preliminary version of EXPLORER was implemented on a NOW consisting of eight DEC Alpha workstations connected through an Ethernet. The Pthread package was used to support multithreading. Experiments on synthetic loops showed speedups of up to 6.5 in DOACROSS loops and 7 in DOALL Loops.

Keywords: DOACROSS loops, inspector/executor, multithreading, networks of workstations, run-time parallelization

1 Introduction

A recent trend in high performance computing is to use networks of workstations (NOW) [1] as a parallel computation engine. In such systems off-the-shelf workstations are connected through standard or proprietary networks and cooperate to solve one large problem. Each workstation typically runs the full operating system and communicate with each other by passing messages using standard protocols such as TCP/IP.Unfortunately, the communication overhead in such system is often two or three order of magnitude higher than the computation. This thus makes NOWs ideal only for coarse-grain regular computations.

For computations whose data dependence relations can only be determined at runtime, parallelization is often done at runtime, which induces a large amount of messages for dependence detection, data accesses, and computation scheduling. With the high communication overhead, there is hardly any performance

---

* This work was supported in part by the National Science Council under Grant NSC86-2213-E-007-043 and NCHC-86-08-024.

gain by running such programs on NOWs without special hardware or software supports. It follows that only few attempts have been made so far for runtime parallelization on distributed memory systems, not to mention on NOWs. Furthermore, only restricted cases have been solved [3, 4, 5].

This paper reports a new attempt to runtime parallelization, with an emphasis on NOWs without special hardware or system software supports. Our approach, called *EXPLORER*, can handle loops with and without loop-carried dependences. To address the problem of high communication overhead in NOWs, we employ the latency hiding strategy and realize it through thread programming, which is commonly supported in current workstation systems. The basic idea is to use one thread to exploit data dependencies and parallelism in the loop and to schedule ready iterations for execution by computation threads. Transferring dependent data and checking for data availability can be performed by additional threads. EXPLORER has been implemented a NOW consisting of eight DEC Alpha workstations connected through an Ethernet. Multithreading in EXPLORER is supported through the Pthread package and communication between nodes is accomplished through TCP/IP. Since there is no system specific components in EXPLORER, its portability is very high. Preliminary performance results showed that the proposed scheme is feasible and an almost linear speedup can be obtained.

2 Preliminaries

2.1 Loop Model

Fig. 1 shows the model of the loop that will be used to demonstrate our runtime parallelization method. Assume that the content of the array b is known only at runtime, and that the loop contains no loop-carried output dependence. The latter implies that in the loop in Fig. 1(b), the array elements of f are distinct.

$$
\begin{array}{ll}
\text{for } (i = 1; i < N; i++) \; \{ & \qquad \text{for } (i = 1; i < N; i++) \; \{ \\
\quad a[i] = \ldots ; & \qquad\quad a[f[i]] = \ldots ; \\
\quad \ldots = \ldots a[b[i]] \ldots ; & \qquad\quad \ldots = \ldots a[b[i]] \ldots ; \\
\} & \qquad \} \\
\qquad\quad \text{(a)} & \qquad\qquad\quad \text{(b)}
\end{array}
$$

Fig. 1. The loop model

Now, consider the loop in Fig. 1(a). If $b[i] = i$, then there is no dependence entering or out of this iteration and it can be executed independently of other iterations. When $b[i] < i$, there is a true dependence between the iteration that writes to $b[i]$ and the current iteration. The former must be executed before the latter. Finally, when $b[i] > i$, there is an anti-dependence between the two involved iterations. Such a dependence can be removed by renaming.

In Fig. 1(b), each element of array a in the loop is written at most once, based on the assumption of no output dependence. It is then possible to find an inverse array of f to indicate which element is to be written in which iteration. Given the inverse array, the loop in Fig. 1(b) can be handled in the same way as that in Fig. 1(a). Thus, we will mainly consider the latter in the following discussions.

2.2 Previous Approaches

The CHAOS/PARTI system [3] provides a set of library routines to identify data access patterns at runtime and to optimize data distribution and communication based on the inspector-executor model. The routines hide low-level messaging passing details of a distributed memory system from the intended irregular applications. Unfortunately, the system only handles loops without loop-carried dependences.

The RAPID system [4, 5] is another runtime parallelization system. It parallelizes loops by first extracting a task dependence graph from the data access patterns of the given loop and then executing the loop according to an efficient scheduling of the graph on the target machine. The system was implemented on the distributed-memory multicomputer, Meiko CS-2, and utilized its Remote Memory Access (RMA) mechanism for low-cost communications. It is not clear whether the proposed method can be applied efficiently to general networks of workstations which have no special hardware supports.

3 EXPLORER for Runtime Parallelization

The organization of our EXPLORER runtime system is shown in Fig. 2. It consists of five primitive threads: *explorer, worker, scheduler, input*, and *output*. Their functions are described below. For simplicity of presentation, we assume that every node has the full copy of both arrays a and b.

The explorer thread in each node runs through every iteration in the loop. It first checks to see if the iteration is "owned" by the local node. The *owner* of an iteration is responsible for performing all the computations of that iteration, including updating the array element $a[i]$. In EXPLORER the iterations are assigned to the nodes in a cyclic fashion. We will discuss why this is the case later.

Next, the explorer checks whether $a[b[i]]$ is available. An array element $a[b[i]]$ is *available* to an iteration i if either of the following two conditions are satisfied:

- owner($b[i]$)=owner(i) and, if $b[i] < i$, the write to $a[b[i]]$ has completed.
- owner($b[i]$)$\neq$owner(i) and $a[b[i]]$ has been received from owner($b[i]$).

The explorer thread then determines whether the iteration can be executed or it needs to wait for data availability. The complete pseudo code of the explorer thread is shown in Fig. 3.

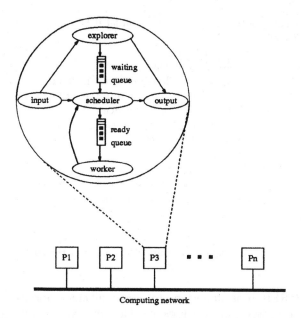

Fig. 2. The organization of the EXPLORER run-time system

The scheduler thread is responsible for maintaining the waiting queue. It receives signals from the worker and the input thread. The signals indicate that a particular array element is available. The scheduler thread then removes the iterations whose right-hand side, i.e., $a[b[i]]$, matches the available element. If the owner of the removed iteration is the local node, then the iteration is put into the ready queue for the worker thread to execute. Otherwise, the local node must be the owner of iteration $b[i]$. The scheduler then informs the output thread to send $a[b[i]]$ to the owner of the removed iteration.

The worker thread fetches an iteration from the ready queue and performs the computations contained in that iteration until the ready queue is empty. When it finishes executing an iteration, it informs immediately the scheduler of the availability of the updated array element.

The output thread is responsible for sending data to other nodes. It is invoked by the worker and the scheduler threads. The output thread may aggregate the messages if they are sent to the same node.

The input thread is responsible for receiving data from other nodes. When data arrives, it informs the explorer and the scheduler threads that the received data is available. Note that we separate message sending and receiving operations into two threads in order to avoid possible deadlock situations.

As mentioned above, the EXPLORER system allocates iterations to the nodes in a cyclic fashion. We could have allocated the iterations in blocks. However, if all iterations in node P_x depend on the last iteration in P_y, then all iterations in P_x may have to wait for the last iteration in Py. Thus, a cyclic allocation is more appropriate.

```
1    duplicate another copy of array a and let it be tmp_a;
2    for (i = 0; i < N ; i + +)
3        if (owner(i) == myid)
4            if (i > b[i]) /* flow dependence */
5                if (a[b[i]] is available) execute the iteration;
6                else put the iteration into to the waiting queue;
7            else if (i < b[i]) /* anti-dependence */
8                    execute the iteration using tmp_a[b[i]];
9            else execute the iteration;
10       else /* not owner */
11           if (owner(b[i]) == myid)
12               if (b[i] is available) inform the output thread to send a[b[i]] to owner(i);
13               else put the iteration into the waiting queue;
```

Fig. 3. The pseudo code of the explorer thread

In EXPLORER it is important to assign different priorities to different threads, if such an assignment is possible. The input and output threads transmit data between nodes. If they are given a higher priority, then data may be moved between nodes more swiftly and processor idleness becomes less likely. On the other hand, the scheduler and worker threads handle iterations upon which other iterations depend. Thus they may be given the middle priority, while the explorer thread has the lowest priority.

4   Performance Evaluation

4.1   The Experimental Environment

We has implemented EXPLORER on a network of workstations consisting of eight DEC Alpha3000 workstations connected by a 10 Mbps Ethernet. Each workstation ran DEC OSF1 operating system. Pthread was used to support multithreading. Communications among workstations in EXPLORER were implemented using UNIX sockets through TCP/IP. which supported the Pthread package. The communications are designed to be implemented using UNIX sockets on top of TCP/IP protocols. In the experiments we used a synthetic loop, which is shown in Fig. 4 and modified from that in [2]. The iteration count N is set to half of the size of array a. The parameter r is the number of references to array a per iteration, and W is the workload per iteration. The base value of W was taken to be one fourth of the message startup time in our system. The array $INDEX$ controls the locations of elements $a[b[i]]$ in array a. Its contents are prepared according to a parameter Dep_{per}, which denotes the probability that a reference will cause a dependence. If a reference $a[b[i]]$ will cause a dependence, then $a[b[i]]$ is chosen randomly from $a[0]...a[N-1]$. Otherwise, it is chosen from

$a[N]...a[2N-1]$. It follows that a DOALL loop is one with $Dep_{per} = 0\%$, a "mostly-parallel" loop has $Dep_{per} = 10\%$, and a "mostly-serial" loop is one with $Dep_{per} = 90\%$.

```
for (i = 0; i < N; i + +) {
    for (j = 0; j < r; j + +) {
        tmp = INDEX[i][j];
        A[i]+ = A[tmp];
    }
    for (k = 0; k < W; k + +)
        /* workload of the iteration */
}
```

Fig. 4. The benchmark loop for performance evaluation

4.2 Performance Results

In this section, we examine the speedups obtained using the EXPLORER method. Intuitively, a better speedup can be obtained if the workload of the loop (W) is large and the number of references (r) is small. In this case, the loop has a low communication to computation ratio. Fig. 5 shows the resultant speedups of executing the synthetic loop.

- The degree of dependence Dep_{per} : Fig. 5(a)–(c) show the speedup of executing the synthetic loop when the loop is DOALL i.e., $Dep_{per} = 0$. Since EXPLORER does not generate any data communication for DOALL loops, we can obtain very good speedups. From Fig. 5 we can also see that the performance improves when Dep_{per} is small. This is because a small Dep_{per} implies looser dependences among the iterations and fewer remote data accesses.
- The number of nodes P: From Fig. 5 we can see that the speedup of the synthetic loop obtained using EXPLORER is close to linear in most cases. Of course when the number of nodes increases, the reduction rate in computation is faster than that of communication. In other words, the communication becomes relatively more important and dominating. Thus, increasing the node count degrades the speedup.
- The workload W: From Fig. 5 we can see that a larger workload, i.e., a coarser computation leads to a better speedup, because the system has enough computation to hide the communication. As a result, the slopes of the curves in the figures are larger when the workload is heavier.
- The number of references r: When there are more references per iteration, i.e., r is larger, the communication cost increases and the speedup drops.

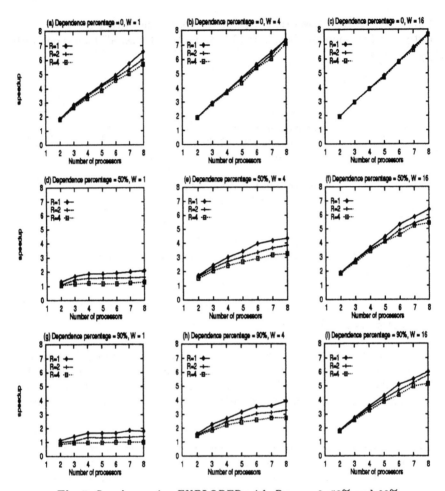

Fig. 5. Speedups using EXPLORER with $Dep_{per}=0$, 50% and 90%

In Fig. 5(g)–(i) we can also find that, for a given number of nodes, a larger workload causes a wider gap between different values of r. The reason is that when Dep_{per} and/or r are large, then the amount of communications is very high . Hence, we need more computations to hide the communication.

5 Conclusions

The EXPLORER system explores parallelism in loops whose parallelism cannot be analyzed at compile time. It targets at general networks of workstations without requiring special hardware or system software supports. It overcomes the high communication overhead in such environments using the latency hiding strategy through thread programming. Preliminary experiments showed that the proposed scheme is feasible and can achieve good performance. As far as we know, there is currently no other work on runtime parallelization of DOACROSS loops on general NOWs.

EXPLORER works best for loops whose runtime schedules cannot be reused. This is in contrast to most previous works, which assumed that the schedule is reusable in order to amortize the cost of the inspector. Of course, EXPLORER can also be applied to the latter case. However, we might need to find out the dependence graph after processing the loop for the first time. In the future, we will study how this can be done. We also plan to study how to aggregate messages in order to reduce the communication overhead and in the mean time not to delay computations too much.

References

1. T. E. Anderson, D. E. Culler, D. A. Patterson, "A case for networks of workstations : NOW," *IEEE Micro 94*.
2. D. K. Chen, J. Torrelas, and P. C. Yew, "An efficient algorithm for the run-time parallelization of DOACROSS loops," *Proc. of Supercomputing 1994*, pp.518-527, November 1994.
3. R. Das, M. Uysal, J. Saltz, and Y. S. Hwang, "Communication optimizations for irregular scientific computations on distributed memory architectures," *Jorunal of Parallel and Distributed Computing*, 22(3), pp.462-479, September 1994.
4. C. Fu and T. Yang, "Run-time compilation for parallel sparse matrix computations," Proc. of the 10th ACM International Conference on Supercomputing, pp.237-244, May 1996.
5. C. Fu and T. Yang, "Efficient run-time support for irregular task computations with Mixed granularities," *Proc. of 10th International Parallel Processing Symposium - IPPS'96*, pp.823-830, April 1996.

Engineering Diffusive Load Balancing Algorithms Using Experiments

Ralf Diekmann[1] and S. Muthukrishnan[2] and Madhu V. Nayakkankuppam[3]

[1] University of Paderborn, Germany. diek@uni-paderborn.de
[2] Information Sciences Center, Bell Labs, USA. muthu@research.bell-labs.com
[3] Courant Institute, New York University, USA. madhu@cs.nyu.edu

Abstract. We study a distributed load balancing problem on arbitrary graphs. *First Order* (FO) and *Second Order* (SO) schemes are popular local diffusive schedules for this problem. To use them, several parameters have to be chosen carefully. Determining the "optimal" parameters analytically is difficult, and on a practical level, despite the widespread use of these schemes, little is known on how relevant parameters must be set. We employ systematic experiments to engineer the choice of relevant parameters in first and second order schemes. We present a centralized polynomial time algorithm for choosing the "optimal" FO scheme based on semidefinite programming. Based on the empirical evidence from our implementation of this algorithm, we pose conjectures on the closed-form solution of optimal FO schemes for various graphs. We also present a heuristic algorithm to locally estimate relevant parameters in the FO and SO schemes; our estimates are fairly accurate compared to those based on expensive global communication. Finally, we show that the FO and SO schemes that use approximate values rather than the optimal parameters, can be improved using a new iterative scheme that we introduce here; this scheme is of independent interest. The software we have developed for our implementations is available freely, and can serve as a platform for experimental research in this area. Our methods are being included in PadFEM, the Paderborn Finite Element Library [1].

"I recommend this method to you for imitation. ... The indirect [iterative] method can be done while half asleep, or while thinking about other things."
– Carl Friedrich Gauss (1823), on iterative methods for solving systems of linear equations (page 325 in [2]).

1 Introduction

We consider the following abstract distributed *dynamic load balancing problem*. We are given an arbitrary, undirected, connected graph $G = (V, E)$ in which node $v_i \in V$ has *weight* w_i of unit-sized tokens. Our goal is to determine a schedule to move tokens in each step across edges so that, finally, the weight on each node is (approximately) equal. In each step we are allowed to move any number of tokens from a node to each of its neighbors in G. This problem models load balancing in synchronous distributed processor networks and parallel machines when we associate each node with a processor, each edge with a communication link of unbounded capacity between two processors, and the tokens with identical, independent tasks [3]. This also models load balancing in parallel finite element methods where a geometric space, discretized using a mesh, is partitioned into sub–regions and the computation proceeds on mesh points in each sub–region independently; here we associate each node with a mesh region, each edge with the geometric adjacency between two regions, and the tokens with the mesh points in each region. It has been extensively studied – see the book [4] for an excellent selection of applications, case studies and references.

Popular schedules for this problem are *diffusive*. Here the weight moved along each edge is dependent on the weight–gradient across the edge. Diffusive schedules are simple and they degrade gracefully in the presence of asynchrony and faults, where edges/nodes fail/reappear arbitrarily [5]. Furthermore, they are *local*, that is, each step of these schedules can be implemented in a local manner by each node consulting only its neighbors (thereby avoiding expensive global communication in distributed applications). In this paper, we focus on diffusive schedules only.

There are two general diffusive load balancing schemes known, namely, the *First Order* (FO) Scheme (FOS), and the *Second Order* (SO) Scheme (SOS). Most existing load balancers use FOSs, while SOSs were proposed recently as a direction to speed up the convergence of FOS's [6]. Both these schemes are parameterized by several variables which must be chosen carefully. Optimal choice of these parameters can often be phrased as constrained eigenvalue minimization problems. However, current mathematical theories of eigenvalue minimization lack the sophistication needed to resolve these problems, especially those arising from SO schemes; therefore, no analytical solutions have been found for these problems. On a practical level, despite the widespread use of diffusive schemes, little work has been done on systematically choosing relevant parameters, or on understanding the effect of any particular choice of parameters on their performance. Typically, the parameters are simply set to intuitive values.

Our main goal is to employ systematic experiments to engineer the choice of relevant parameters in first and second order schemes, and in the process, gain mathematical insight into diffusive load balancing. In the rest of this section, we describe FO and SO schemes and highlight outstanding questions on choice of optimal parameters for them. Following that, we describe our contributions in detail.

1.1 FO and SO Schemes

The *potential* after t iterations, denoted ϕ_t, is $\sum_{i \in V} (w_i - \overline{w})^2$ where the average weight $\overline{w} = (\sum_i w_i)/|V|$; the *initial* potential is denoted ϕ_0. Note that $\phi_t = 0$ if and only if $w_i = \overline{w}$ for all i and $\phi_t > 0$ otherwise. Thus the larger ϕ_t is, larger the imbalance. Also, throughout this paper, we will treat the weights as nonnegative real numbers; therefore, they can be specified to any arbitrary precision. For the load balancing scenario, they ought to be nonnegative integers – see [6] for details on how this condition can be incorporated without significant loss of performance.

FOS*s*. Each iteration of FOS's can be modeled as $\mathbf{w}^{t+1} = M\mathbf{w}^t$ where $M = [\alpha_{ij}]$ is a *diffusion matrix*, that is, M is nonnegative, real, symmetric and doubly stochastic (each of the rows and columns sum to 1). Formally, $0 < \alpha_{ij} < 1$, $\alpha_{ij} = \alpha_{ji}$, and $\alpha_{ii} = 1 - \sum_{j \neq i} \alpha_{ij} \geq 0$. This behavior can be simulated by local movement of weight on the edges of G. If l_{ij}^{t+1} is the amount of load moved between i and j, $(i, j) \in E$ in $(t+1)$st iteration, then $l_{ij}^{t+1} = \alpha_{ij}(w_i^t - w_j^t)$, where w_i^t is the weight on node i after t iterations.

Lemma 1. *Consider* $\mathbf{w}^{t+1} = M\mathbf{w}^t$, *where* M *is real, symmetric and doubly stochastic. Suppose* $\gamma(M)$ *is the second largest eigenvalue of* M *in absolute value. Let* ϕ_0 *and* ϕ_t *be the potentials at the beginning and after* t *iterations respectively. We have,* $\phi_t \leq \gamma(M)^{2t}\phi_0$. *Furthermore, let* μ *be any eigenvalue of* M. *Then there exist initial weight distributions* $\mathbf{w}^0$ *such that* $\phi_t = \mu^{2t}\phi_0$. ∎

When $\gamma(M) < 1$ in the lemma above, we say the schedule *converges*; else, it *diverges*.

SOS*s*. Here we first assume that the weights are arbitrary real numbers; therefore, they may be negative as well. Each iteration in a SOS is modeled as follows: $\mathbf{w}^{k+1} = M\mathbf{w}^k$ if $k = 0$ and $\mathbf{w}^{k+1} = \beta M\mathbf{w}^k + (1 - \beta)\mathbf{w}^{k-1}$ if $k \geq 1$. Here, β is independent of iteration number k and M is a diffusion matrix. Again, this iteration can be simulated by local movement of load between the neighbors in G. Let l_{ij}^k be the weight moved from i to j in iteration k. We have,

$$l_{ij}^{k+1} = \begin{cases} \alpha_{ij}\left(w_i^k - w_j^k\right) & \text{if } k = 0 \\ (\beta - 1)l_{ij}^k + \beta\alpha_{ij}\left(w_i^k - w_j^k\right) & \text{if } k \geq 1 \end{cases}$$

Lemma 2. *If* β *lies in the intervals* $(-\infty, 0)$ *or* $(2, +\infty)$, *the SO iteration does not converge [7]. If* $0 < \beta < 1$ *independent of* M, *then the SOS converges slower than the FOS in the worst case [6]. There exists a value of* β, $1 \leq \beta \leq 2$, *such that the SOS converges faster than the FOS. The* β, *denoted* β_M, *for which the SOS converges the fastest amongst all those based on* M *is* $\frac{2}{1+\sqrt{1-\gamma(M)^2}}$ *[8].* ∎

In [6], the authors showed that there are weight distributions, graphs, M's and β_M's such that $\sum_{j \neq i} l_{ij}^k > w_i^k$ during some iteration k. Thus the weight at node i is not sufficient to meet the demand on the load to be moved on its incident edges as per the SO schedule. In order to ensure that the weights are nonnegative, the authors suggested sending as much as can be sent with the current load at node i, and keeping track of *I Owe You* (IOU_{ij}) on edges (i, j) which will be met as the iteration proceeds (see [6] for details). We remark that both FOS and SOSs are inspired by the mature literature on iterative solution of linear systems (see [9, 7, 10] etc.). In particular, they are related to the stationary iterative methods.

1.2 Outstanding Questions

In order to employ either a FOS or a SOS with a given G, appropriate parameters must be chosen. For the FOS's the parameters are the entries of M; for the SOS's, the additional parameter is β. In this section, we list a sample of outstanding questions regarding the appropriate choice of FOSs and SOSs that have remained open. The following questions are conceptual.

• *Determining Optimal FOS's and SOS's:* For a given G, what is the optimal FOS, that is, what is the M that converges the fastest in the worst case over all input weight distributions? Similarly, for a given G, what is the optimal SOS?

• *Determining Parameter Ranges:* For what ranges of β does the SOS outperform the FOS? For what choices of β's do IOUs vanish rapidly on the edges?

•*Finding Alternative Schemes.* Are there alternative simple iterative schemes that outperform FOSs without introducing IOUs? Are there simple iterative systems that use the knowledge of the input weight distribution to outperform FOSs and SOSs significantly?

The following are technical questions mainly motivated by the fact that local algorithms are preferable in distributed load balancing scenarios.

• *Choosing Efficient Schemes Locally:* What is the most efficient FOS or SOS that can be chosen locally for a given G?

•*Robust Schemes:* Since theoretical results show that β_M is dependent on the global properties of M, local computation of β will be only an approximation to β_M. Given such an approximate choice for β_M, can alternative, local, iterative schemes outperform the SOSs?

• *Engineering Guidelines for Choosing Parameters:* Can useful guidelines be developed for choosing parameters for FOSs and SOSs over a broad range of input problem instances based on suitable benchmarks?

These questions have not been resolved in the literature on matrix iterative methods. In the absence of analytical solutions, our work focuses on developing a strong experimental basis to understand these problems.

1.3 Our Contributions

We have performed extensive computer simulations of diffusive schemes on Matlab[4]. Our test cases included synthetic inputs as well as examples from real life applications. In order to clearly convey our insights from these experiments, we focus on the following few issues in this paper.

• We show that the optimal FOS for any graph can be computed in polynomial time by solving a semidefinite program (SDP). Since even small topologies result in large SDPs, and since algorithms for solving SDPs are memory-intensive and time-consuming, we can analyze only small-sized graphs experimentally. Nevertheless, we are able to make conjectures on optimal M's for hypercubes, linear arrays, star graphs etc. These results are in Section 3.

• We present a simple, local heuristic for estimating M and β for SOSs in linear

---

[4] Matlab is a registered trademark of The MathWorks Inc.

number of rounds on any graph G. Although there exist graphs on which this heuristic proves to be a bad approximation, on almost all the graphs in our extensive test suite, the approximation is remarkably effective. See Section 4.

• We introduce a new stationary iterative method called the *generalized* SOS (GSOS). Each iteration $k + 1$ of GSOS can be modeled as: $\mathbf{w}^{k+1} = M\mathbf{w}^k$ if $k = 0$ and

$$\mathbf{w}^{k+1} = \beta M\mathbf{w}^k + (1 - \beta)N\mathbf{w}^{k-1}$$

if $k \geq 1$. Setting $N = I$ in GSOS yields the standard SOS. We show that when β used in the SOS is only an approximation to β_M, GSOSs can be used with an appropriate "perturbation" of I to outperform such SOSs. See Section 5 for details. Curiously, such a natural generalization of the stationary iterative methods does not seem to have been investigated in the literature. Our investigation is in a preliminary state, and is continuing.

• We have developed a set of library routines in Matlab to implement and study these load balancing schemes, and collate relevant performance statistics. These routines could aid researchers and practitioners in experimental investigations in this area, or could act as building blocks in a more comprehensive package on load balancing; these routines are freely available.[5] All our methods will also be included in PadFEM, the Paderborn Finite Element Library [1].

2 The Test Suite

In this section, we describe the gamut of graphs we included in our test suite.

Graphs. The graphs we considered include several standard topologies, random graphs of various edge densities, and three graphs drawn from application areas. Standard topologies include Hypercubes, two-dimensional grids, linear arrays, cylinders (vertically wrapped grids), tori (vertically and horizontally wrapped grids) and Xgrids (grids with additional diagonal edges). The three graphs from applications were *AF1*, *WH3*, and *CR*: these are quotient graphs of the 2-dimensional finite-element meshes *airfoil1* (AF1), *whitaker3* (WH3), and *crack* (CR) respectively. They arise from applications in computational fluid dynamics (*airfoil1*), earthquake wave propagation (*whitaker3*) and structural mechanics (*crack*). All three are frequently used as test-cases for partitioning and load balancing algorithms [11]. The quotient graphs were obtained using *Unbalanced Recursive Bisection (URB)* in Par^2 [11]. Table 1 summarizes the test suite of graphs and their properties. In the table, λ_1 and λ_{n-1} are the largest and second smallest eigenvalues of the Laplacian matrix $L = D - A$ associated with graph G (here A is the adjacency matrix of G and D is the diagonal matrix of the degrees of the nodes). They are shown in closed form wherever possible.

Input weight distributions. The weight distributions used included: *Spike(k)*: A weight of kn tokens placed on node 1, where n is the number of nodes in the graph; *Scattered(k)*: A weight of $2k$ units on the even numbered nodes (all our test cases have an even number of nodes); *Random(k)*: A random weight drawn

---

[5] See http://www.cs.nyu.edu/phd_students/madhu/loadbal/papers.html.

from a uniform distribution on $[0, 2k]$ on each node. For the test cases AF1, WH3 and CR, the input weight distributions were obtained from the applications.

Experiments. The termination criterion for the load balancing algorithms was $\max_{i,j \in V} |w_i - w_j| < 0.1$. The number of iterations needed for load balancing was obtained by averaging the iteration count for eighteen *Random* distributions and one each of the other two; iterations counts for AF1, WH3 and CR were obtained for the weight distribution from the application scenarios.

3 Optimal Design of First Order Schemes

For a given G, the *optimal* diffusion matrix M^* is the one with minimum μ_2. An ϵ-optimal diffusion matrix M for G satisfies $\mu_2(M) \leq \mu_2(M^*) + O(\epsilon)$, where M^* is the optimal diffusion matrix for G. We can compute an ϵ-optimal diffusion matrix M for any G using semidefinite programming.

Theorem 3.1 *For any graph $G = (V, E)$ with $n = |V|$ and $e = |E|$, an ϵ-optimal diffusion matrix can be determined in time polynomial in n and $|\log \epsilon|$.*

Proof. The problem of determining the optimal M for the given G can be formalized as: $\min \{\mu_2(M) : M = [\alpha_{ij}] \in \mathcal{D}^n, \alpha_{ij} = 0 \; \forall (i,j) \notin E\}$, where $\mathcal{D}^n$ is the set of symmetric, doubly stochastic matrices, *i.e.* each row and each column sums to 1 and all entries are nonnegative. Since $\mu_1(M) = 1$ for any $M \in \mathcal{D}^n$, we can replace the objective function by $\mu_1(M) + \mu_2(M)$.

It is well known that minimizing the sum of the k absolutely largest eigenvalues of a symmetric matrix subject to linear constraints can be formulated [12] as a semidefinite program (SDP). Solving this SDP, an ϵ-optimal M can be obtained in time $O(n^6 \sqrt{e} |\log \epsilon|)$; the details are omitted. ∎

Implementation. We implemented this solution in Matlab, using the state-of-the-art code for SDP from [13]. However, even modest sized graphs lead to large semidefinite programs (the 4×4 mesh in Figure 1 an SDP with 424 constraints in a 138×138 matrix variable) taking up several Megabytes of memory. We limited ourselves to computing diffusion matrices for graphs with at most 50 nodes (say, a 7×7 mesh or torus); these took a few hours and used at less than the 640 MB of memory that was available on our SGI R10000 workstation.

Observation 1. We computed optimal diffusion matrices M for hypercubes, star graphs and linear arrays. Based on our solutions, we make the following conjectures (entries of M not specified below can be inferred using its doubly stochastic property).

Conjectures. For the d-dimensional Hypercube, $\alpha_{ij} = 1/(d+1)$ for $(i,j) \in E$, and $\alpha_{ij} = 0$ for $(i,j) \notin E$ when $i \neq j$. For a linear array on n nodes, $\alpha_{ij} = \frac{1}{2}$ for $(i,j) \in E$. For a star graph on n nodes, $\alpha_{ij} = \frac{1}{n-1}$ for $(i,j) \in E$.

Surprisingly, in all the three cases above, all the α_{ij}'s are identical for $(i,j) \in E$! Of the many such diffusion matrices that have been considered in the literature (for example with $\alpha_{ij} = 1/d$ for hypercubes in [3], $\alpha_{ij} = 1/n$ for star graphs, etc.), our conjecture identifies the correct one. Proving such conjectures appears to be nontrivial. The entry α_{ii} corresponds to the fraction of the load retained by node i. For the d-dimensional hypercube, α_{ii} is $1/(d+1)$, while for the star, it is 0 at the center. So, there appears to be no general principle to determine the fraction retained at the nodes during this diffusive process.

Observation 2. Not all graphs yield an optimal diffusion matrix with identical α_{ij}'s. The optimal M's for grids and cylinders are shown in Figure 1; we have not been able to identify closed-form expressions for the optimal diffusion matrices for these types of graphs.

4 Local Estimation of M and β

In what follows, we use "$\widehat{}$" to denote quantities estimated heuristically. For simplicity, we assume that all the off-diagonal entries in M are equal to α. In this case, $M = I - \alpha L$ (where L is the Laplacian induced by M), and it is well known that $\alpha = \frac{2}{\lambda_{n-1}(L) + \lambda_1(L)}$ yields an M with minimum $\gamma(M)$. (If $\alpha > \frac{1}{deg_i}$ for a node i, then $\alpha_{ii} < 0$. Therefore, $\widehat{\alpha}_{ij}$ is set to $\max\{\widehat{\alpha}, \frac{1}{\max\{deg_i, deg_j\}}\}$ in $\widehat{M}$. We omit this detail here.) For this M, $\gamma(M) = 1 - \frac{2 \cdot \lambda_{n-1}(L)}{\lambda_{n-1}(L) + \lambda_1(L)}$, and the best choice of β is given by

$$\beta = \frac{2}{1 + \sqrt{1 - \gamma^2(M)}} = \frac{2}{1 + 2 \cdot \sqrt{\frac{\lambda_{n-1}(L)\lambda_1(L)}{(\lambda_{n-1}(L) + \lambda_1(L))^2}}}.$$

Since eigenvalues of general graphs are not easy to compute especially in a distributed setting, we explore the possibility of estimating them locally.

Estimating $\lambda_1(L)$: It is known that $\lambda_1(L) \leq 2d$, where d is the maximum degree of the graph. In fact, for regular graphs, $\lambda_1(L) = 2d$, and, usually, for irregular graphs, $\lambda_1(L) \ll 2d$. For the graphs in our test suite (Table 1), we observed that $\widehat{\lambda}_1(L) \sim d + \bar{d}$, where $\bar{d}$ is the average degree of the graph, and we used this as an estimate of $\lambda_1(L)$.

Estimating $\lambda_{n-1}(L)$: Our approach to estimate $\lambda_{n-1}(L)$ is via the isoperimetric constant of the graph. For a finite graph $G = (V, E)$, the *isoperimetric constant* $h(G) = \min\left\{\frac{|\partial \bar{V}|}{|V|}; \bar{V} \subset V, |\bar{V}| \leq \frac{1}{2}|V|\right\}$, where $\partial \bar{V} = \{\{i, j\} \in E; i \in \bar{V}, j \in V \setminus \bar{V}\}$ is the set of boundary edges of $\bar{V}$. It is known [14] that the quantity $h(H)$ can be used to approximate $\lambda_{n-1}(L)$. In particular, we have $\frac{h^2(G)}{2d} \leq \lambda_{n-1}(L) \leq 2h(G)$, where d is the maximum degree of G, and L is the Laplacian of G. However, it is difficult to compute $h(G)$ efficiently. Instead we estimate $h_i(G)$ for each node i by doing a breadth-first-search (BFS) from i until approximately $n/2$ nodes are visited, and use the definition of the isoperimetric constant. Then, we compute $h'(G) = \min_i h_i(G)$. Determining $\widehat{\lambda}_{n-1}$ from $h'(G)$ is a subtle issue from the formula above since the range there is wide. Our experiments on the test suite (cf. Table 1) suggest that $\frac{1}{2}h'(G)$ is a good estimation for λ_{n-1}. In summary, we use $\widehat{\lambda}_1 = d + \bar{d}$ and $\widehat{\lambda}_{n-1} = \frac{1}{2}h'(G)$, and choose $\widehat{\alpha}$ and $\widehat{\beta}$ as described earlier.

Experimental Results: We call the FO and SO schemes in which α and β are computed using the exact spectrum as the *exact FOS* and the *exact SOS* respectively; in contrast, when these quantities are estimated by our heuristic above, the schemes are called the *heuristic FOS* and the *heuristic SOS* respectively. Table 2 compares the heuristic and the exact schemes. For a SOS, $w^t = r_t(M)w^0$ where $r_t(M)$ is a matrix polynomial in M (see [6]). The table shows the *effective γ per iteration*, defined as $\gamma(SO) = (\gamma(r_t(M))^{1/t}$, after $t = 10$ iterations. Our observations are as follows.

The heuristic FOS is remarkably close to the exact FOS; most of the time, the heuristic is within 10% of the exact value. In several cases, the heuristic SOS outperforms the exact SOS! See, for instance, the rows corresponding to XG(8), G(8) and St(64), for which at the end of 10 iterations, the potential for the heuristic SOS will be lower than that of the exact SOS by a factor of approximately 0.48, 0.49, and 0.76 respectively in the worst case. This clearly demonstrates that choosing an optimal SOS is not equivalent to finding the optimal FOS followed by choosing the best β for this M, *i.e.* $\beta = \beta_M$. For random graphs, the performance of the heuristic SOS deteriorates significantly with increase in edge density (see rows R1 through R10). The heuristic choice of $\widehat{\lambda}_{n-1}$ is typically an underestimate, especially if the graphs are regular. The estimation of $\widehat{\lambda}_1$ is nearly exact for the standard topologies such as grids and hypercubes. In this case, even a small under-estimation of λ_{n-1} leads to $\widehat{\alpha}$'s much larger than α (the effect amplifies with decreasing λ_{n-1}). On linear arrays, grids and cylinders, this is compensated, because most of the $\widehat{\alpha}_{ij}$'s are set to $\frac{1}{\max\{deg_i,deg_j\}}$ which is the best α for them. On tori and hypercubes however, this over-estimation of α is too severe, and that explains the table entries for these graphs. In general, our choice seems best suited for a variety of graphs. However, for the special case of regular graphs, other estimates may be more suitable.

Remark. In [6] it was shown experimentally that although IOUs can get large, they vanish rapidly as the iteration progresses. From our experiments, we concluded that the same holds for the heuristic SOSs as well. As an example, consider XG(8) in Figure 2 where the initial weight distribution was *Spike*(100). As shown there, the sum of the IOUs disappears after only a few iterations. Thus the heuristic SOS is a suitable load balancing scheme in practice.

5 Generalized Second Order Schemes

The GSOS is an iterative method in which each iteration can be modeled as $w^1 = Mw^0$; $w^{k+1} = \beta Mw^k + (1 - \beta)Nw^{k-1}$, $(k \geq 1)$, where M and N are both diffusion matrices for the same graph G. Observe that for the case $N = I$, this reduces to the standard SOS.

Lemma 3. *Let l_{ij}^k be the amount of load shifted over edge (i, j) during iteration k. Further, let $M = [\alpha_{ij}]$ and $N = [\bar{\alpha}_{ij}]$. Then,*

$$l_{ij}^{k+1} = \begin{cases} \alpha_{ij}(w_i^k - w_j^k) & \text{if } k = 0 \\ (\beta - 1)l_{ij}^k + \beta\alpha_{ij}(w_i^k - w_j^k) - (1 - \beta)\bar{\alpha}_{ij}(w_i^{k-1} - w_j^{k-1}) & \text{if } k > 0 \end{cases}$$

Proof. Straightforward. ∎

Consider a SOS using a β that is an approximation of β_M. By means of an experiment, we show that a carefully chosen GSOS can improve on the performance of this SOS. We choose a GSOS with an N that satisfies $\max_{i,j} |N_{ij} - I_{ij}|$ is $O(\epsilon)$ for some small $\epsilon \ll 1$, *i.e.* N is a small perturbation of the identity matrix.

Experiment. In our experiment, β was varied in the vicinity of β_M. For each value of β, $[\epsilon_{\min}, \epsilon_{\max}]$ indicates the range of ϵ in which GSOS performed as well or better than SOS, at the end of 10 iterations. ϵ_{best} denotes the value of ϵ in this range for which GSOS achieved greatest improvement over the SOS, and γ_{best} denotes the effective γ of this GSOS. For each graph, we see a substantial range of β where GSOS improves on SOS (see Table 3). We provide another example of this phenomenon, but this time, we fix β at the heuristic local estimate $\widehat{\beta}$ of Section 4 instead. As before, there is a visible improvement of the GSOS over the SOS (see Table 4).

Remark. To see the behavior of GSOSs on the realistic load balancing problem, we simulated a GSOS on XG(8) with IOUs on the initial weight distribution *Spike(100)*. Figure 2 shows that the GSOS has the desirable property of all the IOUs disappearing after a few iterations.

6   Concluding Remarks

We have experimentally addressed the question of choosing various parameters in FO and SO schemes for distributed load balancing on arbitrary graphs. Based on our experiments, we conjectured closed-form expressions for the optimal M's in FOSs. Proving these conjectures is open. Other theoretical issues that are open include analysis of the new iterative load balancing method we introduced in this paper, namely, the GSO scheme, and developing efficient algorithms to determine the optimal SOSs. We are continuing further experiments on this. For example, we are fine-tuning our heuristic methods to estimate FO and SO parameters for different classes of graphs. Several issues in GSOS (for example, the choice of N, its effect on the convergence of the scheme, etc.) remain open and are the focus of some of our current investigations. Moreover, further experiments on larger topologies and actual parallel implementations of the schemes proposed here are necessary to assess their practical viability. These questions will be addressed in a journal version of this paper. Our methods are being incorporated in to PadFEM [1], a library of routines to perform finite element simulations on MPPs using DD-preconditioned Conjugate Gradient methods on adaptively changing meshes. We hope to use the benchmark applications there to develop broad guidelines for choice of various parameters in diffusive load balancing strategies.

Acknowledgments.
This work was supported by the DFG-Sonderforschungsbereich 376 "Massive

Parallelität: Algorithmen, Entwurfsmethoden, Anwendungen" and the EC ES-PRIT Long Term Research Project 20244 (ALCOM-IT). The first author also thanks Sergej Bezrukov and Burkhard Monien for many helpful discussions.

References

1. R. Diekmann, U. Dralle, F. Neugebauer, and T. Römke. PadFem: A portable parallel FEM-tool. In *Proceedings of the International Conference on High-Performance Computing and Networking (HPCN-Europe '96)*, volume 1067 of *Lecture Notes in Computer Science*, pages 580–585. Springer Verlag, 1996.
2. N. Higham. *Accuracy and Stability of Numerical Algorithms*. SIAM, Philadelphia, 1996.
3. G. Cybenko. Dynamic load balancing for distributed memory multiprocessors. *Journal of Parallel and Distributed Computing*, 7:279–301, 1989.
4. G. Fox, R. Williams, and P. Messina. *Parallel Computing Works!* Morgan Kaufmann Publishers, 1994.
5. W. Aiello, B. Awerbuch, B. Maggs, and S. Rao. Approximate load balancing on dynamic and asynchronous networks. In *Proceedings of the 26th ACM Symposium on Theory of Computing*, pages 632–641, 1993.
6. S. Muthukrishnan, B. Ghosh, and M. Shultz. First and second order diffusive methods for rapid, coarse, distributed load balancing. In *Jounal version*, 1996. A preliminary version appeared in the Proceedings of the 8th Annual ACM Symposium on Parallel Algorithms and Architectures.
7. R. Varga. *Matrix Iterative Analysis*. Prentice–Hall, Englewood Cliffs, New Jersey, 1962.
8. G. H. Golub and R. Varga. Chebyshev semi–iterative methods, successive over-relaxation iterative methods and second order richardson iterative methods. *Numerische Mathematik*, 3:147–156, 1961.
9. L. A. Hageman and D. M. Young. *Applied Iterative Methods*. Academic Press, 1981.
10. R. Barrett, M. Berry, T. Chan, J. Demmel, J. Donato, J. Dongarra, V. Eijkhout, R. Pozo, C. Romine, and H. van der Vorst. *Templates for the Solution of Linear Systems: Building Blocks for Iterative Methods*. SIAM, Philadelphia, 1993. URL: http://netlib2.cs.utk.edu/linalg/html_templates/Templates.html.
11. R. Diekmann, D. Meyer, and B. Monien. Parallel decomposition of unstructured FEM–meshes. *Concurrency: Practice and Experience*, 1997. (to appear).
12. F. Alizadeh. Interior–point methods in semidefinite programming with applications to combinatorial optimization. *SIAM Journal on Optimization*, 5(1):13–51, February 1995.
13. F. Alizadeh, J.-P. A. Haeberly, M. V. Nayakkankuppam, and M. L. Overton. *SDP-pack User's Guide (Version 0.8 BETA)*. NYU Computer Science Department Technical Report #734, March 1997.
14. B. Mohar. Isoperimetric number of graphs. *Journal of Combinatorial Theory (Series B)*, 47:274–291, 1989.

Name	Description	$|V|$	$|E|$	Degree min	avg	max	$\lambda_{n-1}(L)$	$\lambda_1(L)$
$L(n)$	lin. array	n	$n-1$	1	$2-\frac{2}{n}$	2	$4\sin^2(\frac{\pi}{2n})$	$4\cos^2(\frac{\pi}{2n})$
$G(n)$	$n\times n$-grid	n^2	$2n^2-2n$	2	$4-\frac{4}{n}$	4	$4\sin^2(\frac{\pi}{2n})$	$8\cos^2(\frac{\pi}{2n})$
$C(n)$	$n\times n$-cylinder	n^2	$2n^2-n$	3	$4-\frac{2}{n}$	4	$n=8:\ 0.152$	7.848
$T(n)$	$n\times n$-torus	n^2	$2n^2$	4	4	4	$4-4\cos^2(\frac{\pi}{n})$	8
$XG(n)$	$n\times n$-Xgrid	n^2	$4n^2-6n+2$	3	$8-\frac{12}{n}+\frac{4}{n^2}$	8	$n=8:\ 0.416$	11.391
$Q(n)$	Hypercube	2^n	$n2^{n-1}$	n	n	n	2	$2n$
$ST(n)$	Star-graph	n	$n-1$	1	$2-\frac{2}{n}$	$n-1$	1	n
R1	rand. graph	100	266	1	5.3	9	0.809	11.900
R3	rand. graph	100	1356	11	27.1	43	10.593	45.560
R4	rand. graph	100	1663	14	33.3	53	13.521	55.331
R5	rand. graph	100	2227	21	44.5	65	20.466	67.184
R10	rand. graph	100	3737	43	74.7	98	42.485	99.032
AF1	qu.-gr. (CFD)	64	167	2	5.2	9	0.150	10.655
WH3	qu.-gr. (FEM)	128	337	2	5.3	8	0.063	10.611
CR	qu.-gr. (FEM)	128	357	2	5.6	8	0.107	9.782

Table 1. The set of test graphs.

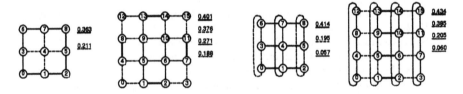

Fig. 1. The optimal edge weights for G(3), G(4), C(3) and C(4).

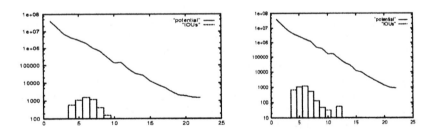

Fig. 2. Potential and sum of absolute values of all IOUs on XG(8) for $Spike(100)$. Left plot: SOS with heuristic $\widehat{\alpha}$ and $\widehat{\beta}$; Right plot: GSOS with ε_{best} from Table 4.

Name	α,β with $\lambda_{n-1}(L), \lambda_1(L)$				λ_{n-1}, λ_1 heuristic						$\frac{\gamma(FO)}{\widehat{\gamma}(FO)}$	$\frac{\gamma(SO)}{\widehat{\gamma}(SO)}$
	α	β	$\gamma(FO)$	$\gamma(SO)$	$\widehat{\lambda}_{n-1}$	$\widehat{\lambda}_1$	$\widehat{\alpha}$	$\widehat{\beta}$	$\widehat{\gamma}(FO)$	$\widehat{\gamma}(SO)$		
L(32)	0.500	1.824	0.995	0.962	0.016	3.938	0.506	1.777	0.995	0.966	1.000	1.004
G(8)	0.255	1.568	0.962	0.827	0.102	7.500	0.263	1.627	0.961	0.771	0.999	0.932
C(8)	0.250	1.571	0.962	0.826	0.125	7.750	0.254	1.600	0.962	0.804	1.000	0.973
T(8)	0.233	1.329	0.864	0.628	0.438	8.000	0.237	1.386	0.896	0.677	1.037	1.546
XG(8)	0.169	1.461	0.946	0.820	0.186	14.563	0.136	1.636	0.947	0.763	1.001	0.930
Q(6)	0.143	1.177	0.716	0.447	0.938	12.000	0.155	1.317	0.860	0.633	1.201	1.416
St(64)	0.031	1.605	0.984	0.929	0.500	64.969	0.031	1.703	0.984	0.901	1.000	0.973
R1	0.157	1.344	0.882	0.677	0.667	14.320	0.134	1.416	0.894	0.609	1.014	0.900
R3	0.036	1.122	0.676	0.483	5.500	70.120	0.026	1.316	0.725	0.512	1.072	1.060
R4	0.029	1.115	0.676	0.495	7.000	86.260	0.021	1.310	0.721	0.507	1.067	1.024
R5	0.023	1.083	0.607	0.436	10.500	109.540	0.017	1.278	0.654	0.469	1.077	1.076
R10	0.014	1.043	0.499	0.358	18.580	172.740	0.011	1.256	0.541	0.447	1.084	1.249
AF1	0.185	1.621	0.978	0.893	0.080	14.219	0.140	1.740	0.980	0.839	1.002	0.940
WH3	0.187	1.743	0.991	0.941	0.060	13.266	0.150	1.765	0.991	0.937	1.000	0.996
CR	0.202	1.657	0.985	0.921	0.100	13.578	0.146	1.709	0.985	0.904	1.000	0.982

Table 2. Comparison of (global) best choice for α and β with (local) heuristic estimates for these parameters for FO and SO schedules. $\gamma(SO)$ is the effective γ of the SO method after 10 iterations.

Name	β	ε_{min}	ε_{max}	ε_{best}	$\gamma(SO)$	γ_{best}	$\frac{\gamma_{best}}{\gamma(SO)}$
Q(6)	1.5	0.0	0.0944	0.0930	0.6544	0.6482	0.991
	1.7	0.0	0.1061	0.1047	0.7983	0.7872	0.986
	1.9	0.0	0.1119	0.1107	0.9372	0.9260	0.988
XG(8)	1.7	0.0	0.2197	0.2114	0.8157	0.8022	0.983
	1.9	0.0	0.2376	0.2343	0.9377	0.9237	0.985
	2.0	0.0	0.2389	0.2314	0.9990	0.9707	0.972

Table 3. Generalized Second Order Schedule: Ranges of ε where GSOS performs at least as well as SOS. γ's are the effective values after 10 iterations. $\alpha = \frac{2}{\lambda_2(L)+\lambda_n(L)}$ is fixed, β is varied.

Name	ε_{min}	ε_{max}	ε_{best}	$\widehat{\gamma}(SO)$	$\widehat{\gamma}_{best}$	$\frac{\gamma_{best}}{\gamma(SO)}$
G8	0.0	0.0126	0.0093	0.77121	0.76872	0.997
XG(8)	0.0	0.1240	0.0442	0.76259	0.74946	0.983
R3	0.0	0.0515	0.0488	0.51229	0.49634	0.967
R4	0.0	0.0393	0.0381	0.50665	0.48928	0.966
R5	0.0	0.0338	0.0333	0.46945	0.46185	0.984

Table 4. Generalized Second Order Schedule: Ranges of ε where GSOS performs at least as well as SOS. γ's are the effective values after 10 iterations. $\widehat{\alpha}$ and $\widehat{\beta}$ are fixed local, heuristic estimates (cf. Table 2).

Comparative Study of Static Scheduling with Task Duplication for Distributed Systems[*]

Gyung-Leen Park[1], Behrooz Shirazi[1], and Jeff Marquis[2]

[1]University of Texas at Arlington, Dept. of CSE, Arlington, TX 76019, USA
{gpark, shirazi}@cse.uta.edu
[2]Prism Parallel Technologies, Inc., 2000 N. Plano Rd, Richardson, Texas 75082

Abstract. Efficient scheduling of parallel tasks onto processing elements of concurrent computer systems has been an important research issue for decades. The communication overhead often limits the speedup of parallel programs in distributed systems. Duplication Based Scheduling (DBS) has been proposed to reduce the communication overhead by duplicating remote parent tasks on local processing elements. The DBS algorithms need task selection heuristics which decide the order of tasks to be considered for scheduling. This paper explores the speedup obtained by employing task duplication in the scheduling process and also investigates the effect of different task selection heuristics on a DBS algorithm. Our simulation results show that employing task duplication achieves considerable improvement in the application parallel execution time, but employing different task selection heuristics does not affect the speedup significantly.

1 Introduction

Efficient scheduling of parallel programs, represented as a task graph, onto processing elements of parallel and distributed systems has been under investigation for decades [14-17, 20, 22]. The typical objective of a scheduling algorithm is to map the program tasks onto the processing elements of the target architecture in such a way that the program parallel execution time is minimized. Since it has been shown that the multiprocessor scheduling problem is NP-complete in general forms, many researchers have proposed scheduling algorithms based on heuristics. The scheduling methods can be classified into two general categories of duplication-based and non-duplication algorithms. Task duplication algorithms attempt to reduce communication overhead by duplicating tasks that would otherwise require interprocessor communications if the tasks were not duplicated.

A majority of the non-duplication scheduling methods are based on the *list scheduling* algorithm [1]. A list scheduling algorithm repeatedly carries out the following steps: (1) Tasks ready to be assigned to a processor are put onto a queue, based on some priority criteria. A task becomes ready for assignment when all of its parents are scheduled. (2) A suitable "Processing Element" (PE) is selected for assignment. Typically, a suitable PE is one that can execute the task the earliest. (3) The task at the head of the priority queue is assigned to the suitable PE.

---

[*] This work has in part been supported by grants from NSF (CDA-9531535 and MIPS-9622593) and state of Texas ATP 003656-087.

Duplication Based Scheduling (DBS) is a relatively new approach to the scheduling problem [1, 3, 5, 6, 7, 9, 17]. The DBS algorithms reduce the communication overhead by duplicating remote parent tasks on local processing elements. Similar to non-duplication algorithms, DBS methods have been shown to be NP-complete [10]. Thus, many of the proposed DBS algorithms are based on heuristics.

This paper explores the advantages of employing a task duplication method in the scheduling process and also investigates the effect of different node selection heuristics on a DBS algorithm. Duplication First and Reduction Next (DFRN) algorithm is used as the task duplication method while the list scheduling algorithms Heavy Node First (HNF), High Level First with Estimated Times (HLFET), and Decisive Path Scheduling (DPS) are used as node selection heuristics. Other scheduling algorithms are also considered for comparison purposes.

Our simulation study shows that employing task duplication achieves considerable performance improvement over existing algorithms without task duplication, especially when the *Communication to Computation Ratio* is increased. However, employing different node selection heuristics does not affect the performance significantly.

The remainder of this paper is organized as follows. Section 2 presents the system model and the problem definition. Section 3 covers the related algorithms which are used in our performance study. The comparative study is conducted in Section 4. Finally, Section 5 concludes this paper.

2 The System model and Problem Definition

A parallel program is usually represented by a *Directed Acyclic Graph (DAG)*, which is also called a *task graph*. As defined in [6], a DAG consists of a tuple (V, E, T, C), where V, E, T, and C are the set of task nodes, the set of communication edges, the set of computation costs associated with the task nodes, and the set of communication costs associated with the edges, respectively. $T(V_i)$ is a computation cost for task V_i and $C(V_i, V_j)$ is the communication cost for edge $E(V_i, V_j)$ which connects task V_i and V_j. The edge $E(V_i, V_j)$ represents the precedence constraint between the node V_i and V_j. In other words, task V_j can start the execution only after the output of V_i is available to V_j. When the two tasks, V_i and V_j, are assigned to the same processor, $C(V_i, V_j)$ is assumed to be zero since intra-processor communication cost is negligible compared with the interprocessor communication cost. The weights associated with nodes and edges are obtained by estimation [20].

This paper defines two relations for precedence constraints. The $V_i \Rightarrow V_j$ relation indicates the strong precedence relation between V_i and V_j. That is, V_i is an immediate parent of V_j and V_j is an immediate child of V_i. The terms *iparent* and *ichild* are used to represent immediate parent and immediate child, respectively. The $V_i \rightarrow V_j$ relation indicates the weak precedence relation between V_i and V_j. That is, V_i is a parent of V_j but not necessarily the immediate one. $V_i \rightarrow V_j$ and $V_j \rightarrow V_k$ imply $V_i \rightarrow V_k$. $V_i \Rightarrow V_j$ and $V_j \Rightarrow V_k$ do not imply $V_i \Rightarrow V_k$, but imply $V_i \rightarrow V_k$. The relation $\rightarrow$ is transitive, and the relation $\Rightarrow$ is not. A node without any parent is called an *entry node* and a node without any child is called an *exit node*.

Without loss of generality, we assume that there is only one entry node and one exit node in the DAG. Any DAG can be easily transformed to this type of DAG by adding a dummy node for each entry node and exit node, where the communication cost for an edge connecting a dummy node is zero.

Graphically, a node is represented as a circle with a dividing line in the middle. The number in the upper portion of the circle represents the node ID number and the number in the lower portion of the circle represents the computation cost for the node. For example, for the sample DAG in Figure 1, the entry node is V_1 which has a computation cost of 10. In the graph representation of a DAG, the communication cost for each edge is written on the edge itself. For each node, *incoming degree* is the number of input edges and *outgoing degree* is the number of output edges. For example, in Figure 1, the incoming and outgoing degrees for V_5 are 2 and 1, respectively. A few terms are defined here for a more clear presentation.

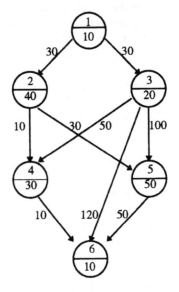

Figure 1. A sample DAG

Definition 1: A node is called a *fork node* if its outgoing degree is greater than 1.
Definition 2: A node is called a *join node* if its incoming degree is greater than 1.

Note that the fork node and the join node are not exclusive terms, which means that one node can be both a fork and also a join node; i.e., both of the node's incoming and outgoing degrees are greater than one. Similarly, a node can be neither a fork nor a join node; i.e., both of the node's incoming and outgoing degrees are one. In the sample DAG, V_1, V_2, and V_3 are fork nodes while nodes V_4, V_5, and V_6 are join nodes.

Definition 3: The *Earliest Start Time, EST(V_i, P_k)*, and *Earliest Completion Time, ECT(V_i, P_k)*, are the times that a task V_i can start and finish its execution on processor P_k, respectively.

Definition 4: The *Message Arriving Time (MAT)* from V_i to V_j, or $MAT(V_i, V_j)$, is the time that the message from V_i arrives at V_j. If V_i and V_j are scheduled on the same processor P_k, $MAT(V_i, V_j)$ becomes $ECT(V_i, P_k)$.

Definition 5: An *iparent* of a join node is called its *primary iparent* if it provides the largest MAT to the join node. The primary iparent is denoted as $V_i = PIP(V_j)$ if V_i is the primary iparent of V_j. More formally, $V_i = PIP(V_j)$ if and only if $MAT(V_i, V_j) >$ $MAT(V_k, V_j)$, for all V_k, $V_k \Rightarrow V_j$, $V_i \Rightarrow V_j$, $i \neq k$. The $MAT(V_i, V_j)$ is $ECT(V_i, P_k) +$ $C(V_i, V_j)$ where $ECT(V_i, P_k) = EST(V_i, P_k) + T(V_i)$ for the processor P_k where V_i is scheduled. If there are more than one iparent providing the same largest MAT, PIP is chosen randomly.

Definition 6: The processor to which $PIP(V_i)$ is assigned is called the *primary processor* of V_i.

Definition 7: The level of a node is recursively defined as follows. The level of an entry node, V_r, is one $(Lv(V_r) = 1)$. Let $Lv(V_i)$ be the level of V_i. Then, $Lv(V_j) =$ $Lv(V_i) + 1$, $V_i \Rightarrow V_j$, for non-join node V_j. $Lv(V_j) = Max(Lv(V_i)) + 1$, $V_i \Rightarrow V_j$, for join node V_j. For example, in Figure 1, the levels of nodes V_1, V_2, V_3, V_4, V_5, and V_6 are 1, 2, 2, 3, 3, and 4, respectively. Even though there is an edge from node 3 to 6, the level of node 6 is still 3 not 2 since $Lv(V_6) = Max(Lv(V_i)) + 1$, $V_i \Rightarrow V_6$, for join node V_6.

Definition 8: The *Decisive Path (DP)* to V_i is the path which decides the distance of V_i from an entry node. For example, in Figure 1, the decisive path to V_6, $DP(V_6)$, is the path through V_1, V_3, V_5, and V_6 since the path decides the distance of the node V_6. The decisive path is defined for every node in the DAG while the critical path is defined only for an exit node. Thus, the critical path becomes a special case of the decisive path. From Figure 1, $DP(V_5)$ is the path through V_1, V_3, and V_5.

Similar to the existing DBS algorithms, we assume a complete graph as a system architecture with unbounded number of processors. Thus, the multiprocessor scheduling process becomes a mapping of the parallel tasks in the input DAG to the processors in the target system with the goal of minimizing the execution time of the entire program. The execution time of the entire program is called its *parallel time,* to be distinguished from the completion time of an individual task node.

3 The Related Work

The critical issue in list scheduling algorithms is the method by which the priorities of the nodes or edges of the input DAG are decided. Most of the these scheduling algorithms use certain properties of the input DAG for deciding the priorities. This section covers typical scheduling algorithms which are used later for performance comparison in Section 4.

3.1 Heavy Node First (HNF) Algorithm

The HNF algorithm [13] assigns the nodes in a DAG to the processors, level by level. At each level, the scheduler selects the eligible nodes for scheduling in descending order based on computational weights, with the heaviest node (i.e. the node which has the largest computation cost) selected first. The node is selected arbitrarily if multiple nodes at the same level have the same computation cost. The selected node is assigned to a processor which provides the earliest start time for the node.

3.2 High Level First with Estimated Time (HLFET) Algorithm

The HLFET [1] algorithm also assigns the nodes in a DAG to the processors level by level. At each level, the scheduler assigns a higher priority to a node with a larger bottom distance. The bottom distance is the length of the path from the node being considered to an exit node. The node with the highest priority is assigned to a processor which provides the earliest start time for the node.

3.3 Linear Clustering (LC) Algorithm

The LC algorithm [8] is a traditional critical path based scheduling method. The scheduler identifies the critical path, removes the nodes in the path from the DAG, and assigns them to a linear cluster. The process is repeated until there are no task nodes remaining in the DAG. The clusters are then scheduled onto the processors.

3.4 Dominant Sequence Clustering (DSC) Algorithm

The *dominant sequence* is a dynamic version of the critical path. The dominant sequence is the longest path of the task graph for un-scheduled nodes [21]. Initially, the dominant sequence is the same as the critical path for the original input DAG. At each step, the scheduler selects one edge in the dominant sequence and eliminates it if this would reduce the length of the dominant sequence. The scheduler identifies a new dominant sequence since an edge elimination may change the longest path. The operations are repeatedly carried out until all the edges are examined.

3.5 Decisive Path Scheduling (DPS) Algorithm

The DPS algorithm [12] first identifies the decisive paths for all the nodes in the input DAG. Since we are assuming an input DAG with one entry node and one exit node, the decisive path to the unique exit node becomes the critical path. The DPS algorithm inserts the nodes on the critical path one by one into the task queue (parent first) if the node does not have any parent which is not in the task queue yet. If there are some parents which are not in the task queue, DPS algorithm inserts the parent belonging to the decisive path to the node first. The node is put into the task queue after all the parents are inserted into the queue according to the decisive path concept. The DPS algorithm can be implemented as either a list scheduling algorithm or a node selection algorithm for a DBS algorithm. Either case is implemented and compared in this paper.

3.6 Scalable Task Duplication based Scheduling (STDS) algorithm

The STDS [7] algorithm first calculates the start time and the completion time of each node by traversing the input DAG. The algorithm then generates clusters by performing depth first search starting from the exit node. While performing the task assignment process, only critical tasks which are essential to establish a path from a particular node to the entry node are duplicated. The algorithm has a small complexity because of the limited duplication. If the number of processors available is less than that needed, the algorithm executes the processor reduction procedure. In this paper, unbounded number of processors are used for STDS for performance comparison in Section 4.

3.7 Critical Path Fast Duplication (CPFD) algorithm

The CPFD algorithm [2] classifies the nodes in a DAG into three categories: Critical Path Node (CPN), In-Branch Node (IBN), and Out-Branch Node (OBN). A CPN is a node on the critical path. An IBN is a node from which there is a path to a CPN. An OBN is a node which is neither a CPN nor an IBN. CPFD tries to schedule CPNs first. If there is any unscheduled IBN for a CPN, CPFD traces the IBN and schedules it first. OBNs are scheduled after all the IBNs and CPNs have been scheduled. The motivation behind CPFD is that the parallel time will likely be reduced by trying to schedule CPNs first. Performance comparisons shows that CPFD outperforms earlier DBS algorithms in most cases [2].

3.8 Duplication First Reduction Next (DFRN) algorithm

Unlike other DBS algorithms, the DFRN algorithm [11] first duplicates any duplicable parent without estimating the effect of the duplication. Then the duplicated tasks are deleted if the tasks do not meet certain conditions. Another characteristic of the DFRN algorithm is that it applies the duplication only for the primary processor. These two differences enable DFRN algorithm to achieve high performance with a short running time to generate the schedule.

In this paper we use DFRN as the bases for investigating the effect of task duplication in the scheduling process. The effect of node selection heuristics for the purpose of duplication is studied using DFRN as well. The details of the DFRN algorithm are presented in [11]. A high level description of the algorithm is as follows:

1. Use a heuristic to prioritize the task nodes in the input task graph.
2. Take the next unscheduled task with the highest priority:
    2.1. If the task is a not a join node, duplicate its parents; obviously, if the task is neither a fork nor a join node, then no duplication is needed and the task is scheduled after its parent.
    2.2. If the task is a join node:
        2.2.1. Identify the task's primary processor.
        2.2.2. Duplicate all parents of the join node onto the primary processor.

2.2.3. Delete the duplicated tasks if the start time of the join node is increased due to the duplication.

For a DAG with V nodes, the complexities of steps 2.2.1, 2.2.2, and 2.2.3 are $O(V)$, $O(V^2)$, and $O(V^2)$, respectively. Thus, the complexity of the entire algorithm becomes $O(V^3)$. The DFRN algorithm is generic in the sense that any heuristic can be used in step 1 as a node selection heuristic. The effects of different node selection heuristics combined with the DFRN algorithm are investigated in Section 4.

3.9 Comparison

The time complexity and the priority criteria heuristic for the aforementioned algorithms are summarized in Table I.

Table I. Characteristics of scheduling algorithms

Algorithm	Classification	Priority Criteria Heuristic	Complexity
HNF	List Scheduling	Level and Node Weight	$O(V \log V)$
HLFET	List Scheduling	Level and Bottom Distance	$O(V^2)$
LC	Clustering	Critical Path	$O(V^3)$
DSC	Clustering	Dominant Sequence	$O(V^2)$
DPS	List Scheduling	Decisive Path	$O(V^3)$
STDS	DBS	Topological Order	$O(V^2)$
DFRN	DBS	any heuristic	$O(V^3)$
CPFD	DBS	Critical Path	$O(V^4)$

Figure 2 presents the schedules for the sample DAG in Fugure 1 obtained by schedulers covered in this section. In the schedules, for a node V_i assigned to processor P_k, the three elements in each bracket represent $EST(V_i, P_k)$, the node number of V_i (which is i), and $ECT(V_i, P_k)$, from left to right.

P1: [0, 1, 10] [10, 2, 50] [110, 4, 140]
P2: [40, 3, 60] [80, 5, 130] [150, 6, 160]
      (1) Schedule by HNF (PT = 160)

P1: [0, 1, 10] [10, 2, 50] [110, 4, 140]
P2: [40, 3, 60] [80, 5, 130] [150, 6, 160]
      (2) Schedule by HLFET (PT = 160)

P1: [0, 1, 10] [10, 3, 30] [110, 5, 160] [160, 6, 170]
P2: [40, 2, 80] [80, 4, 110]
      (3) Schedule by LC (PT = 170)

P1: [0, 1, 10] [10, 3, 30] [110, 5, 160] [160, 6, 170]
P2: [40, 2, 80] [80, 4, 110]
      (4) Schedule by DSC (PT = 170)

P1: [0, 1, 10] [10, 3, 30] [30, 2, 70] [70, 5, 120] [120, 6, 130]
P2: [80, 4, 110]
> (5) Schedule by DPS (PT = 130)

P1: [0, 1, 10] [10, 2, 50]
P2: [0, 1, 10][10, 3, 30] [80, 5, 130] [130, 6, 140]
P3: [0, 1, 10] [10, 3, 30] [70, 4, 90]

> (6) Schedule by STDS (PT = 140)

P1: [0, 1, 10] [10, 2, 50]
P2: [0, 1, 10][10, 3, 30] [30, 2, 70][70, 5, 120] [120, 6, 130]
P3: [0, 1, 10] [10, 3, 30] [70, 4, 90]
> (7) Schedule by HNF-DFRN (PT = 130)

P1: [0, 1, 10] [10, 2, 50]
P2: [0, 1, 10][10, 3, 30] [30, 2, 70][70, 5, 120] [120, 6, 130]
P3: [0, 1, 10] [10, 3, 30] [70, 4, 90]
> (8) Schedule by HLFET-DFRN (PT = 130)

P1: [0, 1, 10] [10, 3, 30] [30, 2, 70] [70, 5, 120] [120, 6, 130]
P2: [0, 1, 10] [10, 2, 50]
P3: [0, 1, 10] [10, 3, 30] [60, 4, 90]
> (9) Schedule by DPS-DFRN (PT = 130)

P1: [0, 1, 10] [10, 3, 30] [30, 2, 70] [70, 5, 120] [120, 6, 130]
P2: [0, 1, 10] [10, 2, 50] [50, 3, 70] [70, 4, 90]
> (10) Schedule by CPFD (PT = 130)

Figure 2. Schedules obtained by various schedulers

4 Performance Comparison

We generated 1000 random DAGs to compare the performance of HNF-DFRN, HLFET-DFRN, and DPS-DFRN with the existing scheduling algorithms. We used three parameters the effects of which we were interested to investigate: the number of nodes, CCR (average Communication to Computation Ratio), and the average degree (defined as the ratio of the number of edges to the number of nodes in the DAG). The numbers of nodes used are 20, 40, 60, 80, and 100 while CCR values used are 0.1, 0.5, 1.0, 5.0, and 10.0. Forty DAGs are generated for each case of the 25 combinations, which makes 1000 DAGs.

From the existing schedulers, we selected three list scheduling algorithms, two clustering algorithms, and two DBS algorithms for performance comparison. HNF, HLFET, and DPS are chosen from the list scheduling algorithms. Since all the three list scheduling algorithms are also chosen as the node selection methods in DFRN, the effect of task duplication can be easily seen by comparing HNF with HNF-DFRN,

HLFET with HLFET-DFRN, and DPS with DPS-DFRN, respectively. Also, we can compare the significance of the node selection criteria in the DFRN algorithm. The LC and DSC algorithms are chosen from clustering algorithms. The STDS and CPFD algorithms are chosen from the DBS algorithms. STDS has a low complexity and requires a minimal time to generate the schedule, while CPFD has a high complexity and takes several orders of magnitude longer to produce a schedule [11].

For performance comparison, we define a *normalized* performance measure named *Relative Parallel Time (RPT)*. RPT is the ratio of the parallel time generated by a schedule to the program execution time lower bound which is the sum of the computation costs of the nodes in the critical path. For example, if the parallel time obtained by DPS-DFRN is 200 and the execution time lower bound is 100, RPT of DPS-DFRN is 2.0. A smaller RPT value is indicative of a shorter parallel time. The RPT of any scheduling algorithm can not be lower than one.

Figures 3, 4, and 5 depict the performance comparison results with respect to N (the number of DAG nodes), CCR, and the average degree, respectively. Each case in Figure 3 is an average of 200 runs with varying CCR and average degree parameters. The average CCR value is 3.3 and the average degree value is 3.8. As shown in this Figure, the number of nodes does not significantly affect the relative performance of scheduling algorithms. In other words, the performance comparison shows similar patterns regardless of N. In the pattern, HNF-DFRN, HLFET-DFRN, and DPS-DFRN algorithms show much shorter parallel time than existing algorithms with equal or lower time complexity while they show a comparable performance to CPFD (which has a higher time complexity compared to DFRN).

CCR is a critical parameter, determining the grain of a task. As CCR is increased, the performance gap becomes larger as shown in Figure 4. The difference among ten algorithms was negligible until CCR becomes 1. But when CCR becomes greater than 1, the differences in RPTs become significant. As expected, duplication-based scheduling algorithms show considerable performance improvement for a DAG with high CCR values.

As shown in Figure 5, it seems the number of communication links among the DAG node, represented by the average degree parameter in our study, do not significantly impact the performance of the schedulers. In this Figure, the change in average degree affects the parallel times of the DAGs, but does not change the relative performance the algorithm with respect to each other.

One interesting observation is that a node selection heuristic, such as HNF, HLFET, and DPS, does not play a significant role in a DBS algorithm (DFRN). However, the same heuristics make significant differences in list scheduling algorithms. For example, when CCR is 5, the parallel times obtained by HNF, HLFET, and DPS are 3.38, 3.12, and 2.70, respectively; while the parallel times of HNF-DFRN, HLFET-DFRN, and DPS-DFRN are all 1.67. It is because the duplication process reduces the effect of different node selection heuristics, as already shown in Figure 2. In this Figure, the parallel times of HNF, HLFET, and DPS are larger or equal to the parallel times of the DFRN when using the respective node selection heuristic. Thus, in DBS algorithms, the design of an efficient node selection algorithm is not critical; rather the effort should be focused on the design of the duplication method with a low complexity. A simple heuristic such as HNF would be sufficient as a node selection heuristic for a DBS algorithm such as DFRN.

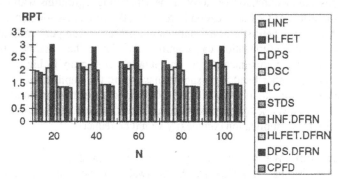

Figure 3. Comparison with respect to N.

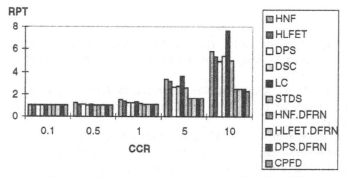

Figure 4. Comparison with respect to CCR (N=100).

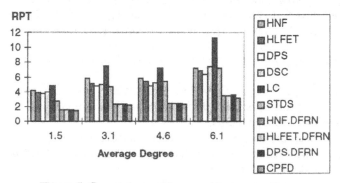

Figure 5. Comparison with respect to average degree

5 Conclusion

This paper presented a comparative study of the scheduling algorithms with respect to two issues: the effect of task duplication in reducing the communication costs and the effect of task selection heuristic in the duplication process. The performance results clearly show that task duplication can significantly reduce the parallel time of an application when the communication to computation cost ratio is high (fine-grain computing). On the other hand, it is shown that during the duplication process, the task selection heuristic (to determine the next task to consider for duplication) is immaterial. Thus, one is better off using a simple and fast heuristic for node selection so that the overall task duplication complexity would be reduced.

Acknowledgment. We would like to express our appreciation to Drs. Ishfaq Ahmad, Sekhar Darbha, and Dharma P. Agrawal and their research groups for providing the source code for the CPFD and STDS schedulers which were used in our performance comparison study.

References

1. T. L. Adam, K. Chandy, and J. Dickson, "A Comparison of List Scheduling for Parallel Processing System," *Communication of the ACM*, vol. 17, no. 12, Dec. 1974, pp. 685-690.
2. I. Ahmad and Y. K. Kwok, "A New Approach to Scheduling Parallel Program Using Task Duplication," *Proc. of Int'l Conf. on Parallel Processing*, vol. II, Aug. 1994, pp. 47-51.
3. H. Chen, B. Shirazi, and J. Marquis, "Performance Evaluation of A Novel Scheduling Method: Linear Clustering with Task Duplication," *Proc. of Int'l Conf. on Parallel and Distributed Systems*, Dec. 1993, pp. 270-275.
4. Y. C. Chung and S. Ranka, "Application and Performance Analysis of a Compile-Time Optimization Approach for List Scheduling Algorithms on Distributed-Memory Multiprocessors," *Proc. of Supercomputing'92*, Nov. 1992, pp. 512-521.
5. J. Y. Colin and P. Chretienne, "C.P.M. Scheduling with Small Communication Delays and Task Duplication," *Operations Research*, 1991, pp. 680-684.
6. S. Darbha and D. P. Agrawal, "SDBS: A task duplication based optimal scheduling algorithm," *Proc. of Scalable High Performance Computing Conf.*, May 1994, pp. 756-763.
7. S. Darbha and D. P. Agrawal, "A Fast and Scalable Scheduling Algorithm for Distributed Memory Systems," *Proc. of Symp. On Parallel and Distributed Processing*, Oct. 1995, pp. 60-63.
8. Kim, S. J., and Browne, J. C., "A general approach to mapping of parallel computation upon multiprocessor architectures," *Proc. of Int'l Conf. on Parallel Processing*, vol III, 1988, pp. 1-8.
9. B. Kruatrachue and T. G. Lewis, "Grain Size Determination for parallel processing," *IEEE Software*, Jan. 1988, pp. 23-32

10. C. H. Papadimitriou and M. Yannakakis, "Towards an architecture-independent analysis of parallel algorithms," *ACM Proc. of Symp. on Theory of Computing (STOC)*, 1988, pp. 510-513.

11. G.-L. Park, B. Shirazi, and J. Marquis, "DFRN: A New Approach on Duplication Based Scheduling for Distributed Memory Multiprocessor Systems," *Proc. of Int'l Parallel Processing Symp.*, April. 1994, pp. 157-166.

12. G.-L. Park, B. Shirazi, and J. Marquis, "Decisive Path Scheduling: A New List Scheduling Method," Tech. Report, Dept. of Computer Science and Engineering, Univ. of Texas at Arlington, 1997.

13. B. Shirazi, M. Wang, and G. Pathak, "Analysis and Evaluation of Heuristic Methods for Static Task Scheduling," *Journal of Parallel and Distributed Computing*, vol. 10, No. 3, 1990, pp. 222 -232.

14. B. Shirazi, A.R. Hurson, "Scheduling and Load Balancing: Guest Editors' Introduction," *Journal of Parallel and Distributed Computing*, Dec. 1992, pp. 271-275.

15. B. Shirazi, A.R. Hurson, "A Mini-track on Scheduling and Load Balancing: Track Coordinator's Introduction," *Hawaii Int'l Conf. on System Sciences (HICSS-26)*, January 1993, pp. 484-486.

16. B. Shirazi, A.R. Hurson, K. Kavi, "Scheduling & Load Balancing," *IEEE Press*, 1995.

17. B. Shirazi, H.-B. Chen, and J. Marquis, "Comparative Study of Task Duplication Static Scheduling versus Clustering and Non-Clustering Techniques," *Concurrency:Practice and Experience*, vol. 7(5), August 1995, pp. 371-389.

19. M. Y. Wu and D. D. Gajski, "Hypertool: A Programming Aid for Message-Passing Systems," *IEEE Trans. on Parallel and Distributed Systems*, vol. 1, no. 3, Jul. 1990, pp. 330-340.

20. M.Y. Wu, A dedicated track on "Program Partitioning and Scheduling in Parallel and Distributed Systems," in the *Hawaii Int'l Conference on Systems Sciences*, January 1994.

21. T. Yang and A. Gerasoulis, "DSC: Scheduling Parallel tasks on an Unbounded Number of Processors," *IEEE Trans. On Parallel and Distributed Systems*, vol. 5, no. 9, pp. 951-967, Sep. 1994.

22. T. Yang and A. Gerasoulis, A dedicated track on "Partitioning and Scheduling for Parallel and Distributed Computation," in the *Hawaii Int'l Conference on Systems Sciences*, January 1995.

A New Approximation Algorithm for the Register Allocation Problem

Klaus Jansen[1], Joachim Reiter[2]

[1] Fachbereich IV, Mathematik, Universität Trier, 54 286 Trier, Germany,
Email: jansen@dm3.uni-trier.de
[2] Fachbereich IV, Wirtschaftsinformatik, Universität Trier, 54 286 Trier, Germany,
Email: jr@wiinfo.uni-trier.de

Abstract. In this paper we study the problem of register allocation in the presence of parallel conditional branches with a given branching depth d. We start from a scheduled flow graph and the goal is to find an assignment of the variables in the flow graph to a minimum number of registers. This problem can be solved by coloring the corresponding conflict graph $G = (V, E)$. We describe a new approximation algorithm with constant worst case rate for flow graphs with constant branching depth. The algorithm works in two steps. In the first step, the lifetimes are enlarged such that the lifetimes form one unique interval across the different possible execution paths for each variable. We prove that the conflict graph $\bar{G}$ with enlarged lifetimes satisfies $\omega(\bar{G}) \leq (2d + 1)\omega(G)$ where $\omega(G)$ is the cardinality of a maximum clique in G. In the second step, we propose an algorithm with approximation bound $(d + 1)\chi(\bar{G})$ for $\bar{G}$, using its specific structure.

Keywords: register allocation, approximation, branching flow graphs, compiling

1 Introduction

High level synthesis starts with a behavioral specification of a digital system and produces a register transfer level structure that realizes the behavior [5]. The two major optimization tasks in high level synthesis are scheduling and hardware allocation. Hardware allocation is an assignment of functional units to arithmetic and logical operations, of registers to variables, and of interconnections between registers and functional units. The goal of hardware allocation is to minimize the total amount of hardware elements. Branching flow graphs (acyclic directed graphs with a conditional branching structure) are used to model the behavior of a program [7]. For an overview about different methods to solve the scheduling and allocation problems, we refer to [3, 6, 14]. In this paper we study the problem of register allocation. A value of a variable which is generated in one control step and used later must be stored in a register. In order to minimize the number of registers, the possibility of register sharing is used. This problem can be solved using graph theoretical methods [1].

The branching structure of the flow graph is given by a set $\mathcal{B}$ of *conditional blocks*. Each block $B \in \mathcal{B}$ has m_B ($m_B \geq 2$) different execution paths and is equivalent to a *case-statement* with m_B cases. We draw a block as a closed rectangle where the rectangle is separated by $(m_B - 1)$ vertical dashed lines into m_B parts. Each part is one *execution path* or *branch* of the conditional block. Let m be the maximum number of execution paths in blocks. A block B can lie beside (parallel) to other blocks or in one execution path of another block called the *father block* $f(B)$. The set of conditional blocks forms possibly a *nested* branching structure with allowed parallel blocks. The maximum number of nested blocks is called the branching depth d. The flow graph in Figure 1 contains two nested conditional blocks, each with two branches. The "program execution" determines which path in each reached conditional block is really executed during the program. This depends on the value of the input variables.

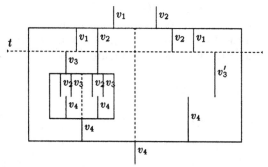

Fig. 1. A flow graph with two nested conditional blocks

To a variable v *lifetime intervals* $[t_a(v), t_e(v))$ are given, which are drawn as vertical line (in Figure 1 labeled with the variable names). The real lifetime depends on the program execution. The starting time $t_a(v)$ of v is the control step at which a value of the variable is computed for the first time and the ending time $t_e(v)$ the control step at which the value is probably used the last time. There are several possible ways in which registers can be shared by variables. For example, the variables v_3 and v_4 can share the same register since their lifetimes do not overlap. On the other hand, the variables v_3 and v_3' in different branches can share the same register even if their lifetimes overlap since only one of them is used during a program execution.

The register allocation problem can be solved by coloring the vertices of the graph $G = (V, E)$ where $\{v, w\}$ is in E if and only if the variables v and w are in conflict [4]. For flow graphs with one conditional block, the register allocation problem is already NP-complete even if the lifetimes form unique intervals for each variable [10]. Several algorithms are proposed for the the register allocation problem for flow graphs without conditional blocks [8, 16, 17, 18]. If the flow graph contains many conditional blocks, the results by these algorithms are less satisfactory. The system HIS [2] uses an analysis of all possible program executions and a coloring method for the optimization of the hardware resources. The complexity of this approach depends on the number of possible program executions which may be exponential. Two other heuristics which handle conditional register sharing are proposed by Kurdahi and Parker [13] and by Park, Kim and Liu [15].

Treating one variable as two separate variables will require a register transfer operation when the variables are assigned to different registers. In this case a complex control logic can select the register assignment at run time. We study here the case as in [2, 13, 15] that all value assignments to a variable are stored in the same register to avoid additional transfer operations and to obtain a simpler control logic.

We denote with $\chi(G)$ the number of colors in a minimum coloring of the conflict graph G. Given a heuristic H, $\chi_H(G)$ is the number of colors generated by H. For a flow graph with branching depth d and at most m execution paths in a conditional block, the heuristics by Kurdahi, Parker and by Park, Kim and Liu have worst case bound $m^d \cdot \chi(G)$ (see [11]). For flow graphs with conditional blocks, but without parallel blocks, Kannan and Proebsting [12] have proposed an algorithm with ratio $2\chi(G)$. This ratio gets rapidly worse if parallel conditional blocks appear in the flow graph as Figure 2 illustrates.

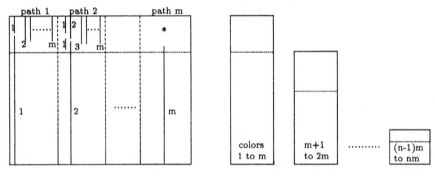

a) 1 block, colors denoted for the heuristic

b) n parallel blocks of the same structure as in a), colors denoted for the heuristic

Fig. 2. Worst-case example for the heuristic of Kannan and Proebsting

The heuristic needs because of its successive method in a) m colors, which can be transferred and extended to $m \cdot n$ colors in b). In the optimum all variables lying behind the dashed horizontal line in a) would get color 1 and m colors would be used above the dashed line (sufficient for all execution paths). Consequently, in b) $m + n$ colors would be sufficient, because in each block i only colors with number i would appear after the respective horizontal dashed line ($i \in \{1, \ldots, n\}$) and for the areas above the dashed lines one needs the color numbers from $m+1, \ldots, m+n$. In a recursive construction with branching depth d, we will reach $m^d \cdot n$ colors for the heuristic and $m^d + n$ in the optimum. With $n = m^d$, we will reach a worst case bound of $m^d \cdot \chi(G)$.

We propose an approximation algorithm with constant worst case bound for flow graphs with parallel conditional blocks and constant branching depth d. Our algorithm consists of two steps. In the first step, called equalization, the lifetimes are enlarged such that the lifetimes form an unique interval across all possible program executions for each variable. Let $\omega(G)$ be the cardinality of a maximum clique in G. We prove that the conflict graph $\bar{G} = (V, \bar{E})$ with enlarged lifetimes satisfies $\omega(\bar{G}) \leq (2d + 1) \cdot \omega(G)$ where d is the branching depth of the flow graph. In the second step we propose an algorithm with approximation value $(d + 1) \cdot \chi(\bar{G})$ for conflict graphs $\bar{G}$ with equalized variables.

2 Definitions

The *stage* $d(B)$ of a block B is the depth of the block in the nested branching structure. If a block B does not lie in another block, then $d(B) = 1$; otherwise $d(B) = d(f(B)) + 1$ (see again Figure 1).

Let $T_a(v)$ and $T_e(v)$ be the minimum starting and maximum ending time of variable v. In the further, we may assume that the lifetime structure for a variable is connected. This means, that for each time step $t \in [T_a(v), T_e(v))$ there exists at least one program execution such that v is alive at step t (otherwise the lifetimes are separated and we can use different variable names).

Two variables v and w are *in conflict* ($v \not\sim w$) if there is at least one program execution such that v and w are alive at one time step t. The graph $G = (V, E)$ with edges $\{v, w\}$ for two variables that are in conflict is called the *conflict graph*. In this definition we implicitly make the assumption that parallel blocks are processed independently. The paths in the case-statements are chosen by evaluation of arithmetical or logical expressions.

The start and end times of block B in the flow graph are denoted by $T_a(B)$ and $T_e(B)$. At the start and end of a conditional block B_i we have an input and output set of variables I_i and O_i. We assume that each input variable v is alive at step $T_a(B_i)$ for all program executions that reach B_i. This means that v must get a value before B_i is reached. Therefore, the set of input variables I_i (and analogously the set of output variables O_i) forms a clique in the conflict graph.

3 Equalization of variables

We distinguish between global and local variables. A variable v is *local with respect to block B* if its lifetime interval occurs only inside one execution path ℓ and does not occur in the blocks nested in line ℓ of B. Such a variable is called a local variable at stage $d(B)$. Variables where all lifetime intervals are outside of all blocks are called global variables at stage 0. In contrast, the other variables cross at least one block. A variable is said *global with respect to block B* if the lifetime intervals lie both inside and outside of B. If a variable v occurs inside of B, inside of the father block $f(B)$ and not outside of $f(B)$, then v is called a global variable at stage $d(B)$. Variables that lie inside a block B at stage 1 and outside of all blocks are called global variables at stage 1. We may assume that variables that occur only in different execution paths have different names; otherwise we can rename them. In Figure 3, a is global at stage 0, y global at stage 1, z global at stage 2 and b local at stage 1.

Our approximation algorithm works in two steps. In this step, the lifetime intervals of the variables in the flow graph are enlarged such that $t_a(v) = T_a(v)$ and $t_e(v) = T_e(v)$ for all program executions and for all variables v (for an example see Figure 3). Thereby, the conflict graph $G = (V, E)$ is transformed into a simpler structured graph $\bar{G} = (V, \bar{E})$ with $E \subset \bar{E}$. The size of a maximum clique in $\bar{G}$ can be increased. Therefore, we prove an upper bound for the new clique size $\omega(\bar{G})$. Later in Section 4, we generate an approximate coloring of $\bar{G}$ using an iterative path search. This coloring can be used also as a coloring of G.

The main difficulty in register allocation is that variables can have different lifetimes across different execution paths. We specify the entire flow graph as an unconditional block B_F with one execution path. Consider a variable $v \in V$. Then, there exists an unique block $B(v)$ at minimal stage such that all lifetime intervals lie completely in one execution path ℓ of $B(v)$. Equalization of v means that the lifetimes of v form now one interval $[T_a(v), T_e(v))$ in execution path ℓ of block $B(v)$. If v is a global variable at stage 0 or 1, then we get one lifetime interval in B_F. The other global variables at stage $h + 1$ are transformed into local variables at stage h (for $h \geq 1$). For example, variable z in Figure 3 will become local at stage 1.

Let $\bar{G} = (V, \bar{E})$ be the conflict graph for the enlarged intervals. A maximum clique in $\bar{G}$ can be computed in polynomial time [9]. Moreover, for each clique $\bar{C}$ in $\bar{G}$ there is at least one time step t and one program execution such that all variables $v \in \bar{C}$ are alive at time t. We note that this in general does not hold for the original conflict graph $G = (V, E)$.

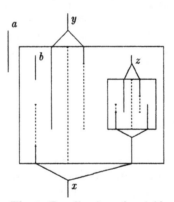

Fig. 3. Equalization of variables

Another interpretation of equalization of a variable $v \in V$ is the following. We denote with $I(v, B)$ the intersection of the lifetime interval $[T_a(v), T_e(v))$ and the block interval $[T_a(B), T_e(B))$. Consider a block B nested in $B(v)$ such that $I(v, B) \neq \emptyset$ and assume that v is not alive in execution path ℓ of B during a set of intervals within $I(v, B)$. Then, equalization does the same as including this set of lifetime intervals in execution path ℓ of B. These lifetime intervals are called pseudo intervals and are drawn as vertical dashed lines. Conflicts between two pseudo intervals and between pseudo and real intervals are called pseudo conflicts. In Figure 3, we have a pseudo conflict between x and y.

Theorem 1. *Let G be the conflict graph for a flow graph with branching depth d. Then, the conflict graph $\bar{G}$ generated by equalization of variables in V satisfies:*

$$\omega(\bar{G}) \leq (2 \cdot d + 1) \cdot \omega(G)$$

Proof. **No parallel structure**: Let $\bar{C}$ be a clique in $\bar{G}$. Then, all variables in $\bar{C}$ are alive at one time step t in one program execution. Since we have no parallel structure, there is an unique block B_h at stage h, $h \in \{1, \ldots, d\}$ that contains $\bar{C}$. Let C be the real variables at time t in $\bar{C}$. Pseudo intervals in $\bar{C}$ can be obtained only by equalization of variables that cross block B_h or a block B_i at stage $i = h - 1, \ldots, 1$ that contain B_h. Note, that other blocks at stage i ended before B_i starts generate no further pseudo intervals. Let $A_i \subset I_i$ (and $E_i \subset O_i$) be sets of the input (and output) variables of block B_i, $1 \leq i \leq h$ at stage i that generate pseudo intervals in $\bar{C}$. Then, $|\bar{C}| \leq |C \cup A_1 \cup E_1 \cup \ldots A_h \cup E_h|$ and since $A_i \subset I_i$ and $E_i \subset O_i$ are cliques in G:

$$|\bar{C}| \leq |C| + \sum_{i=1}^{h} |A_i| + |E_i| \leq (2 \cdot h + 1)\omega(G).$$

Parallel structure: Again, $\bar{C}$ can be found at a given time step t for one program execution. Let us assume that thereby n parallel execution paths are active at step t. Notice, that the path of the unconditional block B_F at stage 0 is possibly one of them. Each of these parallel paths lies in an unique block $B_{h_j}^{(j)}$ at stage $h_j \in \{0, \ldots, d\}$, $1 \leq j \leq n$. We denote with $\bar{C}_j$ the set of variables in execution path j at step t and with C_j the real variables in $\bar{C}_j$. For $h_j = 0$, we have $C_j = \bar{C}_j$. Let us assume that $h_j > 0$. Then, pseudo intervals in $\bar{C}_j$ can be obtained only by equalization of variables that cross $B_{h_j}^{(j)}$ or blocks $B_i^{(j)}$ at stage $i = h_j - 1, \ldots, 1$ that contain $B_{h_j}^{(j)}$. If $A_{ij} \subset I_i^{(j)}$ (and $E_{ij} \subset O_i^{(j)}$) are the sets of input (and output) variables of $B_i^{(j)}$ at stage i that generate pseudo intervals in $\bar{C}_j$, then

$$|\bar{C}_j| \leq |C_j \cup A_{1j} \cup E_{1j} \cup \ldots \cup A_{h_j j} \cup E_{h_j j}|.$$

Denote $X_{ij} = A_{ij} \cup E_{ij}$, $C = C_1 \cup \ldots \cup C_n$ and $Y_i = X_{i1} \cup \ldots \cup X_{in}$ (not defined sets X_{ij} are empty sets). Then, we get:

$$|\cup_{j=1}^{n} \bar{C}_j| = |(C_1 \cup X_{11} \cup \ldots X_{h_1 1}) \cup \ldots \cup (C_n \cup X_{1n} \cup \ldots X_{h_n n})|$$
$$\leq |(C_1 \cup \ldots \cup C_n) \cup (X_{11} \cup \ldots \cup X_{1n}) \cup \ldots \cup (X_{d1} \cup \ldots \cup X_{dn})|$$
$$\leq |C| + \sum_{i=1}^{d} |Y_i|.$$

Clearly, the set C of real variables is a clique in G. Let us consider one set Y_i. Then, for all $j \in \{1, \ldots, n\}$ and each element $x \in X_{ij}$ there is a real lifetime interval in block $B_i^{(j)}$ that contains time step t (since x is a global variable at stage i that generates a pseudo conflict). Using the parallelism, for $x \in X_{ij}$ and $y \in X_{ij'}$ with $j \neq j'$ and $X_{ij} \neq X_{ij'}$ we have $\{x, y\} \in E$. We prove this as follows. $B_i^{(j)}$ and $B_i^{(j')}$ are parallel blocks at stage i. Therefore, the program execution reaches both blocks together. For x (and y) we have a real lifetime interval in $B_i^{(j)}$ (and in $B_i^{(j')}$) that contains step t. Since the blocks are parallel, we can construct a program execution that reaches both real lifetime intervals. This implies that $|Y_i| \leq 2 \cdot \omega(G)$ for $1 \leq i \leq d$ and

$$|\bar{C}| = |\cup_{j=1}^n \bar{C}_j| \leq (2 \cdot d + 1)\omega(G).$$

An example to show $\omega(\bar{G}) = (2d+1)\omega(G)$ is given in [11]. $\square$

4 Approximation algorithm for equalized variables

The main idea of our algorithm is to find iteratively a set K of variables in the flow graph such that the chromatic number $\chi(\bar{G}[K])$ is bounded by a constant and that the maximum clique size $\omega(\bar{G}) > \omega(\bar{G}[V \setminus K])$. Later, we observe that $\chi(\bar{G}[K])$ depends on the branching depth of the flow graph. Let $B^{(\ell)}$ be the subflow graph in path ℓ of B and let $S(B^{(\ell)})$ be the set of conditional blocks in $B^{(\ell)}$. The set of all blocks nested in $S(B^{(\ell)})$ is denoted by $\bar{S}(B^{(\ell)})$.

Using the equalization in Section 3, we have only local variables at stage $1, \ldots, d$ or global variables at stage 0. For each local or global variable v we replace v by an unconditional block B_v with $[T_a(B_v), T_e(B_v)) = [T_a(v), T_e(v))$. The starting (ending) time $T_a(B)$ $(T_e(B))$ of an (un-)conditional block is the earliest starting (latest ending) time of all variables in B. We assume that there exists at least one variable in B that is alive at step t for each time step $t \in [T_a(B), T_e(B))$; otherwise we split B into a sequence of blocks with this property.

4.1 Pathsearch in one execution line

First, we consider one execution path ℓ in a conditional block B. Let D be the set of conditional and unconditional blocks in this path. The following algorithm computes a set $X \subset D$ of blocks in B with width at most two (this means with at most two blocks at one time step). We define $T_a(D) = min_{B' \in D} T_a(B')$ and $T_e(D) = max_{B' \in D} T_e(B')$.

Algorithm 1
(1) choose a block $B \in D$ with earliest starting time and set $X = \{B\}$, $D = D \setminus \{B\}$,
(2) let $Y = \{B' \in D : T_e(B') > T_e(B)\}$,
(3) if $Y = \emptyset$ then stop; otherwise take a block $B' \in Y$ with earliest starting time and set $B = B'$, $X = X \cup \{B'\}$, $D = D \setminus \{B'\}$,
(4) if there three blocks in X at one time step then
　　(4.1) determine a block $B^* \in X$, $B^* \neq B$ with earliest starting time such that $[T_a(B^*), T_e(B^*)) \cap [T_a(B), T_e(B)) \neq \emptyset$,
　　(4.2) set $X = X \setminus \{B' \in X | B' \neq B, B' \neq B^*, T_a(B') \geq T_a(B^*)\}$,
(5) goto (2).

If there are three blocks at one time step t in step (4), then exactly one block B' must be deleted. We note that the computed set X is a chain of blocks from $T_a(D)$ to $T_e(D)$ in D. Using step (4) in the algorithm, for each time step $t \in [T_a(D), T_e(D))$ there are at most two blocks at step t. We can use this algorithm for each execution path of a conditional block B. Then, the union $\bar{X}$ of the computed sets forms a set of blocks without time gaps. This implies that there is at least one block $B' \in \bar{X}$ with $t \in [T_a(B'), T_e(B'))$ for each $t \in [T_a(D), T_e(D))$.

4.2 Recursive pathsearch

The idea to compute a closed chain of blocks in one conditional block can be extended to nested blocks. The goal is to find a set K of intervals such that $\omega(\bar{G}) > \omega(\bar{G}[V \setminus K])$. In the first step, we have parallel blocks $B_1, \ldots, B_k$ at stage 1 in the flow graph. Each of these blocks B_i is either an unconditional block (a global variable at stage 0) or a conditional block. Recursively, we compute the collection K as follows.

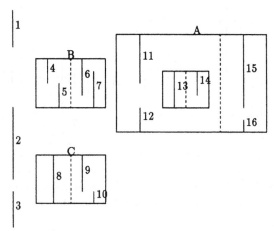

Fig. 4. Recursive pathsearch in blocks

Algorithm 2
given: a set of parallel blocks $\mathcal{B} = \{B_1, \ldots, B_k\}$,
output: a set K of intervals in $\mathcal{B}$ and a set of blocks $\bar{\mathcal{B}}$ (either blocks in $\mathcal{B}$ or blocks $B \in \bar{S}(B_i^{(\ell)})$ with $B_i \in \mathcal{B}$),
(1) apply algorithm 1 to $B_1, \ldots, B_k$ and obtain a set X of blocks with width at most two,
(2) set $\bar{\mathcal{B}} = X$ and $K = \{B \in X | B$ is an unconditional block$\}$,
(3) for each conditional block $B \in X$ and each execution path ℓ in B do
  (3.1) let $D = S(B^{(\ell)})$,
  (3.2) apply algorithm 2 to D and obtain a set K' of variables and a set D' of blocks,
  (3.3) set $K = K \cup K', \bar{\mathcal{B}} = \bar{\mathcal{B}} \cup D'$.

Let us consider the example in Figure 4. Using algorithm 1 a sequence of blocks $X = \{1, A, 2, C, 3\}$ with width at most two is computed. Since A and C are conditional blocks, we compute in each path of A and C again a sequence of blocks. In total, the collection K of variables is given by $\{1, 2, 3, 8, \ldots 16\}$.

Lemma 2. *Algorithm 2 computes a set K of variables such that the maximum clique size $\omega(\bar{G}) > \omega(\bar{G}[V \setminus K])$.*

Proof. It is sufficient to reduce the clique size $\omega(\bar{G}_t)$ by at least one for each time step t with $\omega(\bar{G}_t) > 0$. We prove this only for $d = 1$, the general case is proved by induction. The algorithm finds in step (1) at least one block $B \in X$ with $t \in [T_a(B), T_e(B))$. If B is an unconditional block with variable v, then $v \in K$ and $\omega(\bar{G}_t[V \setminus K]) \leq \omega(\bar{G}_t) - 1$. If B is a conditional block, then algorithm 2 computes for each path ℓ in B at least one variable alive at step t (if there is at least one variable in path ℓ at step t). In this case, this variable set is a subset of K and reduces again the maximum clique size at step t by at least one. □

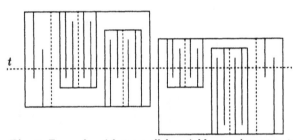

Fig. 5. Example with 8 parallel variables at time step t

On the other hand, for a flow graph with branching depth d it is possible that the set K contains a clique C with 2^{d+1} variables (for $d = 2$ see time step t in Figure 5).

4.3 Deletion algorithm

Next, we describe an algorithm to reduce the maximum number of parallel intervals in K. Using algorithm 2 (step (1)) we have computed a chain of blocks $B_1, \ldots, B_k$ at stage 1. We may assume that these blocks are ordered by their starting times. For $i = 2, \ldots, k$ we consider pairs of blocks B_{i-1}, B_i and we delete some blocks in B_{i-1} that are generated by algorithm 2. Let y be the starting time of block B_i. If $y \in [T_a(B_{i-1}), T_e(B_{i-1}))$, then we delete the following blocks in the $\bar{B}$:

(1) delete in each execution line of B_{i-1} all blocks $B \in \bar{B}$ with $T_a(B) \geq y$,

(2) if there are two blocks $B, B' \in \bar{B}$ in one execution path of B_{i-1} at step y, then delete one block with later starting time,

(3) if there is one conditional block $B \in \bar{B}$ in one execution path of B_{i-1} at step y, then apply steps (1) and (2) recursively to each path of B.

After this reduction for block pairs at stage 1, we apply the same algorithm recursively to each remaining block in $\bar{B}$ at stages $2, \ldots, d$. For our example in Figure 4, the variables $10, 12, 16$ are deleted.

Algorithm 3

(1) apply algorithm 2 to $B = \{B_1, \ldots, B_k\}$ and obtain a set K of variables and a set $\bar{B}$ of blocks,

(2) apply the deletion algorithm to $\bar{B}$ and obtain a set $\tilde{B}$ of remaining blocks,

(3) set $K' = \{B \in \tilde{B} | B$ is an unconditional block $\}$,

(4) apply an algorithm to color K as given in Lemma 3.

Lemma 3. *Given a flow graph with depth d, algorithm 3 produces a set of variables such that the maximum clique size $\omega(\bar{G})$ is reduced by at least one and that the variables can be colored with at most $d + 2$ colors.*

Proof. First, the maximum clique size is reduced by one. We consider the case $d = 1$; the general case can be proved by induction. Again, it is sufficient to prove that the clique size $\omega(\bar{G}_t)$ is reduced by one for each time step t with $\omega(\bar{G}_t) > 0$. Step (1) deletes only variables v in B_{i-1} with $T_a(v) \geq T_a(B_i)$. This implies that the clique size $\omega(\bar{G}_t)$ for $t \in [T_a(B_{i-1}), T_e(B_{i-1})) \cap [T_a(B_i), T_e(B_i))$ is reduced by variables in B_i. Using step (2), only one of two variables with the later starting time is deleted. This implies that the remaining variables in B_{i-1} reduce the clique size $\omega(\bar{G}_t)$ for each $t \in [T_a(B_{i-1}), T_a(B_i))$.

Next, the set of variables K can be colored with $d+2$. Again, we consider only the case $d = 1$ and left the induction proof to the reader. Consider the sequence of blocks $B_1, \ldots, B_\ell$ at stage 1 produced by algorithm 3 and ordered by their starting times. For each time step t there are at most two blocks B_{i-1}, B_i and at most two chosen intervals in each execution path of these blocks. Furthermore, for each $t \in [T_a(B_i), T_e(B_{i-1}))$ we have at most one variable alive at t in each path of B_{i-1}. We start with B_ℓ and color all chosen variables in B_ℓ with colors $1, 2$. In $B_{\ell-1}$ the latest interval in each path is colored with 3. Then, we continue the coloring of $B_{\ell-1}$ with colors $2, 3$. By induction on the length ℓ, the chosen intervals in $B_\ell, \ldots, B_1$ can be colored with 3 colors. $\square$

Using algorithm 3 we can compute iteratively sets of variables such that each variable set can be colored with at most $d+2$ colors. The number of iterations is bounded by $\omega(\bar{G})$ and, therefore, such an algorithm needs at most $(d+2)\omega(\bar{G})$ colors. This can be improved using the following ideas.

Again, let $B_1, \ldots, B_k$ be the conditional blocks at stage 1 and let V^{glo} be the set of global variables at stage 0. First, compute for each block B_i and time step $t \in [T_a(B_i), T_e(B_i)) \cap \mathbb{IN}$ the size $\omega(G[V(B_i)_t])$ of a maximum clique at time t. Furthermore, we set $\omega(G[V(B_i)_t]) = 0$ for $t \leq T_a(B_i) - 1$ and $t \geq T_e(B_i)$.

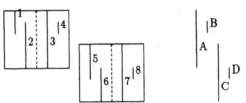

Fig. 6. The construction of the interval graph G^*

Next, we construct an interval graph G^* as follows. For each global variable $v \in V^{glo}$ we take an interval $[T_a(v), T_e(v))$. For each block B_i, integer $k \in \mathbb{N}$ and pair $u, v \in [T_a(B_i), T_e(B_i)) \cap \mathbb{N}$, $u < v$ with $\omega(G[V(B_i)_x]) \geq k$ for $x \in [u, v)$ and $\omega(G[V(B_i)_y]) < k$ for $y \in \{u - 1, v\}$ we take one interval $[u, v)$ (called an interval in B_i of width k). By the construction, we have $\omega(G^*) = \omega(\bar{G})$. Now, we compute a minimum coloring f for the interval graph G^*.

The idea is to extend this coloring for the variables in $\bar{G}$. A global variable $v \in V^{glo}$ gets the color of the corresponding interval $[T_a(v), T_e(v))$. For a block B_i we consider a sequence of intervals $[u_1, v_1), \ldots, [u_\ell, v_\ell)$ of width k with $T_a(B_i) \leq u_1 < v_1 < u_2 < \ldots < v_\ell \leq T_e(B_i)$. In the first step, we consider the case $k = 1$. For each execution path ℓ of B_i we compute a set of variables K_ℓ using algorithm 3 applied on $S(B_i^{(\ell)})$. By Lemma 3, the set $K = \bigcup_{\ell=1}^{m_{B_i}} K_\ell$ can be colored with $d + 1$ colors. Each lifetime interval $[T_a(v), T_e(v))$ for $v \in K$ lies in exactly one of the intervals $[u_j, v_j)$ of width $k = 1$. For a variable $v \in K$ with $[T_a(v), T_e(v)) \subset [u_j, v_j)$ we use the colors $(f([u_j, v_j)), 1), \ldots, (f([u_j, v_j)), d+1)$. Then, we delete the colored variables in block B_i and use the same algorithm for $k = 2, \ldots, \omega(G[V(B_i)])$ to obtain the colors of the remaining variables in $V(B_i)$. In total, we use at most $(d + 1)\chi(G^*)$ colors.

In Figure 6 we have a flow graph with two conditional blocks and eight variables and the corresponding interval graph G^*. The graph G^* can be colored with two colors $f(A) = 1$, $f(D) = 1$, $f(B) = 2$ and $f(C) = 2$. Using the algorithm above, we obtain the following colors for the variables: $f(1) = (1, 1)$, $f(2) = (1, 2)$, $f(3) = (1, 1)$, $f(4) = (2, 1)$, $f(5) = (2, 1)$, $f(6) = (2, 2)$, $f(7) = (2, 1)$, $f(8) = (1, 1)$.

Theorem 4. *Let G be the conflict graph for a flow graph of depth d. Then, the number of colors generated by our algorithm is bounded by $(d + 1)(2d + 1) \cdot \chi(G)$. If the branching depth is constant, we have an algorithm with constant approximation bound.*

Proof. The conflict graph $\bar{G}$ produced by equalization satisfies $\omega(\bar{G}) \leq (2d + 1)\omega(G) \leq (2d+1)\chi(G)$ and the algorithm above for equalized variables needs at most $(d + 1)\chi(G^*) = (d + 1)\omega(\bar{G})$ colors. $\square$

An example such that the worst case ratio $(d + 1)$ can be achieved asymptotically for large n is given in [11].

5 Conclusion

Further topics to work on consist in:

(1) Generalize the proposed method to flow graphs with loops.
(2) Find better algorithms for flow graphs with small branching depth.
(3) Implementation of the heuristic and its evaluation in comparison with existing algorithms (as e.g. [13], [15]).

References

1. A. Aho, R. Sethi and J. Ullman: Compilers, Principles, Techniques and Tools, Addision Wesley, 1988.
2. R.A. Bergamaschi, R. Camposano and M. Payer: Data-path synthesis using path analysis, 28.th Design Automation Conference (1991), 591 – 595.
3. R. Camposano and W. Wolf: High-Level Synthesis, Kluwer, Dordrecht, 1992.
4. G.J. Chaitin: Register allocation and spilling via graph coloring, Symposium on Compiler Construction (1982), 98-101.
5. M.C. MacFarland, A.C. Parker and R. Camposano: Tutorial on high-level synthesis, 25.th Design Automation Conference (1988), 330 – 336.
6. D. Gajski, N. Dutt, A. Wu and S. Lin: High-Level Synthesis: Introduction to Chip and System Design, Kluwer, Dordrecht, 1992.
7. M.S. Hecht: Flow Analysis of Computer Programs, Elsevier, New York, 1977.
8. C.Y. Huang, Y.S. Chen, Y.L. Lin and Y.C. Hsu: Data path allocation based on bipartite weighted matching, 27.th Design Automation Conference (1990), 499 – 503.
9. K. Jansen: Processor-optimization for flow graphs, Theoretical Computer Science 104 (1992), 285 – 298.
10. K. Jansen: On the complexity of allocation problems in high level synthesis, Integration - the VLSI Journal 17 (1994), 241 – 252.
11. K. Jansen and J. Reiter, Approximation algorithms for register allocation, Universität Trier, Forschungsbericht 13 (1996).
12. S. Kannan and T. Proebsting, Register allocation in structured programs, Symposium on Discrete Algorithms (SODA), 1995, 360-368.
13. F.J. Kurdahi and A.C. Parker: Real: a program for register allocation, 24.th Design Automation Conference (1987), 210 – 215.
14. P. Michel, U. Lauther and P. Duzy: The Synthesis Approach to Digital System Design, Kluwer, Dordrecht, 1992.
15. C. Park, T. Kim and C.L. Liu: Register allocation for general data flow graphs, 2.nd European Design Automation Conference (1993), 232 – 237.
16. P. Pfahler: Automated datapath synthesis: a compilation approach, Microprocessing and Microprogramming 21 (1987), 577 – 584.
17. D.L. Springer and D.E. Thomas: Exploiting the special structure of conflict and compatibility graphs in high-level synthesis, International Conference on Computer Aided Design (1990), 254 – 257.
18. C.J. Tseng and D. Siewiorek: Automated synthesis of data paths in digital systems, IEEE Transactions on CAD 6 (1989), 379-395.

A Parallel Approximation Algorithm for Resource Constrained Scheduling and Bin Packing

Anand Srivastav[*]   Peter Stangier[**]

Abstract

[1] We consider the following classical resource constrained scheduling problem. Given m identical processors, s resources $R_1, \ldots, R_s$ with upper bounds $b_1, \ldots, b_s$, n independent jobs $T_1, \ldots, T_n$ of unit length, where each job requires one processor and an amount $R_i(j) \in \{0, 1\}$ of resource R_i, $i = 1, \ldots, s$, the optimization problem is to schedule the jobs at discrete times in $\{1, \ldots, n\}$ subject to the processor and resource constraints so that the latest scheduling time is minimum. Note that multidimensional bin packing is a special case of this problem. We give for every fixed $\alpha > 1$ the first *parallel* 2α-factor approximation algorithm and show that there cannot exist a polynomial-time approximation algorithm achieving an approximation factor better than $4/3$, unless $P = NP$.

Keywords: resource constrained scheduling, multidimensional bin packing, derandomization, $log^c n$-wise independence, NC-approximation algorithm.

1   Introduction

Resource constrained scheduling is the following problem: The input is

- a set $T = \{T_1, \ldots, T_n\}$ of independent jobs. Each job T_j needs one time unit for its completion.

- a set $\mathcal{P} = \{P_1, \ldots, P_m\}$ of identical processors. Each job needs one processor.

- a set $\mathcal{R} = \{R_1, \ldots, R_s\}$ of limited resources. This means that at any time all resources are available, but the available amount of each resource R_i is bounded by $b_i \in \mathbb{N}$, $i = 1, \ldots, s$.

- For $i = 1, \ldots, s$, $j = 1 \ldots, n$ let $R_i(j)$ be 0/1 resource requirements saying that every job T_j needs $R_i(j)$ units of resource R_i during its processing time. For a job $T_j \in \mathcal{T}$ and a time $z \in \{1, \ldots, n\}$ let x_{jz} be the 0/1 variable which is 1 iff job T_j is scheduled at time z.

- Given a valid schedule let C_{max} be the latest completion time defined by $C_{max} = \max\{z \mid x_{jz} > 0, j = 1, \ldots, n\}$.

The combinatorial optimization problem is:

Definition 1.1 *(Resource Constrained Scheduling)*

(i) *(Integral Problem) Find a schedule, that is a 0/1 assignment for all varibles x_{jz}, subject to the processor and resource constraints such that $\sum_{z=}^{n} x_{jz} = 1$ for all jobs T_j and C_{max} is minimum. Let C_{opt} denote this minimum.*

[*]Institut für Informatik, Humboldt Universität zu Berlin, Unter den Linden 6, 10099 Berlin, Germany; e-mail: srivasta@informatik.hu-berlin.de

[**]Zentrum für Paralleles Rechnen (ZPR), Universität zu Köln, Weyertal 80, 50931 Köln, Germany; e-mail: pstangier@mac.de.

[1]The results of this extended abstract will appear in *Discrete Applied Mathematics (1997)*.

(ii) *(Fractional Problem) Find a fractional schedule, that is an assignment of each x_{jz} to a rational number in the closed interval $[0, 1]$ subject to the processor and resource constraints so that $\sum_{z=1}^{n} x_{jz} = 1$ for all jobs T_j and C_{max} is minimum. Let C denote this minimum. We shall call C_{opt} the (integral) optimal and C the fractional optimal schedule.* [2]

According to the standard notation of scheduling problems the integral problem can be formalized as $P|\text{res} \cdot \cdot 1, p_j = 1|C_{max}$. This notation means: the number of identical processors is part of the input ($P|$), resources are involved (res), the number of resources and the amount of every resource are part of the input (res $\cdot \cdot$), every job needs at most 1 unit of a resource (res $\cdot \cdot 1$), the processing time of all jobs is equal 1 ($p_j = 1$) and the optimization problem is to schedule the jobs as soon as possible ($|C_{max}$). The integral problem is NP-hard in the strong sense even for $m = 3$ [GaJo79], while the fractional problem can be solved by linear programming in polynomial time. An interesting special case of resource constrained scheduling is the following generalized version of the multidimensional bin packing problem.

Definition 1.2 *(Bin Packing Problem $BIN(\vec{l}, d)$) Let $d, n, l_i \in \mathbb{N}$, $i = 1, \ldots, d$, and let $\vec{l} = (l_1, \ldots, l_d)$. Given vectors $\vec{v}_1, \ldots, \vec{v}_n \in [0, 1]^d$, pack* [3] *all vectors in a minimum number of bins such that in each bin B and for each coordinate i, $i = 1, \ldots, d$, $\sum_{\vec{v}_j \in B} v_{ij} \leq l_i$. Define $L_R = \lceil \max_{1 \leq i \leq d} \frac{1}{l_i} \sum_{j=1}^{n} v_{ij} \rceil$. (Observe that L_R is the minimum number of bins, if fractional packing is allowed.)*

$BIN(1, d)$ is the multidimensional bin packing problem, and $BIN(1, 1)$ is the classical bin packing problem. The intention behind the formulation with a bin size vector $\vec{l}$ is to analyse the relationship between bin sizes and polynomial-time approximability of the problem.

Previous Work. The known polynomial-time approximation algorithms for resource constrained scheduling are due to Garey, Graham, Johnson, Yao [GGJY76] and Röck and Schmidt [RS83]. Garey et al. constructed with the First-Fit-Decreasing heuristic a schedule of length C_{FFD} which asymptotically is a $(s + \frac{1}{3})$-factor approximation, i.e. there is a non-negative integer C_0 such that $C_{FFD} \leq C_{opt}(s + \frac{1}{3})$ for all instances with $C_{opt} \geq C_0$. De la Vega and Lueker [VeLu81] improved this result presenting for every $\epsilon > 0$ a linear-time algorithm which achieves an asymptotic approximation factor of $s + \epsilon$. For restricted bin packing and scheduling Błazewicz and Ecker gave a linear time algorithm (see [BESW93]). The first polynomial-time approximation algorithm for resource constrained scheduling with an *constant factor* approximation guarantee has been given in [SrSt97]: For every $\epsilon > 0$, $(1/\epsilon) \in \mathbb{N}$, an integral schedule of size at most $\lceil (1 + \epsilon)C_{opt} \rceil$ can be constructed in strongly polynomial time, provided that all resource bounds b_i are at least $\frac{3(1+\epsilon)}{\epsilon^2} \log(8Cs)$ and the number of processors is at least $\frac{3(1+\epsilon)}{\epsilon^2} \log(8C)$. This approximation guarantee is independent of the number of processors or resources. For $\epsilon = 1$ this gives a sequential 2-factor approximation algorithm for resource constrained scheduling.

Results. In this paper we give a NC-version of the sequential 2-factor approximation algorithm. Applying an extension of the method of $\log^c n$-wise independence to multivalued random variables, we show the following: For every $\tau \geq \frac{1}{\log n}$ and every $\alpha > 1$

---

[2] Note that by definition the fractional optimal schedule C is an integer, only the assignments of jobs to times are rational numbers.

[3] Packing simply means to find a partitioning of vectors in a minimum number of sets so that the vectors in each partition (= bin) satisfy the upper bound conditions of the definition.

there is an NC approximation algorithm for resource constrained scheduling with an approximation guarantee of 2α which requires $O(n(ns)^{\frac{1}{\tau}+1})$ EREW-PRAM processors and runs in $O((\frac{\log(ns)}{\tau})^4))$ time, provided that $m, b_i \geq \alpha(\alpha-1)^{-1}n^{\frac{1}{2}+\tau}\sqrt{\log 3n(s+1)}$ for all $i = 1, \ldots, s$.

Since we assume here that the number of processors and resource constraints is "large", one might wonder, whether this is an interesting problem class. The non-approximability result we prove shows that the problems we consider have complexity-theoretic significance: Let $\tau \in [0, 1/2)$. Even if $m, b_i \in \Omega(n^{1/2+\tau}\sqrt{\log(ns)})$ for all i, then for any $\rho < 4/3$ there cannot exist a polynomial-time ρ-factor approximation algorithm, unless $P = NP$. [4]

Note that both results carry over to the multidimensional bin packing problem: For every $\tau \geq \frac{1}{\log n}$ and $\alpha > 1$ there is an NC approximation algorithm for multidimensional bin packing with an approximation guarantee of 2α which requires $O(n(nd)^{\frac{1}{\tau}+1})$ EREW-PRAM processors and runs in $O((\frac{\log(nd)}{\tau})^4))$ time, provided that $l_i \geq \alpha(\alpha-1)^{-1}n^{\frac{1}{2}+\tau}\sqrt{\log 3nd}$ for all i. And, even if $l_i \in \Omega(n^{1/2+\tau}\sqrt{\log(nd)})$ for all i, then there cannot exist a polynomial-time ρ-factor approximation algorithm for multidimensional bin packing for any $\rho < 4/3$, unless $P = NP$

2 Parallel Scheduling and Bin Packing

For convenience let us consider $s+1$ resources $R_1, \ldots, R_{s+1}$ where resource R_{s+1} represents the processors. There is no obvious way to parallelize the sequential 2-factor approximation algorithm of [SrSt97]. In this section we will show that at least in some special cases there is an NC approximation algorithm. The algorithm is based on the method of $\log^c n$-wise independence.

2.1 The Parallel Scheduling Algorithm

The algorithm has 4 steps.

Step 1: *Finding an optimal fractional scheduling in parallel.* We wish to apply randomized rounding and therefore have to generate an optimal fractional solution. The following integer program is equivalent to resource constrained scheduling.

$$
\begin{aligned}
\min D \\
\sum_j R_i(j)x_{jz} &\leq b_i & \forall R_i \in \mathcal{R}, \\
& & z \in \{1, \ldots, D\} \\
\sum_z x_{jz} &= 1 & \forall T_j \in \mathcal{T} \\
x_{jz} &= 0 & \forall T_j \text{ and } z > D \\
x_{jz} &\in \{0, 1\}.
\end{aligned}
$$

Its optimal fractional solution C is given by $C = \lceil C' \rceil$, where $C' = \max_{1 \leq i \leq s+1} \left\{ \frac{1}{b_i} \sum_{j=1}^n R_i(j) \right\}$. This is easily checked by setting $x_{jz} = 1/C$ for all jobs T_j and all times $z \in \{1, \ldots, C\}$.

---

[4]Note that this result does not contradict the sequential polynomial-time approximation guarantee of $\lceil (1 + \epsilon)C_{opt} \rceil$: For $C_{opt} = 3$ and every $\epsilon \leq 1/3$, $\lceil (1 + \epsilon)C_{opt} \rceil = 4$, thus the sequential approximation algorithm outputs the numbers 3 or 4. Hence this algorithm is not a ρ-factor approximation algorithm for $\rho < 4/3$, because a ρ-factor approximation algorithm would always give us a valid schedule of optimal length 3 (for problems with $C_{opt} = 3$).

Step 2: *Enlarged fractional schedule.* In order to define randomized rounding we enlarge the fractional schedule. The reason for such an enlargement is that for the analysis of the rounding procedure we need some room. Define for every $\alpha > 1$

$$d = 2^{\lceil \log(\alpha C) \rceil}, \tag{1}$$

hence $d \leq 2\alpha C$. (The reason for the above definition of d as a power of 2 is that we will later need the fact that $GF(d)$ is a field) The new fractional assignments of jobs to times are $\tilde{x}_{jz} = 1/d$ for all $j = 1, \ldots, n$, $z = 1, \ldots, d$. Note that this assignment defines a valid fractional schedule, even with tighter resource bound

$$\sum_{j=1}^{n} R_i(j)\tilde{x}_{jz} \leq b_i/\alpha \tag{2}$$

for all $i = 1, \ldots, s+1$, $z = 1, \ldots, d$.

Step 3: *Randomized Rounding.* The sequential randomized rounding procedure is to assign randomly and independently job T_j to time z with uniform probability $1/d$. With the method of limited independence we parallelize and derandomize this rounding scheme ini the next step.

Step 4: *Drandomized Rounding in NC.* We apply the parallel conditional probability method for multivalued random variables (Theorem 2.3).

2.2 Basic Tools for the Analysis

Throughout this subsection let k be an even integer and d be a power of 2. In the next subsection we will fix the values for k and d. (d will become the size of the enlarged fractional schedule)

First, the size C of the optimal fractional schedule can be computed in parallel with standard methods:

Lemma 2.1 *The size C of an optimal fractional schedule can be computed with $O(ns)$ $EREW - PRAM$-processors in $O(\log(ns))$ time.*

For parallel derandomization we wish to apply the method of $\log^c n$-wise independence. Berger and Rompel [BeRo91] have already discussed the case of multi-valued random variables. Let us here briefly point out how their framework can be applied to our problem and fix the work and time bounds.

Let $n = 2^{n'} - 1$ for some $n' \in \mathbb{N}$. A representation of $GF(2^{n'})$ as a n'-dimensional algebra over $GF(2)$ can be explicitly constructed using irreducible polynomials, for example the polynomials given in [Li82], Theorem 1.1.28. Let $b_1, \ldots, b_n$ be the n non-zero elements of $GF(2^{n'})$ in such an irreducible representation and let $B = (b_{ij})$ the following $n \times \lceil \frac{k-1}{2} \rceil$ matrix over $GF(2^{n'})$

$$B = \begin{bmatrix} 1 & b_1 & b_1^3 & \cdots & b_1^{k-1} \\ 1 & b_2 & b_2^3 & & b_2^{k-1} \\ 1 & b_3 & b_3^3 & & b_3^{k-1} \\ \vdots & \vdots & \vdots & \vdots & \vdots \\ 1 & b_n & b_n^3 & & b_n^{k-1} \end{bmatrix}.$$

B can be viewed as a $n \times \ell$ matrix over $GF(2)$ with $\ell = 1 + \lceil \frac{k-1}{2} \rceil \lceil \log(n+1) \rceil = O(k \log n)$. The matrix B is the well-known parity check matrix of binary BCH codes. Note that any k row vectors of B are linearly independent over $GF(2)$ [McSo77]. Alon, Itai and Babai [ABI86] showed that k-wise independent 0/1 random variables can be constructed from mutually independent 0/1 random variables using a BCH-matrix. The extension to multivalued random variables goes as follows.

Let $Y_1, \ldots, Y_\ell$ be independent and uniformly distributed random variables with values in $\Omega = \{0, \ldots, d-1\}$, $d \in \mathbb{N}$. Let Y be the vector $Y = (Y_1, \ldots, Y_\ell)$ and define Ω-valued random variables $X_1, \ldots, X_n$ by $X_i = (BY)_i \bmod d$ for all $i = 1, \ldots, n$. With $X = (X_1, \ldots, X_n)$ we can briefly write $X = BY \bmod d$. The following lemma follows from the extension of Theorem 2.8 in [BeRo91] to multivalued random variables (see section 4.2 of [BeRo91]).

Lemma 2.2 *The random variables $X_1, \ldots, X_n$ are k-wise independent and uniformly distributed.*

We consider the following class of functions F, arising in the analysis of resource constrained scheduling, for which an NC algorithm constructing vectors with F-value less than (resp. greater than) the expected value $\mathbb{E}(F(X_1, \ldots, X_n))$ can be derived.

Let (R_{ij}) be the resource constraint matrix where we consider the processors as a resource, so (R_{ij}) is a $(s+1) \times n$-matrix. To simplify the notation put $s' = s+1$. For integers $x_j, z \in \{1, \ldots, d-1\}$ let x_{jz} be the indicator function which is 1 if $x_j = z$ and 0 otherwise. For $i = 1, \ldots, s'$, $z \in \Omega$ and integers $x_j \in \Omega$, $j = 1, \ldots, n$, define the functions f_{iz} by

$$f_{iz}(x_1, \ldots, x_n) = \sum_{j=1}^{n} R_{ij}\left(x_{jz} - \frac{1}{d}\right), \tag{3}$$

and set

$$F(x_1, \ldots, x_n) = \sum_{iz} |f_{iz}(x_1, \ldots, x_n)|^k. \tag{4}$$

The following lemma analyses rounding with the conditional probability method in NC. Its proof is given in the appendix section.

Lemma 2.3 *Let $X_1, \ldots, X_n$ be k-wise independent random variables with values in $\Omega = \{0, \ldots, d-1\}$ defined as in Lemma 2.2 and let F be as in (4). Then with $O(\max(d^k, n^{k+1}s'))$ parallel processors we can construct integers $x'_1, \ldots, x'_n \in \Omega$ and $\hat{x}_1 \ldots, \hat{x}_n \in \Omega$ in $O(k^4 \log^4 n + k^2 \log n \log d + k \log n \log s')$ time such that*

(i) $F(x'_1, \ldots, x'_n) \geq \mathbb{E}(F(X_1, \ldots, X_n))$

(ii) $F(\hat{x}_1, \ldots, \hat{x}_n) \leq \mathbb{E}(F(X_1, \ldots, X_n))$.

Furthermore, we need estimates of the k-th moments.

Lemma 2.4 *Let $k = \lceil \frac{\log 3n(s+1)}{\tau \log n} \rceil$. Then*

$$\mathbb{E}\left(\sum_{iz} |f_{iz}(X_1, \ldots, X_n)|^k\right)^{1/k} \leq n^{\frac{1}{2}+\tau} \lceil \log 3n(s+1) \rceil^{1/2}.$$

Proof: Using an analog of the Chernoff bound due to Alon and Spencer (Corollary A7, [ASE92]) we can show exactly as in Berger/Rompel [BeRo91], proof of Corollary 2.6,

$$\mathbb{E}\left(\sum_{iz} |f_{iz}(X_1, \ldots, X_n)|^k\right) \leq 2(k/2)! (n/2)^{k/2}. \tag{5}$$

Robbins exact Stirling formula shows the existence of a constant γ_n with $\frac{1}{12n+1} \leq \gamma_n \leq \frac{1}{12n}$ so that $n! = (n/e)^n \sqrt{2\pi n} e^{\gamma_n}$ ([Bol85], page 4), thus $n! \leq 3(n/e)^n \sqrt{n}$. Furthermore, for all $k \geq 2$, $2\sqrt{k}(4e)^{-k/2} \leq 1$. With this bounds the right hand side of inequality (5) becomes $3(k/n)^{1/2}$. The claimed bound now follows by summing over all $i = 1 \ldots, s+1, z = 1 \ldots, d$ and taking the $(1/k)$-th root.

□

2.3 Main Algorithmic Results

Our main algorithmic results are the following theorems.

Theorem 2.5 *Let* $\alpha > 1$, $\tau \geq \frac{1}{\log n}$ *and suppose that* $b_i \geq \alpha(\alpha - 1)^{-1} n^{\frac{1}{2}+\tau} \sqrt{\log 3n(s+1)}$ *for all* $i = 1, \ldots, s+1$. *Then there is an NC-algorithms that runs on* $O(n(ns)^{\frac{1}{\tau}+1})$ *parallel processors and finds in* $O((\frac{\log(ns)}{\tau})^4))$ *time a schedule of size at most* $2^{\lceil \log(\alpha C) \rceil} \leq 2\alpha C$.

Proof: Put $s' = s + 1$, $k = \lceil \frac{\log 3ns'}{\tau \log n} \rceil$ and let d be as defined in (1), so $d \leq 2\alpha C \leq 2\alpha n$. Thus for fixed α we have $d^k = O(n^k)$ (we are not interested in large values of α). By Theorem 2.3 we can construct integers $\hat{x}_1, \ldots, \hat{x}_n$, where $\hat{x}_j \in \{1, \ldots, d\}$ for all j such that

$$F(\hat{x}_1, \ldots, \hat{x}_n) \leq \mathbb{E}(F(X_1, \ldots, X_n)) \tag{6}$$

holds, using

$$O(n^{k+1}s') = O(n(ns)^{\frac{1}{\tau}+1})$$

parallel processors in

$$O(k^4 \log^4 n + k^2 \log n \log d + k \log n \log s') = O((\frac{\log(ns)}{\tau})^4)$$

time. To complete the proof we must show that the vector $(\hat{x}_1, \ldots, \hat{x}_n)$ defines a valid schedule. This is seen as follows. By definition of the function F we have for all resources R_i and $z = 1, \ldots, d$

$$\left| \sum_{j=1}^{n} R_i(j)(x_{jz} - \frac{1}{d}) \right| \leq (F(x_1, \ldots, x_n))^{\frac{1}{k}} \leq (\mathbb{E}(F(X_1, \ldots, X_n))^{\frac{1}{k}}$$

$$\leq n^{\frac{1}{2}+\tau} \lceil \log 3ns' \rceil^{1/2}. \tag{7}$$

(the last inequality follows from Lemma 2.4). Hence, using (7), (2) and the assumption on the b_i's, we get for all resources R_i and all times $z = 1, \ldots, d$

$$\sum_{j=1}^{n} R_i(j)x_{jz} \leq \left| \sum_{j=1}^{n} R_i(j)(x_{jz} - \frac{1}{d}) \right| + \sum_{j=1}^{n} R_i(j)\frac{1}{d}$$

$$\leq n^{\frac{1}{2}+\tau} \lceil \log 3ns' \rceil^{1/2} + b_i/\alpha$$

$$\leq (1 - 1/\alpha)b_i + b_i/\alpha = b_i,$$

and the theorem is proved.

□

For multidimensional bin packing, the above theorem implies a 2α-factor approximation algorithm in NC.

Theorem 2.6 *For every* $\tau \geq \frac{1}{\log n}$ *and* $\alpha > 1$ *there is an NC-algorithm approximation algorithm for multidimensional bin packing with an approximation guarantee of* 2α *which requires* $O(n(nd)^{\frac{1}{\tau}+1})$ *EREW-PRAM processors and runs in* $O((\frac{\log(nd)}{\tau})^4))$ *time, provided that* $l_i \geq \alpha(\alpha - 1)^{-1} n^{\frac{1}{2}+\tau} \sqrt{\log 3nd}$ *for all* i.

Remark 2.7 The reason why we have to assume $b_i = \Omega(n^{\frac{1}{2}+\tau}\sqrt{\log ns})$ is due to the estimation of the k-th moments. Improvements of this method with the goal to show a 2α-factor (or even better) NC-algorithm under the weaker assumption $b_i = \Omega(\log ns)$ would be interesting. This would match the (presently) best sequential approximation guarantees.

3 Non-Approximability

In this section we will show that in general we cannot achieve a polynomial-time approximation better than a factor of 4/3, even if the number of processors and the resource bounds b_i are in $\Omega(n^{1/2+\tau}\sqrt{\log(ns)})$, $\tau < 1/2$. It is known that even if $m, b_i = \Omega(\log(ns))$ for all resource bounds b_i, it is not possible to find a schedule within a factor of $\rho < \frac{3}{2}$, unless $P = NP$. Here we invoke the conditions $m, b_i \in \Omega(n^{1/2+\tau}\sqrt{\log(ns)})$ for all i. We follow the pattern of the proofs of the non-approximability results in [SrSt97].

Theorem 3.1 *Let* $\tau \in [0, \frac{1}{2})$ *and* $\Delta \geq 3$ *be a fixed integer. Under the assumption* $b_i \in \Omega(n^{1/2+\tau}\sqrt{\log(ns)})$ *for all resource bounds, it is* NP-complete *to decide whether or not there exists an integral schedule of size* Δ.

Proof: We give a reduction to the chromatic index problem which is NP-complete [Hol81]. The following is known about the chromatic index $\chi'(G)$ of a graph G. Let $\Delta(G)$ be the maximal vertex degree in G. Then by Vizing's theorem [Viz64] $\Delta(G) \leq \chi'(G) \leq \Delta(G) + 1$ and an edge coloring with $\Delta(G) + 1$ colors can be constructed in polynomial time. But it is NP-complete to decide whether there exists a coloring that uses $\Delta(G)$ colors, even for cubic graphs, i.e. $\Delta(G) = 3$ [Hol81]. Therefore the edge coloring problem is NP-complete for any *fixed* $\Delta \geq 3$.

For simplicity, we show the reduction for $\tau = 1/4$, but with little modifications the proof goes through for every $\tau \in [0, \frac{1}{2}]$. We start with an instance of the chromatic index problem: let $G = (V, E)$ be a graph with $\nu = |V|, \mu = |E|$ and maximum vertex degree Δ. Put $K = c(\mu\Delta)^4$ where c is a large constant to be fixed later. For simplicity assume that K is an integer. For every edge $e \in E$ we consider $2K$ red and $2K(\Delta - 1)$ blue jobs denoted by

$$T_1^r(e), \ldots, T_{2K}^r(e) \quad \text{and} \quad T_1^b(e), \ldots, T_{2(\Delta-1)K}^b(e).$$

Again we consider $2\mu K\Delta$ identical processors. Hence the processor constraint is trivially satisfied. Let T denote the set of all jobs, T_r the set of red jobs, T_b the set of blue job, $T_r(e)$ the set of red jobs corresponding to an edge e, $T_b(e)$ the set of blue jobs corresponding to an edge e and $T(e) = T_r(e) \cup T_b(e)$.

Resource R_v: For every node $v \in V$ let R_v be a resource with bound $2K$ and requirements

$$R_v(j) = \begin{cases} 1 & \text{if } T_j = T_j^r(e) \\ & \text{and edge } e \text{ contains node } v \\ 0 & \text{else.} \end{cases}$$

Resource R_e: This resource ensures that no more than $2K$ jobs corresponding to the same edge e can be scheduled at the same time. Its bound is $2K$ and the requirements are

$$R_e(j) = \begin{cases} 1 & \text{if } T_j = T_i^r(e) \text{ or } T_j = T_i^b(e) \\ 0 & \text{else.} \end{cases}$$

Resource $R_{i,e}$: For every red job $T_i^r(e)$ choose exactly one other red job $g(T_i^r(e))$ corresponding to e as follows. Put $g(T_i^r(e)) = T_{i+1}^r(e)$ if $i < 2K$ and put $g(T_{2K}^r(e)) = T_1^r(e)$. Let us call $g(T_i^r(e))$ the *buddy* of $T_i^r(e)$. For every red job $T_i^r(e)$ let $R_{i,e}$ be a resource with bound K and requirements

$$R_{i,e}(j) = \begin{cases} 1 & \text{if } T_j^r(e) = T_i^r(e) \\ \frac{1}{2} & \text{if } T_j \in T(e) - \{g(T_i^r(e))\} \\ 0 & \text{else} \end{cases}$$

Claim: G has an edge coloring with Δ colors, if and only if the above defined scheduling problem has a schedule of length Δ.

Suppose that the edges of G can be colored with Δ colors taken from the set $Z = \{1,\dots,\Delta\}$. For each $e \in E$ schedule all red jobs $T_i^r(e)$ at the time corresponding to the color of e, say $z_e \in Z$, and schedule the blue jobs $T_i^b(e)$ in packets of $2K$ at the remaining times $z \neq z_e$, $z \in Z$. Set $x_{jz} = 1$, if job T_j is scheduled at time z, and 0 else. It is straightforward to check that this schedlue does not violate the bounds of the resources R_v and R_e. Now let $e \in E$ and $T_{i_0}^r(e) \in T_r(e)$ be arbitrary. Consider the resource $R_{i_0,e}$. Let $z \in Z$. For $z = z_e$ we have

$$\sum_{T_j \in T} R_{i_0,e} x_{jz_e} = \sum_{T_j \in T(e)-\{g(T_{i_0}^r(e))\}} R_{i_0,e} x_{jz_e} = 1 + \frac{1}{2}(2K-2) = K,$$

and for $z \neq z_e$

$$\sum_{T_j \in T} R_{i_0,e} x_{jz} = \sum_{T_j \in T_b(e)} R_{i_0,e} x_{jz} = \frac{1}{2}(2K) = K,$$

and the first part of the claim is proved.

Suppose now that we have a feasible schedule of length Δ. We show that it is impossible to schedule the red jobs corresponding to the same edge at different times. This is crucial, because then we would be able to define an *unique* scheduling time of all red jobs corresponding to an edge as the color of the edge and would have reduced the scheduling problem to the chromatic index problem.

Assume for a moment that there is an edge $e_0 \in E$ and a time z_0 in a schedule of size Δ so that I red jobs corresponding to e_0 with $1 \leq I < 2K$ are scheduled at time z_0. Since due to resource R_{e_0} at every time exactly $2K$ e_0-jobs must be scheduled, exactly $2K - I$ blue e_0-jobs must be scheduled at time z_0. Since there are strictly less than $2K$ red e_0-jobs scheduled at time z_0, there is a (red) job $T_{i_0}^r(e_0)$ scheduled at time z_0, but whose buddy is *not* scheduled at time z_0. Then the requirement for resource R_{i_0,e_0} at time z_0 is

$$\sum_{T_j \in T} R_{i_0,e_0}(j) x_{jz_0} = \sum_{T_j \in T(e_0)} R_{i_0,e_0}(j) x_{jz_0}$$
$$= 1 + \frac{1}{2}((2K-I)+(I-1))$$
$$= K + \frac{1}{2} > K,$$

and the schedule requires more than K units of resource R_{i_0,e_0} in contradiction to the feasibility assumption. Hence all red jobs corresponding to an edge must be scheduled at the same time.

Finally, we show $b_i \in \Omega(n^{3/4}\sqrt{\log(ns)})$ for all resource bounds:

We have introduced $n = 2K\mu\Delta \leq 2c(\mu\Delta)^5$ jobs and $s = \nu+\mu+2K\mu \leq 4\mu^2\Delta$ resources. Thus $ns \leq 8c(\mu\Delta)^7$, and assuming $c \geq 8$ we get $\log(ns) \leq 14\log(\mu\Delta)\log c$. Hence

$$n^{3/4}\sqrt{\log(ns)} \leq (2c)^{3/4}(\mu\Delta)^{15/4}(14\log(\mu\Delta)\log c)^{1/2} \tag{8}$$
$$\leq c(\mu\Delta)^4 = K.$$

The last inequality follows from the inequality $(\mu\Delta)^{1/4} \geq \log(\mu\Delta)$ and choosing the constant c so that $c \geq (2c)^{3/4}(14\log c)^{1/2}$.

$\square$

Theorem 3.2 *Even under the assumption $b_i \in \Omega(n^{3/4}\sqrt{\log(ns)})$ for all resource bounds b_i, there cannot exist a polynomial-time ρ-factor approximation algorithm for resource constrained scheduling for any $\rho < 4/3$, unless $P = NP$.*

$\square$

Proof: Since we can test the existence of a schedule of size 1 in polynomial time, we may assume that $C_{opt} \geq 2$. Assume for a moment that there is a polynomial-time ρ-approximation algorithm for $\rho < \frac{4}{3}$. Such an algorithm outputs the number 3 if and only if $C_{opt} = 3$. Under the hypoyhesis $P \neq NP$, this is a contradiction to Theorem 3.1.

$\square$

Remark 3.3 Improvements of this method with the goal to show a 2-factor (or even better) NC-algorithm under weaker assumptions like $b_i = \Omega(\log ns)$ would be interesting. This would match the presently best sequential approximation guarantees. We hope that our result can motivate further investigations of parallel approximation algorithms in this area.

References

[ABI86] N. Alon, L Babai, A. Itai; *A fast and simple randomized algorithm for the maximal independent set problem.* J. Algo., 7 (1987), 567 - 583.

[ASE92] N. Alon, J. Spencer, P. Erdös; *The probabilistic method.* John Wiley & Sons, Inc. 1992.

[AnVa79] D. Angluin, L.G. Valiant: *Fast probabilistic algorithms for Hamiltonion circuits and matchings.* J. Comp. Sys. Sci., Vol. 18, (1979), 155–193.

[BeRo91] B. Berger, J. Rompel; *Simulating (log^c n)-wise independence in NC.* JACM, 38 (4), (1991), 1026 - 1046.

[BESW93] J. Błazewicz, K. Ecker, G. Schmidt, J. Węglarz; *Scheduling in computer and maufacturing systems.* Springer-Verlag, Berlin (1993).

[Bol85] B. Bollobàs; *Random Graphs.* Academic Press, Orlando (1985).

[GGJY76] M. R. Garey, R. L. Graham, D. S. Johnson, A.C.-C. Yao; *Resource constrained scheduling as generalized bin packing.* JCT Ser. A, 21 (1976), 257 - 298.

[GaJo79] M. R. Garey, D. S. Johnson; *Computers and Intractability*. W. H. Freeman and Company, New York (1979).

[GLS88] M. Grötschel, L. Lovász, A. Schrijver; *Geometric algorithms and combinatorial optimization*. Springer-Verlag (1988).

[Hol81] I. Holyer; *The NP-completeness of edge coloring*. SIAM J.Comp., 10 (4), (1981), 718 - 720.

[Li82] J. H. van Lint; *Introduction to Coding Theory*. Springer Verlag New York, Heidelberg, Berlin (1982).

[McSo77] F.J. MacWilliams, N.J.A. Sloane; *The theory of error correcting codes*. North Holland, Amsterdam, (1977).

[MNN89] R. Motwani, J. Naor, M. Naor; *The probabilistic method yields deterministic parallel algorithms*. Proceedings 30the IEEE Conference on Foundation of Computer Science (FOCS'89), (1989), 8 - 13.

[Ra88] P. Raghavan; *Probabilistic construction of deterministic algorithms: approximating packing integer programs*. J. Comp. Sys. Sci., 37, (1988), 130-143.

[RT87] P. Raghavan, C. D. Thompson; *Randomized rounding: a technique for provably good algorithms and algorithmic proofs*. Combinatorica 7 (4), (1987), 365-374.

[RS83] H. Röck, G. Schmidt; *Machine aggregation heuristics in shop scheduling*. Math. Oper. Res. 45(1983) 303–314.

[SrSt96] A. Srivastav, P. Stangier; *Algorithmic Chernoff-Hoeffding inequalties in integer programming*. Random Structures & Algorithms, Vol. 8, No. 1, (1996), 27 - 58.

[SrSt97] A. Srivastav, P. Stangier; *Tight aproximation for resource constrained scheduling and bin packing*. To appear in *Discrete Applied Math. 1997*.

[Sriv95] A. Srivastav; *Derandomized algorithms in combinatorial optimization*. Habilitation thesis, Insitut für Informatik, Freie Universität Berlin, (1995), 180 pages.

[VeLu81] W. F. de la Vega, C. S. Lueker; *Bin packing can be solved within $1 + \epsilon$ in linear time*. Combinatorica, 1 (1981), 349 - 355.

[Viz64] V. G. Vizing; *On an estimate of the chromatic class of a p-graph*. (Russian), Diskret. Analiz. 3 (1964), 25 - 30.

4 Appendix: Proof of Lemma 2.3

It suffices to prove part (i) of the lemma. Put $\hat{n} = n^{k+1} s'$. Let I be the set of all k tuples $(\beta_1, \ldots, \beta_k)$ with $\beta_j \in \{1, \ldots, n\}$. For $i = 1, \ldots, s'$, $z \in \Omega$ and integers $x_j \in \Omega$, $j = 1, \ldots, n$ define functions $g_j^{(iz)}$ by

$$g_j^{(iz)}(x_1, \ldots, x_n) = R_{ij}(x_{jz} - \frac{1}{d}),$$

and define for a k-tupel $\beta \in I$, the product function $g_\beta^{(iz)}$ by

$$g_\beta^{(iz)}(x_1,\ldots,x_n) = \prod_{j=1}^{k} g_{\beta_j}^{(iz)}(x_1,\ldots,x_n).$$

Then

$$F(x_1,\ldots,x_n) = \sum_{iz}\sum_{\beta\in I} g_\beta^{(iz)}(x_1,\ldots,x_n).$$

Note that this sum has at most $\hat{n}$ terms. The random variables $X_1,\ldots,X_n$ by definition have the form $X_j = (BY)_j \bmod d$. Therefore we may restrict us to the computation of values for the Y_i's. The conditional probability method goes as follows: Suppose that for some $1 \le t \le \ell$ we have computed the values $Y_1 = y_1,\ldots,Y_{t-1} = y_{t-1}$ where $y_j \in \Omega, j = 1,\ldots,t-1$. Then choose for Y_t the value $y_t \in \Omega$ that maximizes the function

$$w \to \mathbb{E}(F(X_1,\ldots,X_n) \mid y_1,\ldots,y_{t-1}, Y_t = w). \tag{9}$$

After ℓ steps this procedure terminates and the output is a vector $y = (y_1,\ldots,y_\ell)$ with $y_j \in \Omega$, $(j = 1,\ldots,\ell)$. Then the vector $\vec{x}' = (x'_1,\ldots,x'_n)$ with components $x'_j = (By)_j \bmod d$ is the desired solution. For $1 \le t \le \ell$ it will be convenient to use the notation: put $\vec{Y}_t := (Y_1,\ldots,Y_t)$, $\vec{y}_t := (y_1,\ldots,y_t)$ and $\vec{Y}_t^* := (Y_{t+1},\ldots,Y_\ell)$, $\vec{y}_t^* := (y_{t+1},\ldots,y_\ell)$. We are done, if we can compute the conditional expectations $\mathbb{E}(F(X_1,\ldots,X_n) \mid \vec{Y}_t = \vec{y}_t)$ within the claimed time and work bounds. By linearity of expectation, it is sufficient to compute for each tripel $(i,z,\beta), 1 \le i \le s', z \in \Omega, \beta \in I$ the conditional expectations

$$\mathbb{E}(g_\beta^{(iz)}(X_1,\ldots,X_n) \mid \vec{Y}_t = \vec{y}_t). \tag{10}$$

Let $\hat{A}$ be the $k \times \ell$ matrix whose rows are the rows of B with row-indices $\beta_1,\ldots,\beta_k$. Let $\hat{A}_1$ resp. $\hat{A}_2$ be the first t resp. last $\ell - t$ columns of $\hat{A}$. Then

$$\mathbb{E}(g_\beta^{(iz)}(X_1,\ldots,X_n) \mid \vec{Y}_t = \vec{y}_t) = \sum_{x\in\Omega^k} g_\beta^{(iz)}(x) \Pr[\hat{A}\vec{Y} = x \mid \vec{Y}_t = \vec{y}_t]. \tag{11}$$

And for every fixed $x \in \Omega^k$

$$
\begin{aligned}
\Pr[\hat{A}\vec{Y} = x \mid \vec{Y}_t = \vec{y}_t] &= \Pr[\hat{A}_2\vec{Y}_t^* = x - \hat{A}_1\vec{y}_t] \\
&= \begin{cases} 2^{-rank(\hat{A}_2)} & \text{if } \hat{A}_2\vec{Y}_t^* = x - \hat{A}_1\vec{y}_t \\ 0 & \text{otherwise.} \end{cases}
\end{aligned}
$$

(The last equality follows from [BeRo91], section 3.1 and 4.2). Now we are able to estimate the running time and work space. For every $\beta \in I$ we can compute $rank(\hat{A}_2)$ (in $GF(d)$) in $O(\ell^3)$ time. For every $x \in \Omega^k$ the solvability of the linear system $\hat{A}_2\vec{Y}_t^* = x - \hat{A}_1\vec{y}_t$ can be tested in $O(\ell^3)$ time. Since we have d^k vectors x, we can compute $\mathbb{E}(g_\beta^{(iz)})$ for every $1 \le t \le \ell$, $1 \le i \le s'$ and $0 \le z \le d-1$ with d^k parallel processors in $O(\ell^3 + \log(d^k)) = O(\ell^3 + k\log d)$ time. Then we compute for every $y_t, 1 \le t \le \ell$, the expectation $\mathbb{E}(F(X_1,\ldots,X_n) \mid \vec{Y}_t = \vec{y}_t)$ in $O(\log \hat{n})$ time using $O(\hat{n})$ parallel processors. Finally, that

$y_t \in \Omega$ which maximizes (9) can be computed finding the maximum of the d conditional expectations in $O(\log d)$ time with $O(d)$ processors. The maximum number of processors used is $O(\max(d^k, \hat{n})) = O(\max(d^k, n^{k+1}s'))$ and the total running time over all ℓ steps is

$$O(\ell(\ell^3 + k \log d + \log \hat{n} + \log d)) = O(k^4 \log^4 n + k^2 \log n \log d + k \log n \log s').$$

□

Virtual Data Space - A Universal Load Balancing Scheme

Thomas Decker*

Department of Mathematics and Computer Science
University of Paderborn, Germany
e-mail: decker@uni-paderborn.de
http://www.uni-paderborn.de/cs/decker.html

Abstract. The _Virtual Data Space_ is a standard C-library which automatically distributes the work-packets generated by parallel applications across the processing nodes. VDS is a universal system offering loadbalancing-mechanisms for applications which incorporate independent load-items and scheduling algorithms for those which comprise precedence-constraints between their different tasks. This paper presents the concepts of VDS and shows some performance results obtained by synthetic benchmark applications.

1 Introduction

The programming of parallel or distributed systems normally raises questions about the distribution of work, the organization of the communication, and the control of the computation. As these problems are typical for distributed applications, intensive research has been done in these basic research fields in the last years which led to a variety of algorithms.

Nowadays, the challenge is to provide this know-how to the user in a comfortable way in order to increase the acceptance of parallel machines. As these problems are quite general, the tools which help the user to treat them have to be general as well. A very popular example are message passing systems like PVM and MPI. Three main aspects led to the high acceptance of these systems which are universality, simplicity, and portability.

These principles form a guideline which should be taken into consideration for the design of tools which should really aid the programmer of parallel applications. In the last years, several tools have been presented which incorporate load-balancing techniques and which support different programming paradigms. A very good overview of the existing tools is given in [FK95].

One of the first tools was the DYNAMO-library [T94] which handles data-objects representing any work to do. The processing nodes request these objects

---

* This work is supported by the "DFG Sonderforschungsbereich 376 : Massive Parallelität - Algorithmen, Entwurfsmethoden, Anwendungen" and by the EU ESPRIT Long Term Research Project 20244 (ALCOM-IT).

from the (distributed) DYNAMO-system, process them and create new objects which are distributed by the system. The ALDY-library [S95] extends this programming model by introducing the *Process-Worker-Task (PWT)*-model which allows more than one (migratable) worker-process on every processing node, each of which consuming and generating tasks independently from each other. By using task-identifiers, specified by the user, it is possible to restrict the set of workers which are able to process a certain task.

A property common to both libraries is the fact that the tasks are independent from each other. However, many applications incorporate precedence-constraints between their tasks which have to be considered by the scheduling algorithm. A large subclass of these applications is the class of tree-structured computations like for example divide-and-conquer-applications. Tools which support tree-precedence-constraints are for example CILk, Mentat, and the Beeblebrox-system [BJKL95+, G93, PP93]. The first tool, CILK, is an extension of the C-language supporting spawning of threads which are able to return values to the caller. Mentat and Beeblebrox are object orientated libraries distributing objects among the processors like DYNAMO and ALDY under consideration of precedence constraints. The latter is specialized for divide-and-conquer applications.

The need for a universally applicable balancing-tool incorporating specific balancing-algorithms for different application-models led to the Virtual Data Space (*VDS*)-library presented in this paper. It is based on the DYNAMO-concept and supports both, dependent and independent objects.

The remaining part of this paper is organized as follows: The next section presents a short description of the basic concepts of VDS, the third one describes different load-balancing techniques exploited by VDS and their application fields, the fourth section shows some experimental results and the last section presents the conclusions.

2 Concepts of VDS

This section describes the significant concepts of VDS. Many details have been left out because of space-constraints. For a full reference manual see [D96].

Similarly to the DYNAMO-concept, in VDS the load distributed among the processors is represented by arbitrary data-structures which describe any work to be done. It is assumed that these structures can be interpreted by every node in the system. Generally, VDS models a global memory area in which the user can insert and from which he can request data-objects. As long as the objects are inside this area, it is the responsibility of VDS to distribute them among the processors.

However, VDS is not comparable to other systems which simulate global shared memory like Linda, Dome [CGMS95, ABLS96+], or other DSM-systems, as the VDS-data space is not searchable and as the contained objects are not directly addressable.

Class type	Load balancing	Selection rule
normal object	qualitative	FIFO
weighted object	qualitative	Priority queue
message	no lb.	FIFO
message with priority	no lb.	Priority queue
thread	quantitative	LIFO

Table 1. *The class type specifies the method VDS uses to handle the objects belonging to this class*

In order to provide a possibility to distinguish different *kinds* of objects, VDS uses the class-concept of object orientated programming. Each object belongs to a certain class which is declared by the user at the beginning of the program. When requesting objects from VDS, it is possible to restrict the set of classes the requested object should belong to. This mechanism offers a convenient way to implement groups of different objects, representing for example different phases of the computation.

VDS incorporates both, balancing and scheduling methods. The user has to select the right method by specifying the *type* of each class he declares. Currently, VDS offers five class-types, each of which representing different programming-paradigms. Note that it is possible to use several class-types inside the same application.

The simplest paradigm, message-passing, requires no load-balancing at all and is supported by VDS in order to provide a mechanism to control distributed applications.

Normal and weighted objects. Normal and weighted objects are balanced according to a load-measure which is specified by the user. The difference between these types is the additional quality measure which is specified for weighted objects which will again influence the selection rule applied by VDS if more than one object a requested class is available (see Table 1).

Weighted objects offer for example a convenient and efficient way to implement best-first-branch and bound.

Thread-objects. Applications with tree-precedence constraints can often be modelled as a *strict multithreaded computation*. Such a computation is composed of a set of threads. During the execution, a thread can spawn other *child*-threads. Additionally, the execution is influenced by data-dependencies, which will occur if a thread depends on a result to be generated by one of its child-threads.

VDS offers a special object type for modelling strict multithreaded computations. Here, the threads are represented by thread-objects which are generated and handled with special functions.

The concept for transmitting results from a child-object to its parent is based on a special result-type VDS_Result. For every result which will be generated by any of the child-objects, a variable of this type has to be declared within the object-data. Each time a child generates a result, VDS transmits it to the variable of the parent-object which had been declared for this result. The parent-object in turn can use this variable to wait for a result, to check whether the result is available, and to read its content (the result itself).

3   Load balancing techniques used by VDS

The factors influencing the techniques depend on the load balancing quality demanded by the application and by the underlying hardware characteristics.

If the total set of load items processed during a distributed computation does not depend on the schedule determined by the balancing layer, i.e. the order and location where the items are processed, the maximum speedup can be reached if the idle-times of the nodes are minimized. As an example consider tree-structured computations like divide-and-conquer applications which decompose the problem to be solved into parts which directly depend on the problem-instance itself.

For this kind of *quantitative* load balancing, VDS uses randomized worksteal-ing which leads to very good results. The principle of workstealing is to let idle processors send steal messages to other processors selected at random. Processors which receive steal messages while they are in an idle state forward the message to other processors which again are selected at random. Others send an appropriate load item to the idle processor which had initiated the steal message.

Especially for multithreaded computations, these methods proved to be very efficient [BL95, FS96]. The calculation is controlled by using execution-stacks for the ready threads on every node. Each processor works on the topmost thread of its stack. If a thread spawns a new one, then the spawning one is put on top of the stack and the processor starts working on the new thread. Stealing is done at the bottom of the execution stacks. Consequently, threads are migrated which occupy high levels in the execution DAG and thus represent the most amount of work if the execution-DAG is balanced.

However, the load-items generated by a distributed computation may also depend significantly on the order they are processed. For example this is the case in many search algorithms used in artificial intelligence and operations research. In best-first branch-and-bound, for example, the processing order is defined by the quality of the objects (partial solutions). When applications of this kind are parallelized, it is not only important to perform quantitative load balancing, but also some form of *qualitative* load balancing will be neccessary to ensure that all processors are working on good partial solutions and thus prevent the processors from doing ineffective work (work not processed by a sequential best-first algorithm).

In contrast to quantitative algorithms, where processors can only have two states: idle or not idle, qualitative algorithms directly take the load-states of the processors into consideration. Based on comparisons of these states, load is migrated from "source" processors with high load to "sink" processors with low load. The various algorithms for this setting differ in the point of time they get

```
1.  divacon (n)
2.      if (simple (n))
3.          return (evaluate (n));
4.      else
5.          P := partition (n, b);        (|P| = b)
6.          for each p_i ∈ P parallel do
7.              s_i := divacon (p_i);
8.          return (combine (s_1, ..., s_b));
```

Function	Benchmark
simple (n)	$n = 1$
$e(n)$	0
$p(n)$	$p \cdot n$
$c(n)$	c
b	$2, \ldots, 64$

(a)   (b)

Fig. 1. (a): A general parallel divide-and-conquer benchmark, (b): parameters of the benchmark used in the test suite.

active, in the strategies used to select the processors which exchange information about their load-state and in the amount of load which is migrated.

In the current version, VDS uses a balancing strategy which is based on a mixed initiation method which means that the processors will only balance the load in regular intervals if their state is high or low. The approach is based on the diffusion-algorithm [C89, XMLL95] in which every processor aquires the load information of all of its neighbours and afterwards calculates the amount of load-items which have to be transferred to each of its neighbours. Empirical experiments using a branch and bound algorithm for the set partitioning problem showed that this method outperformed simple approaches which only select one neighbour for balancing [XTM95].

Later versions of VDS will also incorporate randomized methods like they were suggested in [MD96] which, in order to support completely connected architectures optimally, do not exploit locality.

4   Performance evaluation for multithreaded computation

This section presents some performance results for tree-structured computations. The experiments are based on the general divide-and-conquer benchmark shown in Fig. 1(a). The parameters of this benchmark are (1) the size of the problems which are directly evaluated, (2) the overhead used for partitioning $p(n)$, combining $c(n)$, and evaluating $e(n)$ the solution of a problem with size n, and (3) the number of partitions b generated in step 5.

The problem-instance is an integer-value n which is partitioned into smaller integers $n_1, \ldots, n_b$ with $\sum_{i=1}^{b} n_i = n$ independently at random.

With the setting shown in Fig. 1(b), the parallelism of the application and thus the performance of VDS depends on the influence of the branching factor b and on the computational effort defined by p and c. For example, the quicksort algorithm would imply $b = 2$, $p > 0$ and $c = 0$.

All runs are performed on a problem instance of size $n = 10000$ on a Parsytec GC/el system with 64 Transputers.

4.1 Theoretical speedup

For the evaluation of the theoretical speedup assume that the partitioning is optimal which means that equally sized partitions of size n/b are generated. Further, let $n = b^k$ for some $k \in \mathbf{N}$. In this case, the sequential execution time T_1 is given by

$$T_1 = c \frac{bn-1}{b-1} + pkn \ . \tag{1}$$

The time needed for the execution of a strict multithreaded computation is at least the length T_∞ of the *critical path* which is a path in the computation tree from the root to a leaf with maximum processing time. From another point of view, T_∞ is the minimum execution time needed if an unbounded number of processors is available. Ignoring the communication-overhead, T_∞ is bound by:

$$T_\infty \geq ck + pn \left(b - \frac{1}{(b-1)n} \right) \ . \tag{2}$$

As the speedup is bound by T_1/T_∞, we can derive upper bounds for some intersting special cases. For the case $p = 0, c > 0$ the speedup is $O\left(\frac{n}{\log n}\right)$ and for $p > 0, c = 0$ it is bound by $O\left(\frac{\log n}{b}\right)$. Note, that the scalability for large values of p is quite poor which is also illustrated by the experiments presented in the next section.

4.2 Influence of the computational effort

In addition to the total running time T_P with P processors, the length T_∞ of the critical path of the application is measured. As the reachable speedup may either be bound by P or by T_1/T_∞ we define the *speedup quality* as a measure for the efficiency of the execution:

$$\text{speedup quality} := \frac{\min\left(\frac{T_1}{T_P}, T_\infty\right)}{T_P} \ .$$

In Fig. 2(a) the speedup is shown for applications with different values for c and p. For a small combining-overhead ($c = 10000$ μsec), a reasonable speedup quality can only be achieved if the partioning-overhead p is large enough to compensate the overhead incurred by the search overhead of workstealing.

For $p = 1000$ μsec and for $c = 10000$ μsec, 100000 μsec the theoretical upper bound of the speedup imposed by the critical path is smaller than P. In these cases the bounds are denoted by the horizontal lines. Fig. 2(b) shows that also in these cases qualities better than 80% can be reached in respect to the critical path.

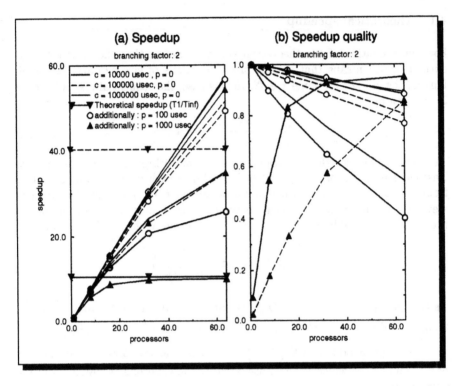

Fig. 2. *Speedup (a) and speedup quality (b) for applications with constant overhead for each thread.*

5 Conclusion

A new tool for balancing the work load in parallel and distributed systems is presented. The Virtual Data Space contains mechanisms for both, quantitative and qualitative load balancing which makes it applicable for a large spectrum of applications starting from tree-structured applications as shown in this paper up to search algorithms like best first branch-and-bound. As VDS automatically balances the load packets generated by the application, the designer of parallel applications is able to concentrate on the problem he wants to solve and does not have to think about how to distribute the load efficiently.

The measurements presented in this paper show that the runtime overhead of VDS is reasonable. The results are nearly optimal as long as the computation contains enough work to be distributed.

The future work on VDS will concentrate on more sophisticated algorithms which are able to adapt themselves to the characteristics of the application and of the architecture [DDLM95].

References

[ABLS96+] J. Arabe, A. Beguelin, B. Lowekamp, E. Seligman, M. Starkey, P. Stephan: *Dome: Parallel Programming in a Heterogeneous Multi-User Environment*, Proc. International Parallel Processing Symposium 1996 (IPPS 96).

[BJKL95+] R. D. Blumofe, C. F. Joerg, B. C. Kuszmaul, C. E. Leiserson, K. H. Randall, Y. Zhou: *Cilk: An Efficient Multithreaded Runtime System*, Proc. of th 5th ACM SIGPLAN Symposium on Principles and Practice of Parallel Programming, (PPOPP '95), pp. 207-216, 1995.

[BL95] R. D. Blumhofe, C. E. Leiserson: *Scheduling Multithreaded Computations by Work Stealing*, Proc. 36th Ann. Symposium on Foundations of Computer Science (FOCS '95), pp. 356-368, 1995.

[C89] G. Cybenko: *Dynamic Load Balancing for Distributed Memory Multiprocessors J. of Parallel and Distributed Computing 7 (1989) pp. 279-301.*

[CGMS95] N. Carriero, D. Gelernter, T. G. Mattson, A. H. Sherman: The Linda Alternative to Message-Passing Systems, *Parallel Computing, 20(4):633-655, 1994.*

[D96] T. Decker: *The Virtual Data Space Reference Manual*, http://www.uni-paderborn.de/SFB376/projects/a2/TP_A2_VDS_ManualE.html.

[DDLM95] T. Decker, R. Diekmann, R.Lüling, B. Monien: *Towards Developing Universal Dynamic Mapping Algorithms*, 7th IEEE Symp. on Parallel and Distributed Processing, SPDP'95, 1995, pp. 456-459.

[FS96] P. Fatourou, P. Spirakis: *Scheduling Algorithms for Strict Multithreaded Computations*, Proc. 7th Annual International Symposium on Algorithms and Computation (ISAAC '96), pp. 407-416, Osaka, Japan, December 1996.

[FK95] B. Freisleben, T. Kielmann: *Approaches to Support Parallel Programming on Workstation Clusters: A Survey*, Informatik Berichte, Fachgruppe Informatik, Universität-GH Siegen, (95-01), 1995.

[G93] A. S. Grimshaw: Easy to Use Object-Oriented Parallel Programming with Mentat, *IEEE Computer, pp. 39-51, May, 1993.*

[MD96] N.R. Mahapatra, S. Dutt: *Random Seeking: A General, Efficient, and Informed Randomized Scheme for Dynamic Load Balancing*, 10th IEEE Parallel Processing Symp., pp. 881-885, 1996.

[PP93] A. J. Piper, R. W. Prager: *Generalized Parallel Programming with Divide-and-Conquer: The Beeblebrox System*, Technical Report, Cambridge University Engeneering Department, CUED/F-INFENG/TR132, 1993.

[S95] T. Schnekenburger: *The ALDY Load Distribution System*, SFB-Bericht 342/11/95 A, Technische Universitaet Muenchen, 1995.

[T94] E. Tärnvik: Dynamo – A Portable Tool for Dynamic Load Balancing on Distributed Memory Multicomputers, *Concurrency: Practice and Experience, 6(8):613-639, 1994.*

[XMLL95] C.-Z. Xu, B. Monien, R. Lüling, F. C. M. Lau: *An Analytical Comparison of Nearest Neighbour Algorithms for Load Balancing in Parallel Computers* Proc. of International Parallel Processing Symposium (IPPS'95), pp. 472-479, 1995.

[XTM95] C.-Z. Xu, S. Tschöke, B. Monien: *Performance Evaluation of Load Distribution Strategies in Parallel Branch and Bound Computations* Proc. 7th Symposium on Parallel and Distributed Processing (SPDP'95), pp. 402-405, 1995.

Improving Cache Performance through Tiling and Data Alignment [*]

Preeti Ranjan Panda[1], Hiroshi Nakamura[2], Nikil D. Dutt[1], and
Alexandru Nicolau[1]

[1] Department of Information and Computer Science,
University of California, Irvine, CA 92697-3425
[2] The University of Tokyo,
Research Center for Advanced Science and Technology,
4-6-1 Komaba, Meguro-ku, Tokyo 153, JAPAN

Abstract. We address the problem of improving the data cache performance of numerical applications – specifically, those with blocked (or tiled) loops. We present DAT, a data alignment technique utilizing array-padding, to improve program performance through minimizing cache conflict misses. We describe algorithms for selecting tile sizes for maximizing data cache utilization, and computing pad sizes for eliminating self-interference conflicts in the chosen tile. We also present a generalization of the technique to handle applications with several tiled arrays. Our experimental results comparing our technique with previous published approaches on machines with different cache configurations show consistently good performance on several benchmark programs, for a variety of problem sizes.

1 Introduction

The growing disparity between processor and memory speeds makes the efficient utilization of cache memory a critical factor in determining program performance. While locality in instruction reference patterns are exploited reasonably well by instruction caches, utilization of data caches is not as effective, especially in numerical applications that typically operate on large sets of data. This inefficient utilization results in performance loss through data cache miss penalties.

Ideally, a fully-associative cache would be the best configuration, but due to the need for fast cache accesses, most commercial processors are equipped with direct-mapped or low-associativity cache memories. Direct-mapped caches have been gaining in popularity in recent processors due to the combined advantages of shorter access times and simplicity of implementation. For example, the HP 735, DEC AXP 3000, SGI Indigo2 and SUN Sparc5 are all equipped with direct-mapped data caches.

Cache misses, which occur when a required data item is absent in the cache and has to be fetched from off-chip memory, are classified into three categories:

---

[*] This work was partially supported by grants from NSF(CDA-9422095) and ONR(N00014-93-1-1348).

compulsory, *capacity* and *conflict* misses[9]. Compulsory cache misses occur when memory data is referenced for the first time. Capacity misses occur when reusable data is evicted from the cache due to cache size limitations. Conflict misses occur when more than one data element competes for the same cache location.

Compiler optimizations to exploit the data cache have been studied and implemented in the past. Compulsory cache misses can be minimized to some extent by software prefetching techniques[2]. Reduction of capacity misses is achieved by *loop blocking* (or *loop tiling*)[13] – a well known compiler optimization that helps improve cache performance by dividing the loop iteration space into smaller *blocks* (or *tiles*). This also results in a logical division of arrays into tiles. The *working set* for a tile is the set of all elements accessed during the computation involving the tile. Reuse of array elements within each tile is maximized by ensuring that the working set for the tile fits into the data cache.

Conflict misses are a consequence of limited-associativity caches. Cache conflicts in the context of tiled loops are categorized into two classes: (1) *self-interference*: cache conflicts among array elements in the same tile; and (2) *cross-interference*: cache conflicts among elements of different tiles (of the same array) or different arrays. Conflict misses can seriously degrade program performance; their reduction is generally addressed during the selection of tile sizes.

Data alignment techniques for reducing cache misses were reported by Lebeck and Wood [7] in a study of the cache performance of the SPEC92 benchmark suite, where they observed significant speedups (upto 3.4X) even on code that was previously tuned using execution-time profilers.

Padding is a data alignment technique that involves the insertion of dummy elements in a data structure for improving cache performance. In this paper, we present DAT, a data alignment technique based on padding of arrays, that ensures a more stable, and in most cases, better cache performance than existing approaches, in addition to maximizing cache utilization, eliminating self-interference, and minimizing cross-interference. Further, while all previous efforts are targetted at programs characterized by the reuse of a single array, we also address the issue of minimizing conflict misses when several tiled arrays are involved. Our experiments on several benchmarks with varying memory access patterns demonstrate the effectiveness of our technique and its stability with respect to changing problem sizes.

The rest of the paper is organized as follows. In Section 2, we discuss previous published work on loop tiling for cache performance enhancement. In Section 3, we demonstrate the significance of the padding technique through a motivating example. In Section 4, we describe our tile size selection and padding strategies, and a generalization of the strategies into the case with several tiled arrays. We describe our experimental results in Section 5, discuss the implications in Section 6, and conclude in Section 7. Appendix A contains details of tile shape derivation for an example benchmark program.

2 Previous Work

Several techniques for exploiting the data cache through loop tiling have been proposed in the past. Loop restructuring techniques to enable tiling are reported in [1, 8, 13, 14], all of which do not address conflict misses that occur in real caches.

Lam, et. al.[6] reported the first work modeling interferences in direct-mapped caches with a study of the cache performance of a matrix multiplication program for different tile sizes. They present an algorithm (LRW) for computing the tile size, which selects the largest *square* tile that does not incur self-interference conflicts. This strategy is effective for reducing cache misses, but is somewhat sensitive to the array size – the tile sizes vary widely with small changes in array sizes. For instance, in the matrix multiplication example, for a 1024-element cache, a 200×200 array results in a 24×24 tile, whereas a 205×205 array results in a 5×5 tile. The 5×5 tile makes inefficient utilization of the 1024-element cache and degrades performance due to loop overheads introduced by tiling.

Essghir[5] presents a tile size selection algorithm (ESS) for one-dimensional tiling, which selects a tile with as many rows of the array, as would fit into the data cache. This algorithm cannot exploit the benefits of two-dimensional tiling, and also does not consider cross-interference among arrays in its computation of tile sizes.

Coleman and McKinley[3] present a technique (TSS) based on the Euclidean G.C.D. computation algorithm for selecting a tile size that attempts to maximize the cache utilization while eliminating self-interferences within the tile. They incorporate the effects of the cache line size, as well as cross-interference between arrays. The tile sizes generated by TSS are, like LRW, also very sensitive to the array dimension. For the matrix multiplication example on a 1024-element cache, a 200×200 array results in a 41×24 tile, whereas, a 205×205 array results in a 4×203 tile. Since the working set in both cases is large, the cache utilization of TSS is good. However, if the working set gets too close to the cache size, cross-interferences, which are handled with a probabilistic estimate in TSS, begin to degrade the performance. We discuss this issue in Section 5.

The data alignment technique proposed by Lebeck and Wood [7] involves padding each record (in an array of records) to the nearest *cache line boundary* to prevent accesses to individual records from fetching an extra cache line from memory, thereby decreasing the number of cache misses. Further, when two or more arrays accessed in a loop have a combined size smaller than the cache, they are laid out in memory so as to map into different regions of the cache. However, their technique does not address programs with tiled loops.

In this work, we present a data alignment technique based on padding of arrays, that ensures a stable cache performance for a variety of problem sizes. Our approach differs from that presented in [6, 5, 3] in that, we are also able to handle the case where several tiled arrays are involved. As in previous work, we focus our attention on two-dimensional arrays.

3 Motivating Example

We illustrate the effect of the padding data alignment technique on the cache performance of the Fast Fourier Transform (FFT) algorithm. Figure 1(a) shows the core loop of FFT[4], highlighting the accesses to array *sigreal*.

Figure 1(b) illustrates the distribution of the accesses in the *sigreal* array, for two iterations of the outer loop (indexed by l), with array size $n = 2048$ words, and an example 512-word direct-mapped cache with 4 words per line. In the first iteration of the outer loop ($l = 0$), we have accesses to the set of pairs (*sigreal*[i], *sigreal*[$i+1024$]), all of which conflict in the cache, because, in general, the pairs (*sigreal*[i], *sigreal*[$i + 512k$]) will map to the same cache location for all integral k. Similarly, we observe similar severe conflicts between the pairs (*sigreal*[i], *sigreal*[$i + 512$]) in the second iteration ($l = 1$).

The pathological conflicts above can be prevented by the judicious padding of the *sigreal* array with dummy elements. Figure 1(d) shows a modification of the storage of the *sigreal* array, with 4 dummy words (size of one cache line) inserted after every 512 words. This ensures that *sigreal*[i] and *sigreal*[$i + 2^n$] never map into the same cache line. We note in Figure 1(d) that for $l = 0$ and 1, *sigreal*[i] and *sigreal*[$i + le$] map into consecutive cache lines. For $l \geq 2$, they map into different regions of the cache (as before), but the padding now avoids the conflicts mentioned before.[3] In order to implement the padding strategy, the array index computation would now change minimally, resulting in the code shown in Figure 1(c). In the computation of i' and le' in the *FFT_Padded* algorithm, NO_LINES (= CACHE_SZ / LINE_SZ) is usually a power of two, and consequently, the respective integer divisions result in shift operations.

A comparison of the execution times of the original FFT algorithm in Figure 1(a) with that of Figure 1(c) shows a speedup of 15% on the SunSparc-5 machine. This shows that the time spent in the extra computation involved in calculating the new array indices (i' and le') is small in comparison to the time saved by preventing the cache misses.

Note that the cache conflict problem identified in the FFT algorithm of Figure 1(a) could be solved by using a 2-way associative cache instead of a direct-mapped one. However, we also need to avoid a similar conflict in array *sigimag* (and between arrays *sigreal* and *sigimag*), so a 4-way associative cache would be required to avoid conflicts in this example. In general, cache access time considerations will not allow the use of a cache with arbitrarily large associativity. The padding technique helps avoid conflicts in a direct-mapped cache, and consequently, also avoids conflicts in any associative cache of the same size. We discuss the the possible effects of the padding data alignment technique on the rest of the program in Section 6.

---

[3] The *sigimag* array, which has an identical access pattern to *sigreal*, is prevented from conflicting with *sigreal* by adjusting the distance between the two arrays, i.e., by inserting padding between them.

Algorithm *FFT_Original*
Input: n, wreal$[n * 2^{n-1}]$,
    wimag$[n * 2^{n-1}]$
Inout: sigreal$[2^n]$, sigimag$[2^n]$
le = 2^n; windex = 1; wptrind = 0;
for (l = 0; l < n; l++) {
    le = le/2;
    for (j=0; j < le; j++) {
        wpr = wreal [wptrind];
        wpi = wimag [wptrind];
        for (i=j; i < 2^n; i += 2*le) {
            xureal = | sigreal [i] |;
            xuimag = sigimag [i];
            xlreal = | sigreal [i + le] |;
            xlimag = sigimag [i + le];
            | sigreal [i] | = xureal + xlreal;
            sigimag [i] = xuimag + xlimag;
            tr = xureal − xlreal;
            ti = xuimag − xlimag;
            | sigreal [i + le] | = tr * wpr −
            ti * wpi;
            sigimag [i + le] = tr * wpi −
            ti * wpr;
        } /* loop i */
        wptrind += windex;
    } /* loop j */
    windex = windex * 2;
} /* loop l */

(a)

Algorithm *FFT_Padded*
Input: n, wimag$[n * 2^{n-1}]$,
    CACHE_SZ, LINE_SZ
Constant: NO_LINES =
    CACHE_SZ / LINE_SZ
Inout: sigreal$[2^n + 2^n/\text{NO\_LINES}]$,
    sigimag$[2^n + 2^n/\text{NO\_LINES}]$
le = 2^n; windex = 1; wptrind = 0;
for (l = 0; l < n; l++) {
    le = le/2;
    le' = le + le / NO_LINES;
    for (j=0; j < le; j++) {
        wpr = wreal [wptrind];
        wpi = wimag [wptrind];
        for (i=j; i < 2^n; i += 2*le) {
            i' = i + i / NO_LINES;
            xureal = | sigreal [i'] |;
            xuimag = sigimag [i'];
            xlreal = | sigreal [i' + le'] |;
            xlimag = sigimag [i' + le'];
            | sigreal [i'] | = xureal + xlreal;
            sigimag [i'] = xuimag + xlimag;
            tr = xureal − xlreal;
            ti = xuimag − xlimag;
            | sigreal [i' + le'] | = tr * wpr −
            ti * wpi;
            sigimag [i' + le'] = tr * wpi −
            ti * wpr;
        } /* loop i */
        wptrind += windex;
    } /* loop j */
    windex = windex * 2;
} /* loop l */

(c)

(b)

(d)

Fig. 1. (a) Original FFT algorithm (b) Distribution of array accesses in data cache for original FFT algorithm (c) Modified FFT algorithm (d) Distribution of array accesses in data cache for modified FFT algorithm

4 Data Alignment Strategy

We now describe our technique, DAT, for data alignment of two-dimensional arrays. In Section 4.1, we describe the procedure for selecting the tile sizes. For the selected tile sizes, we compute the required padding of arrays for avoiding self-interference within the tile in Section 4.2. In Section 4.3, we present a generalization of the technique to handle the case when a tiled loop involves several arrays with closely-related access patterns.

4.1 Tile Size Computation

The tile size selection procedure can be summarized as follows: select the largest tile for which the working set fits into the cache. Note that the working set here could include some elements outside the chosen tile. For computing the working set size, we use the formulation used in [3]. The objective of this strategy is to maximize the cache utilization, but it would normally lead to self-interference among elements of the chosen tile. To prevent this self-interference, we adjust the arrays with an appropriate padding (Section 4.2).

The de-coupling of the two steps (tile size selection and padding) makes the technique both efficient and flexible. The efficiency is derived from the fact that the cache is never under-utilized. The flexibility arises from the ability to eliminate self-interferences independent of the tile size. The important consequence of this flexibility is that the tile size can now be optimized independently, without regard to self-interferences. For instance, if one-dimensional tiling were considered the best choice for an application, it could be easily incorporated into our technique. This cannot be handled in the previous approaches (TSS and LRW).[4] We discuss the de-coupling further in Section 6.

Figure 2 describes procedure *SelectTile* for computing the tile sizes. In addition to the most common case where we select the square tile, we also allow the user to specify the *shape* of the tile in cases where additional information about the application is available. To achieve this, the procedure takes the parameters: *TypeX*, *TypeY*, *ShapeX*, and *ShapeY*. In the common case (case 1: *ShapeX* = *ShapeY* = 0), we choose the largest square tile for which the working set fits into the cache. $ws(x, y)$ gives the size of the working set for a tile with x rows and y columns. We maintain the number of columns to be a multiple of the cache line size to ensure that unnecessary data is not brought into the cache. If *ShapeX* and *ShapeY* have a non-zero value, the user wishes to guide the tile-selection process. *TypeX* and *TypeY* can take the value CONSTANT or VARIABLE. If *TypeX* (*TypeY*) has value CONSTANT and *TypeY* (*TypeX*) has value VARIABLE, as in cases 2 and 3, the user has fixed the number of rows (columns) in the tile as *ShapeX* (*ShapeY*). Procedure *SelectTile* only determines the number of columns (rows). If both *TypeX* and *TypeY* are CONSTANT (case 4), the user has already optimized the tile sizes, and the procedure performs no action. If

---

[4] Note that ESS performs only one-dimensional tiling, and cannot incorporate two-dimensional tiling.

Procedure *SelectTile*
Input:
    TypeX, TypeY, ShapeX, ShapeY,
    Cache Size (C), Associativity (A), Cache Line Size (L)
Output:
    SizeX, SizeY /* No. of rows and columns of selected tile */

    Case 1 (*ShapeX* = *ShapeY* = 0):
        /* Common case: no application-specific info. available */
        $SizeX = SizeY = \max\{k | (k = L \cdot t, t \in I) \wedge (ws(k, k) \le C)\}$
    Case 2 (*TypeX* = CONSTANT **and** *TypeY* = VARIABLE):
        /* Fixed number of rows */
        $SizeX = ShapeX$
        $SizeY = \max\{k | (k = L \cdot t, t \in I) \wedge (ws(SizeX, k) \le C)\}$
    Case 3 (*TypeY* = CONSTANT **and** *TypeX* = VARIABLE):
        /* Fixed number of cols */
        $SizeY = ShapeY$
        $SizeX = \max\{k | (k \in I) \wedge (ws(k, SizeY) \le C)\}$
    Case 4 (*TypeX* = CONSTANT **and** *TypeY* = CONSTANT):
        /* Tile size already determined */
        $SizeX = ShapeX; SizeY = ShapeY$
    Case 5 (*TypeX* = VARIABLE **and** *TypeY* = VARIABLE):
        /* *ShapeX* and *ShapeY* give ratio of number of rows and cols in tile */
        $(SizeX, SizeY) = (ShapeX \cdot t, ShapeY \cdot t)$ such that
        $t = \max\{k | (k \in I) \wedge (ws(ShapeX \cdot k, ShapeY \cdot k) \le C)\}$
end Procedure

Fig. 2. Procedure for selecting tile size

both *TypeX* and *TypeY* are VARIABLE (case 5), the user intends the ratio of rows and columns of the tile to be *ShapeX* : *ShapeY*. The procedure selects the largest tile with the rows and columns in the given ratio for which the working set $\le C$.

Associative Caches Associative caches can be effectively utilized in our tiling procedure to reduce cross interference in a tiled loop. We first determine the *largest allowed working set size* (M) for the tile by considering the possibility of cross-interference. If there is no cross-interference (e.g., there is only one array involved), we set $M = C$ (the cache size). However, there are circumstances where cross-interferences cannot be completely eliminated. For example, in the matrix multiplication example (discussed in Section 5) implementing $P_1 \times P_2 = P_3$, where P_1, P_2, and P_3 are matrices, the tiling algorithm is designed to exploit reuse in matrix P_2 – the other matrices contribute unavoidable cross-interferences with the tile from P_2 in the cache. In this case, if the cache associativity $A > 1$, we set $M = \frac{A-1}{A} \times S$. In other words, in an A-way cache, in any cache entry (with A lines), the *primary* tile (tile of the blocked array) occupies only $(A - 1)$ lines, leaving the remaining one line to be occupied by elements outside the reused tile. This ensures that elements of the primary tile are almost never mistakenly evicted from the cache. [5] After determining M, we use procedure *SelectTile* (Figure 2), replacing the condition $ws(x, y) \leq C$ by $ws(x, y) \leq M$.

4.2 Pad Size Computation

The computation of tile size in Section 4.1 ignores the possibility of self-interferences within the tile. We now formulate the procedure for an appropriate padding of the rows of the array with dummy elements in order to avoid this self-interference. Consider the mapping of a 30 rows $\times$ 30 columns tile in a 256×256 array into a 1024-element cache. Figure 3(a) shows that rows 1 and 5 of the tile cause self-interference because they map into the same cache locations. To overcome this conflict, we can pad each row of the array with 8 dummy elements so that the 5th tile row now occupies the space adjacent to row 1 in the cache (Figure 3(b)). The regularity of the cache mapping ensures that no self-interference occurs among the elements of all the tile rows.

Note that there could be more than one correct pad size that eliminates self-interference within a tile. Clearly, smaller pad sizes are better than larger ones because they reduce the wasted memory space. Algorithm *ComputePad* (Figure 4) outlines the pad size computation procedure.

The inputs to algorithm *ComputePad* are: the cache line size, the cache size, and the dimensions of the array and the tile, and the output is *PadSize*, the smallest pad size that eliminates self-interference within the tile. The initial assignment to *InitPad* ensures that the rows of the padded array are aligned to the cache line size. For different multiples of *LineSize*, from a minimum of 0 to a maximum of *CacheSize*, we test if the resulting padded row of size $(N + PadSize)$ causes self-interference conflicts for the given tile of dimension $Rows \times Cols$. This is done by iterating through all the elements of the tile in steps of *LineSize*, and checking if any two elements map into the same cache line. In Figure 4, element (i, j) of the tile maps into cache location k, i.e., cache line number: $k/LineSize$. For every tile element (i, j), we update $LinesArray[k/LineSize]$ to 1 from 0. If $LinesArray[k/LineSize]$ is already 1, implying that at least one other tile element maps into this cache line, a conflict is flagged, and we repeat the procedure.

4.3 Multiple Tiled Arrays

The padding technique described in Sections 4.1 and 4.2 lends itself to an elegant generalization when computing the tile sizes of an algorithm involving more than

---

[5] Theoretically, this can still occur, e.g., when 2 elements outside the primary tile map into the same entry, but the probability of this occurrence is low because most elements accessed are from the primary tile.

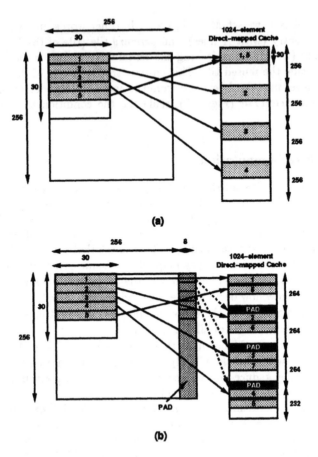

Fig. 3. (a) Self-interference in a 30 × 30 tile: Rows 1 and 5 of tile interfere in cache (b) Avoiding self-interference by padding

one tiled arrays. We assume that the arrays have identical sizes ($S \times R$). Since both arrays are accessed in the same tiled loop, they have identical tile sizes. We choose the tile sizes such that, as before, the working set (which involves elements accessed from all the tiled arrays) fits into the cache. To determine the padding size such that both self-interferences within each tile, as well as cross-interferences across tiles are avoided, we first construct a tile consisting of the smallest rectangular shape (T) enclosing all the tiles. In Figure 5, the new contour formed from the tiles in arrays A and B are shown. We now determine the pad size such that rectangle T in an $S \times R$ array does not have any self-interference, using the method described in Section 4.2. This becomes the pad size of all the tiled arrays. However, the tile size of each array remains $X \times Y$. Finally, we adjust the distance between the starting points of the arrays, so that the tiles are laid out the way they appear in rectangle T, in the first iteration. For example, in Figure 5, arrays A and B (both are now $R \times R'$) are laid out such that $A[0][0]$ and $B[0][0]$ are a distance $((X \cdot R') \bmod C)$ apart in the cache.

Algorithm *ComputePad*
Input:
    Array Dimensions: $N' \times N$; /* Row Size $= N$ */
    Tile Dimensions: *Rows* $\times$ *Cols*
    Cache Size (in words): *CacheSize*
    Cache Line Size (in words): *LineSize*
Output:
    Pad Size for Array: *PadSize*

    if (N mod *LineSize* $== 0$) **then** *InitPad* $= 0$
    else *InitPad* $=$ *LineSize* $- N$ mod *LineSize*
    end if
    for *PadSize* $=$ *InitPad* **to** *CacheSize*
        Status $=$ OK
        Initialize *LinesArray*[i] $= 0$ for all i
        for $i = 0$ **to** *Rows*
            for $j = 0$ **to** *Cols* **step** *Linesize*
                $k = (i \times (N + PadSize) + j)$ mod *CacheSize*
                if (*LinesArray*[k/LineSize] $== 1$) **then** *Status* $=$ CONFLICT
                else *LinesArray*[k/LineSize] $= 1$
                end if
            end for
        end for
        if (*Status* $==$ OK) **then return** *PadSize*
    end for
end Algorithm

Fig. 4. Algorithm for computing pad size for array

5 Experiments

We performed experiments on two common machines widely used today – SUN-Sparc5 (SS5) and SUNSparc10 (SS10). The SS5 has a direct-mapped 8 KB data cache (1024 double precision elements) with a 16-byte cache line (2 elements per line). The SS10 has a 4-way associative 16 KB data cache (2048 elements) with a 32-byte cache line (4 elements per line).

The example programs on which we performed our experiments are: (1) Matrix Multiplication (MM) (with the standard *ijk*-loop nest permuted to *ikj*-order, as in [6]); (2) Successive Over Relaxation (SOR)[3]; (3) L-U Decomposition (LUD)[3]; and (4) Laplace[10]. We did not use the FFT algorithm in the comparisons, because it does not involve tiled loops, and consequently, LRW, TSS, and ESS cannot be applied to it.

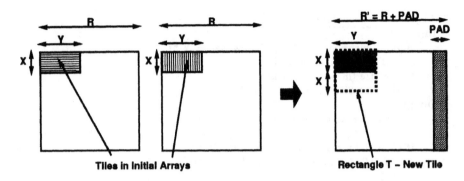

Fig. 5. Multiple Tiled Arrays

Our experimental results were performed both through actual measurement (MFLOPs on SUN SPARC Stations) as well as simulations on the SHADE simulator from SUN Microsystems[11]. Note that execution time is a function of the data cache miss ratio, the number of cache accesses, as well as the number of instructions executed. For tiled algorithms, the total number of instructions executed is smaller for larger tile sizes because the overheads introduced by tiling are smaller. The instruction cache misses are negligible for the above examples, as they all are small enough to fit into the instruction cache. We present performance comparisons of our technique (DAT) versus that of the LRW, ESS, and TSS algorithms in terms of both data cache miss ratio (misses per access) and MFLOPs.

5.1 Uniformity of Cache Performance

Our first experiment was to verify our claim that the padding technique results in a uniformly good performance for a wide variety of array sizes. Figure 6 shows the variation of data cache miss ratios (on the SS5 cache configuration) of four algorithms (LRW, ESS, TSS, and DAT) on the matrix multiplication (MM) example for all integral array sizes between 35 and 350.[6] We observe that the miss ratio of DAT is consistently low, independent of problem size, whereas all other algorithms show some sensitivity to the size. This is attributed to the fact that DAT uses fixed tile dimensions (30 × 30) for the given cache parameters, independent of array size, whereas in the other algorithms, tile dimensions vary widely with array size. Note that although the *number of misses*, and *number of instruction executed*, jointly determine the actual performance, we have plotted the miss ratio here just to show that the miss ratios are consistently low for DAT (low miss ratio is necessary, but not sufficient). Although LRW also has a low miss ratio on an average, it often selects small tile sizes leading to an increase in

---

[6] For small array sizes, all the data fits into the cache, obviating the necessity for tiling; for larger sizes, the simulation times on the commercial simulator, SHADE, were too long to examine every integral data size. Hence the range 35-350.

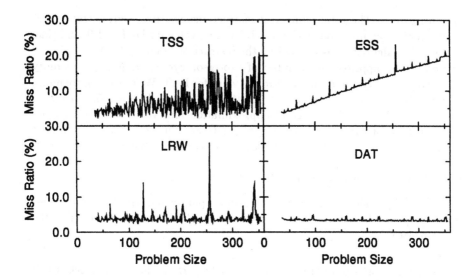

Fig. 6. Matrix multiplication on SS-5: Variation of data cache miss ratio with array dimension

the number of instructions, as described later. Figure 6 demonstrates that our technique DAT exhibits stability of cache performance across the range $35 - 350$ of problem sizes. We expect similar performance for larger array sizes.

5.2 Variation of Cache Performance with Problem Size

We present experiments below on the performance of each technique on the various examples, for several array sizes. We include array sizes of 256, 300, 301, and 550 to enable comparison with TSS [3] (which also presents data on these sizes). Array sizes 256 and 512 are chosen to illustrate the case with pathological cache interference, while 300 and 301 are chosen to demonstrate the widely different tile sizes for small changes in array size. Sizes 384, 640, and 768 illustrate the cases where the array size is a small multiple of a power of 2 ($384 = 128 \times 3; 640 = 128 \times 5; 768 = 256 \times 3$). We also include data for array sizes 700, 800, 900 and 1000.

For each example, we report results of execution on the SS5 and SS10 workstations. The SS5 has an 8KB direct-mapped cache, whereas the SS10 has a 16KB 4-way associative cache, with no second level cache. The results of our experiments on these two platform illustrate the soundness of our technique on direct-mapped as well as associative caches. We present the variation of the following parameters with respect to array size.

Miss Ratio − The data cache miss ratio (misses per data cache access).

Instructions – The total number of instructions executed, normalized to the instruction count of our technique DAT (i.e., value for DAT is 1.0, and that for other algorithms is divided by the instruction count of DAT).

Cycles – The approximate number of processor cycles, assuming a memory latency of 12 cycles, normalized to the value for DAT. We use an approximate formula: *#Cycles = Instruction Count + Latency × Data Cache Misses.* This rough formula is a good approximation for cycle count in a single-instruction stream processor, and takes into account both data cache misses as well as the number of instructions. As explained earlier, the instruction cache misses are negligible.

MFLOPs – The measured MFLOPs rate.

In the graphs shown in Figure 7, the above parameters are along the y-axes of the graphs, whereas the x-axis of all the graphs is the array size (i.e., the problem size).

L-U Decomposition Figure 7 shows the cache performance of the blocked L-U Decomposition program described in [3]. An analysis of the array access patterns in LUD reveals the following relationship for the optimal tile shape:

$$\text{No. of Columns} = \text{No. of Rows} \times \text{Cache Line Size} \qquad (1)$$

We describe the analysis in Appendix A.

Figure 7 shows that DAT and LRW have the lowest cache **miss ratios** (except at array sizes such as $256, 384, 512$, etc., where LRW has considerably higher miss ratios). The lower miss ratio of LRW is, however, often at the expense of instruction count, since the relatively smaller tiles in LRW incur overheads introduced by loop tiling. TSS tends to incur higher miss ratios because it sometimes chooses a comparatively larger tile sizes, leading to the working set size being too close to the cache size, which results in cross-interferences. ESS incurs cache conflicts because it does not account for cross-interferences.

A comparison of the **instruction count** shows that, in general, ESS has a smaller number of executed instructions than DAT and LRW. This is a consequence of the larger tile sizes selected by ESS. The instruction count of TSS fluctuates because it selects widely varying tile sizes.

The processor **cycle count** (with memory latency $= 12$ cycles) shows that DAT has the best overall performance. It is interesting to note that the curves follow roughly the same shape as the cache miss ratio curves, indicating that cache miss ratio is a very important factor in determining performance.

The **MFLOPs** measurements show DAT to perform better than other algorithms. The MFLOPs graph is, actually, the inverse of the cycle count graph, with the difference that the former is a measurement of execution time on the machine, whereas the latter is a computation from the simulation results. The close correlation of the two graphs indicates that the cycle count computations are reliable.

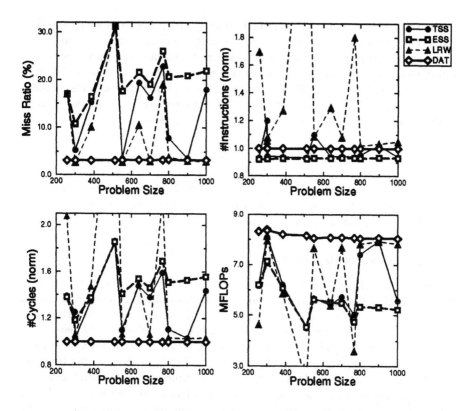

Fig. 7. Performance of L–U Decomposition on SS-5: Variation of (1) Data Cache Miss Ratio, (2) Normalized Instruction Count, (3) Normalized Processor Cycles, and (4) MFLOPs with array size

In summary, Figure 7 shows that our technique DAT consistently incurs lower miss ratios and exhibits better performance than previous techniques for the L–U decomposition example. We observe similar performance results in our experiments using the LUD benchmark on the SS10 machine.

Matrix Multiplication The results for *matrix multiplication* is summarized in Table 1. A comparison of the total number of instructions executed showed that, in general, a larger tile size implies lesser number of instructions due to reduced overheads. We assumed the tile size 30×30 for the SS-5 cache (direct mapped, 1024-element), and 36×36 for SS-10 cache (4-way set-associative, 2048-element – we used the formulation for A-way associative cache discussed in Section 4.1). Note that we only consider multiples of cache line size of candidates for tile size. Here we assumed that no information was available about the characteristic of the application and hence used a square tile.

Successive Over Relaxation (SOR) The results for *SOR* is summarized in Table 1. In this case, an analysis reveals that maximal reuse is obtained by choosing the longest (i.e., maximum number of columns) tile that avoids self-interference. We observed that ESS (which selects as many rows as will fit into the cache) has the same performance as DAT for the smaller array sizes, because both techniques generate the same tile sizes. However, for sizes 550 and greater, DAT achieves better performance because of better reuse.

Laplace The results for *Laplace* is summarized in Table 1. The performance of DAT is better than the other algorithms because the Laplace kernel involves the blocking of multiple arrays – a feature handled by DAT, but not the other techniques. DAT is able to avoid all cross-interference between the arrays, whereas TSS, ESS, and LRW all incur cross-interference.

5.3 Summary of Results

Table 1 summarizes the comparisons of experimental results for all the four examples (MM, LUD, SOR, and Laplace) on the SS5 and SS10 platforms, on the basis of the four metrics identified above. In order to do a fair comparison, the table does not include array sizes such as 512, 640, etc., which would penalize all the other techniques. Thus, the improvements shown in the table are obtained without considering these cases. If these were taken into account, the improvement would be higher. Column 1 gives the example names; Columns 2, 3, and 4 give the improvement in cache miss ratios (MR) of our technique DAT over TSS, ESS, and LRW respectively. Similarly, Columns 5, 6, and 7 compare the normalized instruction counts; Columns 8, 9, and 10 compare the normalized processor cycle counts; and Columns 11, 12, and 13 compare the MFLOPs with respect to DAT, i.e., the speedup of DAT over other techniques. On the SS5, we notice an average speedup of 24.2% over TSS, 38.0% over ESS, and 12.0% over LRW. On the SS10, we observe an average speedup of 7.2% over TSS, 13.8% over ESS, and 4.8% over LRW. The speedup on SS10 is smaller because the SS10 has a larger cache, which reduces the number of cache conflicts.

6 Discussion

The main advantage of DAT, the data alignment technique we presented, is its flexibility – decoupling tile selection from the padding phase allows the tile size to be independently optimized without regard to self-interference conflicts. This allows us to select larger tile sizes to maximize the cache utilization. This also allows us to incorporate a user-supplied tile-shape which might be optimized for a specific application. For instance, we can easily handle one-dimensional tiling, while it cannot be easily incorporated into TSS and LRW.

For a given application and cache size, we choose a fixed tile size, independent of the array size (only the pad size varies with the array size). An important consequence of this choice is the stability of the performance of the resulting tiled

SUN SPARC 5												
Example	MR(x)(%) − MR(DAT)(%)			Inst(x)/ Inst(DAT)			Cycle(x)/ Cycle(DAT)			Speedup (DAT/x)		
	TSS	ESS	LRW	TSS	ESS	LRW	TSS	ESS	LRW	TSS	ESS	LRW
MM	10.21	22.03	3.76	0.99	0.95	1.07	1.29	1.61	1.19	1.30	1.64	1.14
LUD	6.45	14.18	1.59	1.00	0.93	1.14	1.22	1.40	1.17	1.22	1.40	1.13
SOR	8.81	10.28	0.79	1.03	1.00	1.10	1.31	1.34	1.10	1.23	1.26	1.07
Laplace	6.54	6.22	0.35	1.00	1.00	1.07	1.17	1.16	1.06	1.22	1.22	1.14
SUN SPARC 10												
Example	MR(x)(%) − MR(DAT)(%)			Inst(x)/ Inst(DAT)			Cycle(x)/ Cycle(DAT)			Speedup (DAT/x)		
	TSS	ESS	LRW	TSS	ESS	LRW	TSS	ESS	LRW	TSS	ESS	LRW
MM	4.01	9.35	1.39	1.00	0.94	1.03	1.15	1.29	1.08	1.15	1.30	1.06
LUD	2.47	4.54	1.06	0.99	0.96	1.08	1.08	1.13	1.12	1.08	1.13	1.09
SOR	0.72	2.09	1.20	1.05	1.00	1.05	1.07	1.09	1.09	1.02	1.05	1.00
Laplace	1.27	1.88	0.03	1.02	1.00	1.04	1.05	1.05	1.03	1.04	1.07	1.04

Table 1. Summary of performance results on SS5 and SS10. Data for problem sizes 384, 512, 640, and 768 are not included because that would penalize LRW, ESS, and TSS; the cache performance behavior for all examples is similar to that observed in Figure 6.

loop, in comparison to TSS, ESS, and LRW (all of which select tile sizes that are very sensitive to the array size). For instance, in the matrix multiplication example shown in Figure 7, the miss ratio of DAT was in the range [3.11% − 4.87%], in comparison to [2.77% − 25.18%] for LRW, [2.93% − 23.18%] for TSS, and [3.70% − 23.15%] for ESS. Further, Figure 7 shows that all the other techniques suffer from pathological interference, not only for array sizes that are powers of two (256 and 512), but also for small multiples of powers of two (384, 640, and 768). It should be noted, however, that these array sizes are not uncommon in numerical applications. For cache line sizes greater than 1, we observe the same behavior when the sizes are close to one of these numbers, e.g., 254, 255, 385, etc.

Another advantage of our technique is that it is possible to avoid cross-interferences among several tiled arrays, while all the other techniques − LRW, TSS, and ESS are targetted at algorithms in which the reuse of only a single array dominates. While accesses to the several arrays would cause cross-interference in the other approaches, we are able to completely eliminate cache conflicts arising out of this cross-interference using our technique.

Finally, our padding technique can be easily extended to arrays with greater than two dimensions, by serializing the padding procedure over the different dimensions.

One consequence of the padding technique is that the structure of the data arrays is transformed, and references to the array in the rest of the code has to reflect the new structure. This, however, does not lead to any performance penalties. For example, when the row size is updated from R to R', the expression to compute the location for $a[i][j]$ changes from $(A+i \times R+j)$ to $(A+i \times R'+j)$, assuming the array begins at A.

There could be a conflict in padding sizes if the same array is accessed in different tiled loops. In this case, we invoke the padding strategy on the more critical loop, and in the others, use one of the existing methods (LRW or TSS) in the remaining loops, with updated row size for the array. This strategy works because LRW and TSS, which do not modify the data structure, can now use the size of the modified array to generate the tile sizes. However, an examination of existing benchmark programs reveals that such a situation (same array being accessed in two loop nests with different tiling patterns) is rare. It is well known that in most numerical programs, most of the computation tends to be concentrated in a few important kernels.

Another consequence of our proposed data alignment technique is that it cannot be directly incorporated into a library routine, since the array sizes are not known. The optimization is most effective when it is integrated into the program, instead of compiled separately as a library routine. However, the incorporation into a library routine is possible under certain reasonable constraints – for example, if the interface to the tiled array is through file I/O, and the library routine constructs the data structure for the arrays in memory. Consider a program involving matrix multiplication in which the matrices to be multiplied are read from a file. In this case, the library routine (with the file pointer as an argument) would build the array in memory with the appropriate padding. Since the padded array would not affect the performance of any other part of the program in which it is used (as discussed in the previous paragraph), the data alignment strategy can only improve program performance.

One way to work around the problem of incorporating the data alignment technique into a library routine is to copy the arrays into a separate memory space with the padding inserted. The computation is then performed on the new arrays, and the arrays are written back after the computation. This is similar to the *Copy Optimization* technique studied in [6, 12]. However, the overhead of copying the arrays is usually quite significant [12, 3]. In our experiment on the FFT program (Section 3), copying the *sigreal* and *sigimag* arrays into temporary arrays, followed by copying them back after the computation, resulted in a 6% improvement in performance over the original code in Figure 1(a), in contrast to the 15% improvement for Figure 1(c). It is interesting to note that there is a performance improvement of 6% in spite of the overhead incurred due to the copying.

7 Conclusions

We presented a data alignment technique DAT for improving the cache performance of numerical applications. We presented an algorithm for selection of tile sizes and data alignment through padding to maximize the utilization of data caches and minimize cache conflicts. We also presented an extension of this approach to reduce cross-interference in applications with multiple tiled arrays.

Our experiments were performed on machines with different cache configurations (SS-5 and SS-10), using a number of standard numerical benchmarks for a range of array sizes. The results demonstrate that our technique results in consistently low data cache miss ratios and effective cache utilization, leading to good overall performance.

References

1. S. Carr and K. Kennedy, "Compiler blockability of numerical algorithms," Proceedings of Supercomputing '92, Minneapolis, MN, November 1992.
2. D. Callahan, K. Kennedy, and A. Porterfield, "Software Prefetching," International Conference on Architectural Support for Programming Languages and Operating Systems, pp 40-52, April 1991.
3. S. Coleman and K. S. McKinley, "Tile size selection using cache organization and data layout," Proceedings of the SIGPLAN'95 Conference on Programming Language Design and Implementation, La Jolla, CA, June 1995.
4. P. M. Embree and B. Kimble, "C Language Algorithms for Digital Signal Processing," Prentice Hall, Englewood Cliffs, 1991.
5. K. Esseghir, "Improving data locality for caches," Master's thesis, Dept. of Computer Science, Rice University, 1993.
6. M. Lam, E. Rothberg, and M. E. Wolf, "The cache performance and optimizations of blocked algorithms," Proceedings of the Fourth International Conference on Architectural Support for Programming Languages and Operating Systems, Santa Clara, CA April 1991.
7. A. R. Lebeck and D. A. Wood, "Cache profiling and the SPEC benchmarks: A case study," IEEE Computer, Vol. 27, No. 10, October 1994.
8. J. J. Navarro, T. Juan, and T. Lang, "Mob forms: A class of multilevel block algorithms for dense linear algebra operations," Proceedings of the 1994 ACM International Conference on Supercomputing, Manchester, England, June 1994.
9. D. A. Patterson and J. L. Hennessy, "Computer Organization & Design - The Hardware/Software Interface," Morgan Kaufman Publishers, pp. 454-530, 1994.
10. W. H. Press, et. al., "Numerical Recipes in C: The Art of Scientific Computing," Cambridge University Press, 1992.
11. Sun Microsystems Laboratories Inc., "Shade User's Manual," Mountain View, CA, USA, 1993.
12. O. Temam, E. D. Granston, and W. Jalby "to Copy or Not to Copy: A Compile-Time Technique for Assessing When Data Copying Should be Used to Eliminate Cache Conflicts," Proceedings of Supercomputing '93, Portland, OR, USA, November 1993.
13. M. J. Wolfe, "More iteration space tiling," Proceedings of Supercomputing '89, Reno, NV, USA, November 1989.
14. M. E. Wolf and M. Lam, "A data locality optimizing algorithm," Proceedings of the SIGPLAN'91 Conference on Programming Language Design and Implementation, Toronto, Canada, June 1991.

A Tile Shape Computation in L-U Decomposition Example

We present here the procedure for computation of the tile-shape for the L-U Decomposition problem.

Let the matrix size be $M \times M$.

Let L = Cache Line Size.

Let k be the outer loop index and K be the range over which k iterates.

For a tile with a given number of elements (N), we wish to find the number of rows (r) and columns (c) for the tile $(N = r \times c)$, which will minimise the number of cache misses.

Total number of cache misses for a tile $= K \times (r + \frac{c}{L})$, since in each iteration, one row of the tile causes $\frac{c}{L}$ misses and one column causes r misses. Note that K is a constant independent of the tile size.

Total number of tiles $\approx (\frac{M}{r} \times \frac{M}{c})$ (approximation for $\lceil \frac{M}{r} \rceil \times \lceil \frac{M}{c} \rceil$).

Hence, the total number of misses $\approx K \times (\frac{M}{r} \times \frac{M}{c}) \times (r + \frac{c}{L}) \approx K' \times (\frac{1}{c} + \frac{1}{rL})$

Now, since $r \times c = N$, and $c \times rL = NL = constant$, we note that the number of misses is minimised for $c = rL$ (using the fact that for two numbers a and b with constant product ab, the sum $\frac{1}{a} + \frac{1}{b}$ is minimised for $a = b$).

In other words, for minimising the number of cache misses, we have:

$$\text{No. of Columns} = \text{No. of Rows} \times \text{Cache Line Size} \qquad (2)$$

A Support for Non-uniform Parallel Loops and Its Application to a Flame Simulation Code

Salvatore Orlando[1] and Raffaele Perego[2]

[1] Dipartimento di Mat. Appl. ed Informatica - Università Ca' Foscari di Venezia,
Venezia Mestre, 30173 Italy - email: orlando@unive.it.
[2] CNUCE, Consiglio Nazionale delle Ricerche (CNR),
Pisa, 56126 Italy - email: r.perego@cnuce.cnr.it.

Abstract. This paper presents SUPPLE (SUPport for Parallel Loop Execution), an innovative run–time support for parallel loops with regular stencil data references and non–uniform iteration costs. SUPPLE relies upon a static block data distribution to exploit locality, and combines static and dynamic policies for scheduling non–uniform iterations. It adopts, as far as possible, a static scheduling policy derived from the owner computes rule, and moves data and iterations among processors only if a load imbalance actually occurs. SUPPLE always tries to overlap communications with useful computations by reordering loop iterations and prefetching remote ones in the case of workload imbalance. The SUPPLE approach has been validated by many experimental results obtained by running a multi-dimensional flame simulation kernel on a 64–node Cray T3D. We have fed the benchmark code with several synthetic input data sets built on the basis of a load imbalance model, and we have compared our results with those obtained with a CRAFT Fortran implementation of the benchmark.

1 Introduction

Data parallelism is the most common form of parallelism exploited to speed-up scientific applications. In the last few years, high level data parallel languages such as High Performance Fortran (HPF) [2], Vienna Fortran [16] and Fortran D [4] have received considerable interest because they facilitate the expression of data parallel computations by means of a simple programming abstraction. In particular HPF, whose definition is the fruit of several research proposals, has been recommended as a standard by a large Forum of universities and industries. Using HPF, programmers only have to provide a few directives to specify processor and data layouts, and the compiler translates the source program into an SPMD code with explicit interprocessor communications and synchronizations for the target multicomputer. Data parallelism can be expressed above all by using collective operations on arrays, e.g. collective Fortran 90 operators, or by means of parallel loop constructs, i.e. loops in which iterations are declared as independent and can be executed in parallel.

The typical HPF run-time support for parallel loops exploits a static data layout of arrays onto the network of processing nodes, and a static scheduling of iterations which depends on the specific data layout. The adoption of a static policy to map data and computations reduces run-time overheads, because all mapping and scheduling decisions are taken at compile–time. While it produces very efficient implementations for regular concurrent problems, the code produced for *irregular problems* (i.e. problems where some features like non–local data references and iteration costs cannot be predicted until run-time) may be characterized by poor performance. Many researches have been conducted in the field of run-time supports and compilation methods to efficiently implement irregular concurrent problems, and these researches are at the basis of the new proposal of the HPF Forum for HPF2 [3].

In this paper we address particular irregular problems that can be programmed by means of non–uniform parallel loops, i.e. loops where the iteration cost varies considerably and cannot be predicted at compile time. We present a truly innovative support called SUPPLE (SUPport for Parallel Loop Execution) to implement these loops. SUPPLE is not a general support on which we can compile every HPF parallel loop, but only non–uniform (as well as uniform) parallel loops with regular stencil data references. Since stencil references are regular and known at compile time, optimizations such as *message vectorization, coalescing* and *aggregation*, as well as *iteration reordering* can be carried out to reduce overheads and hide communication latencies [5]. SUPPLE is able to initiate loop computation using a statically chosen BLOCK distribution, thus starting iteration scheduling according to the *owner computes rule*. During the execution, however, if a load imbalance actually occurs, SUPPLE adopts a dynamic scheduling strategy that migrates iterations and data from overloaded to underloaded processors in order to increase processor utilization and improve performances. SUPPLE overlaps most overheads deriving from dynamic scheduling, i.e. messages to monitor loads and move data, with useful computations. The main features that allow us to overlap the otherwise large overheads are (1) *prefetching* of remote loop iterations on underloaded processors, and (2) a complete asynchronous and distributed support that does not introduce barriers between consecutive iterations of loops.

SUPPLE has been validated through the implementation of a multi-dimensional flame simulation code [13], which can be programmed by means of parallel loops characterized by regular stencil data reference and non–uniform iteration costs. One example of code that performs a time-dependent multi-dimensional simulation of hydrocarbon flames is described in detail in [12]. It is also included in the HPF–2 draft [3] as one of its motivating application codes. This code is split into two distinct computational phases executed at each time step. The first phase computes fluid *Convection* over a structured mesh, and is expressed by a uniform parallel loop with regular stencil data references. The second phase simulates chemical *Reaction* and subsequent energy releases represented by ordinary differential equations. The solution at each grid point only depends on the value of the point itself, but its computational cost may vary significantly since the chemical combustion proceeds at different rates across the simulation space. It can be thus implemented with a non–uniform parallel loop that refers local data only. The Reaction phase globally requires more than half of the total execution time, and most of this time is spent on a small fraction of the mesh points. Furthermore, the workload distribution evolves as the simulation progresses [12]. The HPF-like code of a simplified two-dimensional flame modeling code is reported in Fig. 1. This is the kernel code we used to validate SUPPLE.

```
C  Loop on the time steps
   DO k= 1,K
C     Convection phase over a structured mesh
C     Nearest neighbor communications - Constant computation time
      FORALL (i= 2:N1-1,  j= 2:N2-1)
        A(i,j) = A(i,j) + F(B(i,j), B(i-1,j), B(i,j-1), B(i+1,j), B(i,j+1), C(i,j))
      END FORALL
      B = A
C     Chemical Reaction phase - No communications - High load imbalance
C     Workload distribution evolves during simulation
      FORALL (i= 1:N1,  j= 1:N2)
        C(i,j) = Reaction(A(i,j))
      END FORALL
   END DO
```

Fig. 1. HPF-like code of a simplified two-dimensional flame modeling code.

The paper is organized as follows. Section 2 gives an overview of our support and presents the SUPPLE implementation of the flame simulation kernel. Assessments of the experiment results are reported and discussed in Section 3, while Section 4 draws the conclusions.

2 SUPPLE and its implementation of the flame simulation code

SUPPLE is a run-time support for the efficient implementation of data parallel loops with regular, *stencil* references, characterized by per-iteration computation times which may be non–uniform, thus generating a workload imbalance.

SUPPLE only supports BLOCK array distributions to exploit locality deriving from stencil computations. Since stencil references are regular and known at compile time, several optimizations such as *message vectorization, coalescing* and *aggregation* [5] can be carried out to reduce communication overheads. These optimizations combine element messages which are then sent to the same processor in one single message without redundant data. Another optimization technique implemented in SUPPLE is *iteration reordering*. The local loops executed by each processor according to the BLOCK distribution and the owner computes rule are split into multiple localized loops: those accessing only local data, and those accessing remote data as well. The loops accessing only local data are scheduled between the asynchronous *sends* and the corresponding *receives* which are used to transmit remote data: in this way the execution of the iterations accessing only local data allows communication latencies to be hidden [5]. As far as our flame simulation benchmark (see Fig. 1), we used this scheme to implement the first Convection loop, which is uniform since the cost of the function $F()$ is constant, and exploits a statically predictable and regular *Five-Point stencil* to access the array B(N,N). For the sake of brevity, we will not discuss in more depth the features of SUPPLE for the support of these kind of uniform and regular loops.

The most innovative feature of SUPPLE is, on the other hand, its non–uniform loop support, which adopts a hybrid (static + dynamic) strategy to schedule loop iterations. In the following of this section we describe this part of the SUPPLE support, and its use for the implementation of the Reaction phase of our benchmark.

2.1 SUPPLE implementation of the Reaction loop

The innovative feature of SUPPLE is its support for non–uniform loops, such as the one that implement the Reaction phase in our flame simulation code. In this case a hybrid (static + dynamic) scheduling strategy is adopted. At the beginning, to reduce overheads, iterations are scheduled statically, by sequentially executing those stored in a queue Q, hereinafter called *local* queue. Once a processor understands that its local queue is *becoming empty*, the dynamic part of the scheduling policy is initiated. Computations are thus migrated at run-time to balance the workload.

Since SUPPLE groups the iterations into *chunks*, the iteration space is statically *tiled*. Chunking is introduced to reduce overhead because SUPPLE migrates chunks instead of single iterations on the basis of the workload distribution, and each processor makes decisions about work migration between the execution of subsequent chunks. To improve locality exploitation [8], SUPPLE adopts tiling also to implement uniform parallel loops.

A further motivation for the adoption of chunking is to avoid synchronizations being introduced between different loop executions. One synchronization might be needed, in fact, if the same array is modified during a previous non–uniform loop, and, due to dynamic work migration, the update of the local array partitions is partially performed by processors that are not the owners of the partitions. SUPPLE avoids waiting for the complete coherence of each data partitions owned by each processor before starting a new loop execution. In fact, it associates a *full/empty-like* flag [1] with each data tile. If a chunk that modifies a given tile is migrated, then the associated flag is set. The flag will be unset when the updated data tile is received back from the remote processors. Flags are checked only when the corresponding tiles need to be read, thus allowing useful computations to be overlapped with communications. As far our flame simulation kernel code, a flag is thus associated with each tile of the array C. In fact, since C is updated during the non–uniform Reaction loop (whose implementation exploits a hybrid scheduling of chunks), and is read during the next execution of the uniform Convection loop, the latter loop must check flags for data coherence.

Work migration requires both chunk identifiers and corresponding data tiles to be transmitted over the interconnection network. Migrated chunks, previously stored in the local queue Q of the owner, are stored by each receiving processor in a queue RQ, called *remote*. Data tiles, once updated remotely, must be returned to owners. In order to overlap computation with communication latency, and to avoid processors becoming idle while waiting for further work, an aggressive *prefetching policy* is adopted.

The size g of chunks is fixed and is decided statically. This size gives only a lower bound on the amount of work migrated at run-time, because the support adopts a heuristics to decide at run-time how many chunks must be moved from an overloaded processor to a underloaded one. On the other hand, since SUPPLE uses a *polling* technique to probe message arrivals, g directly influences the time elapsed between two consecutive *polls*.

Differently from other proposals [10], SUPPLE dynamic scheduling algorithm is fully distributed and asynchronous, and does not introduce bottlenecks which may jeopardize the efficiency when many processors are used. Chunk migration decisions are taken on the basis of local information only. It can be classified as a *receiver initiated* load balancing technique, since it is the *receiver* of remote chunks that asks overloaded processors for work migration [7, 15]. According to the framework proposed by Willebeek-LeMair and Reeves [15], our load balancing technique can be characterized by considering the strategies exploited for *processor load evaluation*, *load balancing profitability determination*, *task migration*, and *task selection*.

Processor Load Evaluation. Each load balancer policy requires a reliable workload estimation to detect load imbalances and take, hopefully, correct decisions about workload migration. Our technique is based on the assumption that each processor can derive the expected, average execution cost of its own chunks stored in the local queue Q. Since the first part of our technique is static, at the beginning each processor only executes chunks belonging to Q. During this phase, the average chunk cost is derived through a non–intrusive code instrumentation. Moreover, when chunks are migrated toward underloaded processors, an estimate of their cost is communicated as well. Each processor is thus aware of the average costs of remote chunks stored in RQ. On the basis of the cost estimates of chunks still stored in Q and RQ, each processor can therefore measure its own *current load*.

Load Balancing Profitability Determination. This strategy is used to evaluate the potential speedup obtainable through load migration weighted against the consequent overheads. In SUPPLE, each processor detects a possible load imbalance, and starts the dynamic part of the scheduling technique on the basis of local information only. It compares the value of its *current load* with a machine-dependent *Threshold*. When the estimated local load becomes lower than *Threshold*, the processor begins to ask other processors for remote chunks. The same comparison of the *current load* with *Threshold* is used to decide whether a chunk migration request should be granted or not. A processor p_j, which receives the migration request from a processor p_i, will grant the request only if its *current load* is higher than *Threshold*. Note that the *Threshold* parameter is chosen high enough to *prefetch* remote chunks, thus avoiding underloaded processors becoming idle while waiting for chunk migrations.

Task Migration Strategy. Sources and destinations for task migration are determined. In our case, since the load balancing technique is receiver initiated, the source (*sender*) of a load migration is determined by the destination (*receiver*) of the chunks. The *receiver* selects the processor to be asked for further chunks by using a *round-robin* policy. This criterion was chosen for its simplicity and also because tests showed that it was the best policy to adopt [7].

SUPPLE uses the same *Threshold* parameter mentioned above to reduce overheads of our task migration strategy. Overheads should derive from requests for remote chunks which cannot be served because the *current load* of the partner that has been asked for is too low (i.e., it is lower than *Threshold*). Thus each processor, when its *current load* (which, in this case, only derives from the chunks in Q) becomes lower than *Threshold*, broadcasts a so-called *termination message*. Our round-robin strategy can thus select a processor from those that have not yet communicated their termination.

Termination messages are also needed to end the execution of a parallel loop, and this is also the reason for the name we have given them. When a processor has already received a termination message from all the other processors, and both its queues Q and RQ are empty, then it can locally terminate the parallel loop.

Task Selection Strategy. Source processors select the most suitable tasks for effectively balancing the load, and send them to the destinations that asked for load migration. In terms of our support, this means choosing the most appropriate number of chunks that must be moved to grant a given migration request. We use a modified *Factoring* scheme to determine this number [6]. *Factoring* is a *Self Scheduling* heuristics formerly proposed to address the efficient implementation of parallel loops on shared–memory multiprocessors. It provides a way to determine the appropriate number of iterations that each processor must fetch from a central queue, and improves other techniques previously proposed [11, 14]. Clearly, the larger this number is, the lower the contention overheads for accessing the shared queue, and the higher the probability of introducing load imbalances when the fetched iterations are actually executed. *Factoring* requires that, if u is the number of remaining unscheduled iterations and P is the number of processors involved, the next P requests for new work are granted with a bunch of $\frac{u}{2 \cdot P}$ iterations. Consequently, at the beginning large bunches of iterations are scheduled, and this number is reduced when the shared queue is going to be emptied. In our case, instead of having a single queue, we have multiple shared queues, one for each processor. When a loop is unbalanced only some of these queues, i.e. those owned by overloaded processors, will be involved in granting remote requests for further work. The other difference regards our scheduling unit, which is a chunk of iterations instead of a single iteration. The modified heuristic we used is thus the following: an overloaded processor replies to a request for further work by sending $\frac{k}{2 \cdot P}$ chunks, where k is the number of chunks currently stored in Q, and P is the number of processors.

3 Experimental results

We tested SUPPLE on a 64 node Cray-T3D by using our flame simulation benchmark. The message passing layer used by SUPPLE is MPI, and for the experiments we exploited the MPI library, release CRI/EPCC 1.3, developed by the Edinburgh EPCC in collaboration with Cray Research Inc. We also compared the SUPPLE implementation of the benchmark with an HPF-style one.

3.1 The synthetic input data sets

We adopted a set of synthetic data sets because of the need to validate SUPPLE as a general support. Conversely, if we tested SUPPLE on a single problem instance, we might have come to the wrong conclusions due to the specific features of the data set chosen for the test. According to the code skeleton shown in Fig. 1, to build the different data sets we made the following assumptions:

- the simulation grid is 1024×1024 points (represented by doubles). Due to the data distribution adopted, a block of 128×128 contiguous points is thus assigned to each of the 64 processing nodes of the target machine;
- one quarter of the total execution time is spent in the Convection phase, while the Reaction phase takes the remaining three quarters;
- experiments were conducted for several values of μ, where μ is the average execution time to process a grid point during the Reaction phase. Note that, because of the previous assumption about the total execution time, the average execution time per point during the Convection phase was $\frac{\mu}{3}$;
- each input data set was built according to a simple model of load imbalance [9] discussed in the following. Note that the imbalance features of the data set are important for the behavior of the Reaction phase, where the per-point execution time depends on the value of the matrix element processed.

Model of load imbalance. Given μ, the average execution time, our model of load imbalance assumes that the workload is not distributed uniformly, and that there exists a region of the simulation grid, hereafter called *loaded region*, in which the average per-point execution time is greater than μ.

The load imbalance model is thus based on two parameters, d and t. The parameter d, $0 < d \leq 1$, is a fraction of the whole simulation grid, and determines the dimension of the loaded region. The parameter t, $0 < t \leq 1$, is a fraction of the total workload T, and determines the whole workload concentrated on the loaded region. Therefore, it follows that $t \geq d$.

Note that the imbalance is directly proportional to t for a given value of d, since larger values of t correspond to larger fractions of T concentrated on the loaded region. Similarly, the imbalance is inversely proportional to d for a given value of t. From these remarks, we can derive F, called *factor of imbalance*, defined as $F = \frac{t}{d}$

Sample data sets. The data sets used for the experiments were characterized by values of F ranging from 1 to 9, where the size of the loaded region was kept fixed $(d = 0.1)$, and $F = 1$ corresponds to the balanced case. All the results reported in this paper were obtained by also keeping fixed the position of the loaded region during all the simulation time steps. We also carried out some tests in which the position of the loaded region is moved at each step. The difference between the results of these experiments and those reported in the paper are not appreciable, because SUPPLE does not exploit any previous knowledge about which the overloaded or the underloaded processors are.

3.2   The experimental parameters

The experiments were conducted on 64 processors for different values of F and μ, and specifically for $F = 1, \cdots, 9$, and $\mu = 0.038, 0.075, 0.15, 0.3$ *msec*. As an example of the imbalance introduced by our synthetic data sets during the Reaction phase, consider that, for $F = 9$ and $\mu = 0.3$ *msec*, the computation of a single point takes 0.033 *msec* if the point itself belongs to the unloaded region, while it takes 2.697 *msec* if it belongs to the loaded region. The external sequential loop of the benchmark was iterated for 10 times in all the experiments.

Given the assumptions above, it is easy to determine the theoretical *Optimal Completion Time* (OCT) when the simulation code is run on 64 processors. The OCT values for $\mu = 0.038, 0.075, 0.15,$ or 0.3 *msec* are 8.2, 16.4, 32.8, or 65.5 *sec*, respectively. This time has been determined without considering any overheads, by symply multiplying the average time required to process a single point of the simulation grid $(\mu + \mu/3)$, times the number of points assigned to each processor (128×128), times the number of simulation time steps (10).

The only parameters that can be modified and tuned in our support are the size g of a chunk of iteration, i.e. our scheduling unit, and the prefetching *Threshold*. The parameter g affects the scheduling overhead, and the polling mechanism used to check for message arrivals. More specifically, a larger g reduces the scheduling overhead, but a larger g may also reduce the ability of our support to effectively balance the workload. A tradeoff must be found, and this depends on both the architecture and the features of the problem.

The tuning of the other parameter, *Threshold*, mainly depends on the specific architecture. It should be larger on architectures where communications are not very efficient. We found, however, that it is also related to the granularity of each chunk, which in turn depends on g but also on the features of the specific problem, i.e. on the parameters μ and F.

Since g and *Threshold* are highly interdependent, one strategy is to fix one and tune the other parameter on the basis of the first setting. Hence, all the experiments were carried out by fixing $g = 21$. We changed the granularity of each chunk by using data sets characterized by different μ. We observed that for larger

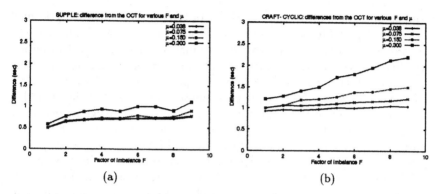

Fig. 2. Differences from the OCT for various μ as a function of the factor of imbalance F: (a) SUPPLE results, (b) CRAFT Fortran results.

μ, it is better to increase the value of *Threshold*. This is not surprising, since for large chunk granularities, overloaded processors become lazier in answering incoming requests due to our polling implementation of input message checking. Thus, for larger μ, work prefetching should be begun earlier to avoid emptying both Q and RQ. To this end, we have to increase the value of *Threshold*. In the exeperiment we used values of *Threshold* ranging from 2 to 40 *msec*, where larger *Threshold* values were used for larger μ.

3.3 Result assessment

Fig. 2.(a) shows several curves. Each curve is related to a distinct μ, and plots the difference between the completion time of the slowest processor and the theoretical OCT for various factors of load imbalance F. The difference from OCT can be considered as the sum of all overheads due to communications, residual workload imbalance, and loop housekeeping. It therefore also includes the time to send/receive the messages that implement the stencil communications of the first Convection loop. Note that there is an overhead of almost 0.5 *sec* even in the balanced case, and the overhead becomes larger when the input data set introduces a more extensive load imbalance. For $F = 9$ and $\mu = 0.15$ *msec*, SUPPLE balances the load with a completion time which is very close to the OCT: the difference from OCT is less than 1 *sec*. Note that, in the absence of load balancing actions, the completion time should have been, at least 229 *sec*, with a difference from the OCT of more than 196 *sec*. If we look at the same results by considering, instead of the absolute differences from OCT, their percentage with respect to OCT, then this percentage ranges from 6% to 9.4% for $\mu = 0.038$, while it ranges between 0.88% and 1.69% for $\mu = 0.3$.

3.4 Comparison with a HPF-like implementation

We implemented the same benchmark with CRAFT Fortran (an HPF-like language), by feeding it with the same synthetic data sets. Since, in this case, the scheduling is static and depends on the data layout, the simulation grid may be distributed CYCLIC, thus losing data locality in favor of a better balance of the processor workloads. Alternatively, we may adopt a BLOCK distribution for the first Convection loop, and a CYCLIC one for the second Reaction one. This solution does not loose data locality exploited by the Convection loop, but would entail using an executable redistribution directive. Although CRAFT Fortran does not support this directive, we simulated it by assigning one array distributed CYCLIC to another distributed BLOCK, and vice versa. Finally, to find a tradeoff, some combinations of BLOCK and CYCLIC may be adopted.

The best results for unbalanced problem instances were obtained by using a pure CYCLIC distribution (see Fig. 2.(b)). One of the reasons of these performance results is the computational weight of the Reaction loop, which is three times that of the first Convection loop. Thus it is more important to try to balance the workload through a pure CYCLIC distribution than to increase locality exploitation with a BLOCK distribution. The differences from the OCT are larger than those obtained by SUPPLE, even for the completely balanced case (nearly a half second more than SUPPLE). These bad results for $F = 1$ are also due to the CYCLIC distribution, which should be turned to BLOCK to take advantage of data locality in the first Convection loop. The results become worse for larger μ and F. For example, for $F = 9$, the CRAFT differences from the optimal times are 1.04, 1.22, 1.50, 2.21 sec for $\mu = 0.038, 0.075, 0.15, 0.3$ $msec$ respectively, against the 0.77, 0.78, 0.90, 1.11 sec obtained on the same tests by SUPPLE. The CYCLIC distribution, besides losing data locality, is not able to perfectly balance the workload, so that, for larger μ, the workload differences between overloaded and underloaded processors increase.

We also conducted other tests with CRAFT Fortran to evaluate combinations of BLOCK and CYCLIC distributions. These results were worse than the pure CYCLIC case. Fig. 3.(a) refers to the case $\mu = 0.038$ $msec$, and shows some curves that plots the differences from the OCT for different values of F. Note that the ordinate scale is logarithmic, and that the pure BLOCK distribution is identified by the label BLOCK_CYCLIC(128).

Finally, we tested the redistribution of the simulation grid between consecutive parallel loop executions. The results, which are shown in Fig. 3.(b) are worse than those obtained by using a pure CYCLIC distribution for both loops. The reason may be the huge volume of data moved simultaneously between all processors, which may cause network congestion problems.

4   Conclusions

We have described SUPPLE, an innovative run–time support, which can be used to compile either uniform or non–uniform parallel loops with regular stencil data references. SUPPLE exploits regular BLOCK distributions of the arrays involved to favor data locality exploitation, and uses the static knowledge of the data layout and of the regular stencil references to perform well-known optimizations such as *message vectorization*, *coalescing* and *aggregation*. *Iteration reordering*

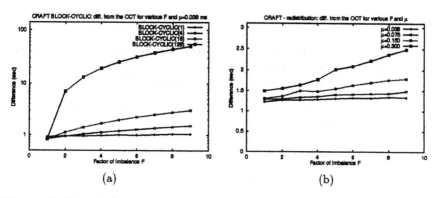

Fig. 3. CRAFT Fortran differences from the OCT: (a) for different data layouts with $\mu = 0.038$ *msec*, (b) for different μ with array redistribution.

is exploited to hide the latencies of the communications needed to fetch remote data. SUPPLE supports two different loop scheduling strategies. Parallel loops characterized by iterations whose costs are uniform, can be scheduled according to a pure static strategy based on the *owner computes rule*. A hybrid scheduling strategy that tries to balance processor workloads can be used instead if the load assigned to the processors according to the owner computes rule is presumed to be non–uniform. In this case loop iterations are initially scheduled statically, and, if a load imbalance is actually detected, an efficient dynamic scheduling, which requires data tiles and iteration indexes to be migrated, is started.

We have shown that SUPPLE hides most dynamic scheduling overheads produced by the need for monitoring the load and moving data, by overlapping them with useful computations. To this end it exploits an aggressive prefetching technique that tries to avoid underloaded processors from becoming idle while waiting for workload migration requests to be satisfied. Nevertheless, since data may be migrated and updated remotely due to dynamic scheduling decisions, and the same data must be kept coherent on the original owner, SUPPLE adopts a fully asynchronous coherence protocol that allows useful computations to be executed while waiting for data to become coherent.

The results of our experiments conducted on a Cray T3D machine by running a kernel code of a flame simulation application, have shown that the performances obtained are very close to the optimal ones. To validate SUPPLE as a general support for non–uniform loops we employed different synthetic data sets built on the basis of a simple load imbalance model. These data sets, fed in input to the flame simulation code, introduce different workload imbalances. For data sets characterized by an average iteration execution times of 0.3 *msec*, the overheads due to communications and the residual load imbalance range between 0.88% and 1.69% of the optimal time. In addition, we have compared the SUPPLE results with those obtained running a CRAFT Fortran implementation of the same benchmark. In all the cases tested, the SUPPLE implementation obtained better performances.

References

1. R. Alverson et al. The Tera computer system. In *Proc. of the 1990 ACM Int. Conf. on Supercomputing*, pages 1–6, 1990.

2. High Performance Fortran Forum. *High Performance Fortran Language Specification*, May 1993. Version 1.0.

3. High Performance Fortran Forum. *HPF-2 Scope of Activities and Motivating Applications*, Nov. 1994. Version 0.8.

4. S. Hiranandani, K. Kennedy, and C. Tseng. Compiling Fortran D for MIMD Distributed-Memory Machines. *Comm. of the ACM*, 35(8):67–80, Aug. 1992.

5. S. Hiranandani, K. Kennedy, and C. Tseng. Evaluating Compiler Optimizations for Fortran D. *J. of Parallel and Distr. Comp.*, 21(1):27–45, April 1994.

6. S.F. Hummel, E. Schonberg, and L.E. Flynn. Factoring: A Method for Scheduling Parallel Loops. *Comm. of the ACM*, 35(8):90–101, Aug. 1992.

7. V. Kumar, A.Y. Grama, and N. Rao Vempaty. Scalable Load Balancing Techniques for Parallel Computers. *J. of Parallel and Distr. Comp.*, 22:60–79, 1994.

8. M. S. Lam, E. E. Rothberg, and M. E. Wolf. The cache performance and optimizations of block algorithms. In *4th Int. Conf. on Architectural Support for Progr. Lang. and Operating Systems*, pages 63–74, Santa Clara, CA, April 1991.

9. S. Orlando and R. Perego. A template for non–uniform parallel loops based on dynamic scheduling and prefetching techniques. In *Proc. of the 1996 ACM Int. Conf. on Supercomputing*, pages 117–124, 1996.

10. O. Plata and F. F. Rivera. Combining static and dynamic scheduling on distributed–memory multiprocessors. In *Proceedings of the 1994 ACM Int. Conf. on Supercomputing*, pages 186–195, 1994.

11. C. Polychronopoulos and D.J. Kuck. Guided Self-Scheduling: A Practical Scheduling Scheme for Parallel Supercomputers. *IEEE Trans. on Computers*, 36(12), Dec. 1987.

12. R. Ponnusamy, J. Saltz, A. Choudary, Y-S Hwang, and G. Fox. Runtime Support and Compilation Methods for User-Specified Irregular Data Distributions. *IEEE Trans. on Parallel and Distr. Systems*, 6(8):815–831, Aug. 1995.

13. J. Saltz et al. Runtime Support and Dynamic Load Balancing Strategies for Structured Adaptive Applications. In *Proc. of the 1995 SIAM Conf on Par. Proc. for Scientific Computing*, Feb. 1995.

14. T.H. Tzen and L.M. Ni. Dynamic Loop Scheduling on Shared-Memory Multiprocessors. In *Proc. of Int. Conf. on Parallel Processing - Vol II*, pages 247–250, 1991.

15. M.H. Willebeek-LeMair and A.P. Reeves. Strategies for Dynamic Load Balancing on Highly Parallel Computers. *IEEE Trans. on Parallel and Distr. Systems*, 4(9):979–993, Sept. 1993.

16. H.S. Zima and B.M. Chapman. Compiling for Distributed-Memory Systems. *Proc. of the IEEE*, pages 264–287, Feb. 1993.

Performance Optimization of Combined Variable-Cost Computations and I/O

Sorin G. Nastea [† 1]   Tarek El-Ghazawi [‡ 1]   Ophir Frieder [‡‡ ‡‡‡ 2]

[1]Department of Electr. Eng. and Comp. Science
The George Washington University
Washington, D.C. 20052
E-mail: {snastea, tarek}@seas.gwu.edu

[2]Department of Computer Science
Florida Institute of Technology
Melbourne, Florida 32901-6975
E-mail: ophir@cs.fit.edu

Abstract

For applications involving large data sets yielding variable-cost computations, achieving both efficient I/O and load balancing may become particularly challenging though performance-critical tasks. In this work, we introduce a data scheduling approach that integrates several optimizing techniques, including dynamic allocation, prefetching, and asynchronous I/O and communications. We show that good scalability is obtained by both hiding the I/O latency and appropriately balancing the workloads. We use a statistical metric for data skewness to further improve the performance by adequately selecting among data-scheduling. We test our approach on sparse benchmark matrices for matrix-vector computations and show experimentally that our method can accurately predict the relative performance of different input/output schemes for a given data set and choose the best technique accordingly.

Keywords: load balancing, parallel I/O, data distribution skewness, sparse matrix computations

1. Introduction

The continually-growing performance gap between the I/O and CPU technologies has constantly challenged authors to propose heuristics for hiding the I/O latency or improving the I/O bandwidth. Most authors focused on I/O solutions alone, and used raw or synthetic data loads in their experiments. Nitzberg and Fineberg [12] presented an overview of raw I/O bandwidth of typical parallel I/O systems using, including the Paragon.

One successful way to enlarge the bandwidth of I/O systems is to access the data before they are actually required by the processing nodes in computations. The technique is generally known as prefetching, and its application is strictly dependent on the application access patterns. Arunachalam, Choudhary, and Rullman [1] described the design and implementation of a prefetching strategy and provided experimental measurements and evaluation of the file system with and without the prefetching capability. They found that, by using prefetching, a maximum speedup of 7.7 could be attained for 8 processing nodes and 8 I/O nodes.

We address the performance optimization of I/O-intensive variable-cost computations. Several aspects differentiate our research from prior work. We focus on studying the interaction between scalable computations and I/O, in the presence of potential load imbalances. We provide integrated solutions for hiding the I/O latency and ensuring the even distribution of workload onto processing units. Such techniques and methods, including prefetching, dynamic data scheduling, and asynchronous I/O and message passing, significantly ease the I/O bottleneck to a limit that can preserve the scalability characteristics of the computations. We use typical computations and variable-cost-yielding data benchmarks to evaluate our techniques. Our research goal is to go beyond just measuring basic system capabilities and test our performance-improving solutions under complex real-life applications.

[†]Author currently on leave of absence from the Department of Control and Computers, the Polytechnic University of Bucharest, Romania.

[‡]Supported by CESDIS/USRA.

[‡‡]This work is supported in part by the National Science Foundation under contract number IRI-9357785.

[‡‡‡]Author currently on leave from the Department of Computer Science, George Mason University, Fairfax, Virginia.

Furthermore, we evaluate and predict performance achievable with various data-scheduling techniques by using statistical information on the data. Previously [6], the *Normalized Coefficient Of Variation*, or *NCOV*, was used as a metric for data skewness to predict the relative performance of data allocations that include *two disjoint phases*: the allocation phase and the computation phase. The method is based on the assumption that a more elaborate allocation produces better load balancing. In other words, the larger overhead of generating better load-balancing may payoff in the computation phase, for adequate computation complexities. In this work, we use address the case in which the data scheduling and the computations *cannot be separated into two disjoint phases* because they overlap. We also use NCOV as a metric for data skewness and, implicitly, for data load-balancing challenges, to predict the performance benefits of our approach and its conditions for optimum-performance usage. The main particularity of this case is that the data-scheduling and the computation phases are not disjoint, but integrated and/or overlapped.

In Section 2, we introduce our load and I/O scheduling technique that makes efficient use of dynamic allocation, prefetching, and asynchronous I/O and message-passing. We compare our data scheduling solution with a typical static data scheduling, that uses node indexes to locate and fetch the data. Furthermore, we use statistical methods for selecting between these two approaches for best performance over an entire range of operational parameters. We apply these proposed solutions to I/O-intensive variable-cost yielding sparse matrix computations. In Section 3, we present our experimental results. We show that our load and I/O scheduling is particularly effective for skewed data distributions that generally raise load-balancing challenges. We end our paper with our concluding remarks.

2. Performance optimization and evaluation

2.1 I/O Latency Hiding

We study the effect on overall performance of I/O-intensive applications with variable-cost computations. Modern MIMD parallel platform are provided with dedicated communication and I/O hardware, which enables and eases the access to data without actually interrupting the compute nodes. Practically, this is achieved through asynchronous I/O and message passing

calls. Therefore, one way of improving the performance is to overlap computation and I/O costs at the maximum extent by using such asynchronous I/O calls. Generally speaking, the overall cost of combined computations and I/O depends on the maximum of the computations and I/O costs. However, it is impractical to access all the data in one I/O transaction because, among other reasons, of memory constraints. Therefore, as several I/O transactions are performed for large data sets, the total execution time depends on the way the two costs overlap for each data set fetched from storage. In (1), we evaluate the overall computation and I/O cost. Let K be the number of read transactions; thus, the overall time becomes:

$$T_{overall} = \sum_{i=1}^{K+1} \max\left\{T_{work_{i-1}}, T_{I/O_i}\right\} \qquad (1)$$

where $T_{work_0} = 0$ and $T_{I/O_{K+1}} = 0$. This delay of computations over I/O is due to the fact that computations always take place on the data set fetched from storage one step ahead. By simplifying, if only one data set is fetched, overlapping cannot occur anymore and the overall execution time becomes:

$$T_{overall} = \max\left\{0, T_{I/O_1}\right\} + \max\left\{T_{work_1}, 0\right\} = T_{I/O_1} + T_{work_1} \quad (2)$$

Thus, another counter argument against large one-time file accesses is the sequentializing of the I/O and computations phases instead of achieving some degree of overlapping. Equations (1) and (2) lead to some interesting generalizations: First, a finer granularity of I/O transactions is likely to produce better overlapping of I/O and computations. However, the I/O read sizes are limited by multi-striping of file contents over physical disks, and a finer granularity than the stripe size will considerably reduce the I/O bandwidth because of an underutilization of the parallel I/O resources. Second, loss of scalability may occur for skewed data distributions. Thus, even if the total computations cost is larger than the total I/O cost, some I/O transactions may not provide enough computations to completely overlap with the I/O. Worse than that, delays at some node may accumulate and ultimately limit the performance.

As outlined by (1), variable-cost I/O-intensive

computations can lead to loss of performance unless additional methods are used. We compare two approaches for allocating onto compute nodes variable computation cost yielding data:

(1) Worker-Worker static allocation

In this simple but typical data scheduling solution, each node generates, based on its node number, the indices of the rows it reads. Thus, the first node reads the first n rows, the second node reads the next n, and so on. At the next read cycle, if there are altogether P processors, the first node reads rows: $P*n, P*n+1, ...,(P+1)*n - 1$. As a rule, node j reads the following rows within read cycle i (if it is the last read cycle, some nodes may not read at all, and one node may read less): $(P*i+j)*n, (P*i+j)*n+1..., (P*i+j+1)*n-1$. Thus, each node reads disjoint areas from the file   We call this approach, which is mainly a static allocation, the worker-worker (W-W) approach. The advantages are the following: the method enables large-size read sessions and the nodes extract the data location based on their node index, by avoiding any overhead. Also, the method attempts some balancing of workloads captured into the fetched data by avoiding reading contiguous blocks, which is not recommended for highly skewed data [7]. The main disadvantage of this method is that it does not attempt to evenly distribute the work onto processing nodes based on the actual distribution of the load. In some cases, imbalances accumulate and have a particularly large negative impact on the overall performance.

(2) Master-Worker dynamic allocation

According to this method, presented in Figure 1, each node informs the master node when it is able to process new data. Conceptually, the method is a typical dynamic allocation, but its novelty consists of the way I/O and control communication are carried out to achieve themaximum of overlapping and performance. As a principle, any worker that is ready to accept new data sends a READY message to the master node. If the master node, that manages the allocation of rows to workers, still has unprocessed rows to allocate, it sends back a GO message. The GO message includes all the necessary information to enable the compute node to fetch the data. Once the EOF is reached, the master node broadcasts a STOP message.

To achieve maximum possible performance, each compute node anticipates the exhaustion of its current workload and requests permission to fetch more data prior to the occurrence of starvation. This minimizes to the complete elimination the wait times in the "infinite"

```
proc Master
    compute the number of I/O reads
    for (all I/O reads)
        receive a READY message from a worker
        identify the node that sent the READY message
        send back GO message with row identifiers
    end for
    for (all processing nodes)
        receive a READY message from a worker
        identify the node that sent the READY message
        send back STOP message with row identifiers
    end for
end proc

proc Worker
    post an asynchronous receive for a STOP message
    post an asynchronous I/O read for the first set of
        data based on node index
    post an asynchronous receive GO message
    send READY message to master node
    Terminate=FALSE
    while not (Terminate)
        for ( ; ; )
            if GO message received from master node
                post a new asynchronous receive GO
                    message
                send READY message to master node
                break from for loop
            end if
            if STOP message received
                cancel GO receive message
                Terminate=TRUE
                break from for loop
            end if
        end for
        wait until previous I/O read ends
        if not (Terminate)
            post a new  asynchronous I/O read
        end if
        perform computations on data previously fetched
            from disk
    end while
end proc
```

Figure 1. Master-Worker (dynamic) data scheduling

for loop. Furthermore, the message passing with the master node takes place while performing I/O and computations. As the I/O calls and the associated computations generally take longer to complete than the short message passing that referees the data allocation, the communication is thus completely carried out in the background and its latency is completely hidden. Whenever a master node is involved in coordinating a parallel application, the concern that the master node may become a serious bottleneck is raised. In this situation, the load associated with the master node has minor cost as compared to the load allocated to compute nodes that deal both with I/O and computational tasks. Therefore, the major node is not impeding on the scalability of the overall execution.

To implement the I/O latency hiding, two memory buffers may be used. At any moment after the start-up, while one buffer is being used to fetch data into it, the other one, already containing data, is used in the computations phase.

One difference between the Static (W-W) and the Dynamic (M-W) approaches is the need for a coordinating master node in the latter approach. In the first approach, the location of the data to be fetched and processed by each compute node can be determined according to the node number of each processor. In the dynamic allocation, a coordinator is necessary to arbitrate requests for more work from compute nodes and keep track of the already allocated and remaining workloads. This avoids any misalignment of allocated data, such as skipping or duplicating the data. As an implementation solution, the master node can be a service node, given its reduced computation attributions.

2.2 Overall performance evaluation and prediction

Besides typical performance evaluation and metrics for parallel applications, some additional issues are raised by this work. One important aspect refers to the data particularities and challenges that would suggest the use of some specific data scheduling for best performance. Also, we would like to anticipate the benefits of using some data scheduling as compared to another approach for some given data set. Furthermore, considering that data allocations provide benefits but incur overheads, as well, it would be expected that, given the implementation particularities, the operation parameter ranges may be divided in optimal-performance domains for each of the approaches. As a key role is played by the data, capturing its particularities and load-balancing challenges is essential. In fact, data skewness is particularly challenging for achieving perfect or almost perfect load-balancing. Therefore, of particular importance is to be able to quantify the data distribution skewness and further use this quantification to predict the load-balancing performance of two data-scheduling techniques with different load-balancing capabilities.

Suppose that the data set involved in the computations can be divided into N workload units, with workloads n_i, $i \in 1,..N$. Also, suppose $M = \frac{1}{N}\sum n_i$ is the average workload. Then, a good metric for data skewness is the *Normalized Coefficient Of Variation* (NCOV) [6], defined as:

$$NCOV = \frac{1}{M\sqrt{N-1}} \sqrt{\frac{1}{N}\sum_{i=1}^{N}(n_i - M)^2} = $$
$$\frac{1}{\sqrt{N-1}} \sqrt{\frac{1}{N}\sum_{i=1}^{N}\frac{n_i^2}{M^2} - 1} \qquad (3)$$

Some interesting properties of NCOV are: (1) NCOV $\in$ [0,1]; (2) reduced influence of the number of workload units on NCOV; and (3) reduced influence of the average workload on NCOV. These properties make the NCOV the best available metric for data skewness.

Previously [6], NCOV was used as a metric for skewness to predict and evaluate the relative performance when the data-allocation and computation phases are disjoint. The method assumes that a more time-consuming allocation produces better load balancing, that helps amortize the initial overhead for adequate computational complexities. Thus, the overhead of the more elaborate allocation is a time overhead. In this work, we address the case in which data scheduling and computations cannot be separated into two disjoint phases, and they overlap in time. We use NCOV as a metric for data skewness and, implicitly, for load-balancing challenges, to predict the performance benefits of the dynamic allocation versus the static one. We base our prediction on existing runtime results with a different data set, and extrapolate prior load-balancing performance for a new data set with load-balancing challenges quantified by NCOV.

An Application to I/O-Intensive Sparse Matrix Computations

To further develop the analysis, let us assume that the data in storage are uncompressed sparse matrices, collected as raw data by a sensor of some type. Therefore, before performing any computations, data has to be retrieved and compressed. Of course, for any subsequent utilization of the matrices, they can be redirected to storage in a compressed representation. Also, some useful operations, such as the matrix-vector multiplication, can be performed with the data. Sparse matrix compression and computations have been well studied [e.g. 8, 9, 10, 11, 13]. In the current work, sparse matrix computations are used to provide irregularly-distributed workloads . Given that each data set fetched from storage contains a random number of non-zero elements, it also involves variable-cost computations. This sets the basis for a demanding testbed to experiment with several implementation solutions that reduce or alleviate the I/O bottleneck while maintaining balanced workload distributions. Proposed solutions can be extrapolated and adapted to other irregularly structured problems with similar requirements.

By adapting to this problem the potential load-imbalance quantification, as described in [6], we obtain the following runtime evaluations for the sparse matrix compression and multiplication, when the data-scheduling approaches are the two previously described techniques:

$$T_{m-w} = a \cdot \frac{n}{P_1} \cdot (Q+d) + e \cdot \frac{N}{P_1} \qquad (4)$$

$$T_{w-w} = a \cdot n \cdot \left[\frac{1}{P_2} + NCOV \cdot \left(\frac{b}{P_2+c} + 1 \right) \right] \cdot (Q+d) + e \cdot \frac{N}{P_2} \qquad (5)$$

where a, b, c, d, and e are constants, Q is the number of multiplying vectors, and $n = \sum n_i = N \cdot M$. The first term in (4) and (5) represents the computations (compression and multiplication) associated with non-zero elements, and the second term represents the compression overhead associated with testing the value of each matrix element.

The performance of the static (W-W) and the dynamic (M-W) data allocations can be compared by considering either the number of compute nodes or the total number of processors used in the implementation. If the total number of processors is a main factor, the dynamic data scheduling requires one controlling processor (the master node) to coordinate the data distribution. This enforces fewer (P-1) processors in the compute pool, while, in the static data scheduling, all P processors can be used for compute purposes. Let $P_1=P-1$ and $P_2=P$; thus, the predicted difference of execution time yielded by the two data-scheduling techniques are:

$$\Delta T = T_{w-w} - T_{m-w} = a \cdot (Q+d) \cdot NCOV \cdot n \cdot \left(\frac{b}{P+c}+1 \right) - \frac{a \cdot n \cdot (Q+d)}{P \cdot (P-1)} - \frac{e \cdot N}{P \cdot (P-1)} \qquad (6)$$

In this case, the overhead of the more elaborate data scheduling technique is a hardware overhead rather a time one. However, for I/O-intensive computations and skewed data sets, it is expected that, for some processor range, better performance to be obtained with the dynamic data scheduling even under such limiting conditions. The goal is to find the number of processors P for which the processor range is divided into two regions with best predicted performance for each of the two approaches.

If the number of compute nodes is considered, the dynamic data scheduling is expected to yield the best performance given its better load-balancing characteristics because the overhead is distributed and mostly hidden by I/O and computations. Therefore, statistic data can primarily be used to evaluate results and predict the performance benefits of the dynamic data scheduling versus the static one for some particular data sets. Given that, in this case, $P_1 = P_2 = P$, the predicted difference of execution time is:

$$\Delta T = T_{w-w} - T_{m-w} = a \cdot (Q+d) \cdot NCOV \cdot n \cdot \left(\frac{b}{P+c}+1 \right) \qquad (7)$$

In both cases, prediction is based on prior run-time results achieved with a different matrix data set, characterized by a different NCOV. Worth noticing is that I/O is transparent to these results because of the assumption that the I/O cost is smaller than the computation cost. However, for larger and larger number of compute nodes, given the reduced hardware scalability of the I/O system, the I/O cost becomes preponderant, making non-essential the selection task of the data scheduling approach.

3. Experiments

3.1 Experimental testbed

In our experiments, we used selected benchmark sparse matrices from the Harwell-Boeing collection [3,4]. We particularly show results based on PSMIGR 1 matrix, with a raw size of almost 40 Mbytes. To further emphasize the impact of the data distribution skewness on overall performance and to create a challenging experimental data set, we generated a new matrix with the same general characteristics with PSMIGR 1 (order: 3140; density: 0.055) but with a different distribution of the non-zero values in the matrix rows. Thus, we used the Zipf distribution, with the *probability density function* described in (8), to generate this new synthetic matrix.

$$p(i) = \begin{cases} \dfrac{C}{i^{1-\theta}}, & for \ i = \{1,2,...,N\} \\ 0, & elsewhere \end{cases}$$

$$where \quad C = \dfrac{1}{\sum\limits_{i=1}^{N} \dfrac{1}{i^{1-\theta}}}, \quad \theta \in [0,1] \qquad (8)$$

In our case, N is the number of rows and $p(i)$ is the probability that some non-zero element is in row i. The data distribution skewness is decisively influenced by the parameter θ. Thus, if θ is 0 or close to 0, the distribution is quite skewed; on the contrary, if $\theta=1$, the distribution is uniform.

We performed our experiments on a 64-node Intel Paragon. The Parallel File System (PFS) is organized as a two-disk system, each disk being a RAID 3. File contents are striped over disks with stripe sizes equal to 64 KBytes. Conceptually, the system combines fine-grained parallelism within each RAID 3 with coarse-grained parallelism at the entire PFS level, seen as a RAID 0.

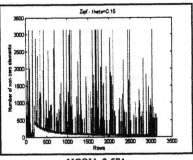

NCOV=3.571

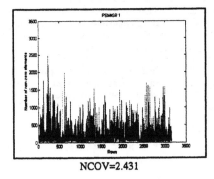

NCOV=2.431

Figure 2. Distribution of non zero elements per rows for PSMIGR 1 and Zipf ($\theta=0.15$) matrices (both matrices have the same order 3140 and density 0.055).

3.2 Experimental results

First, we evaluated the performance of the available PFS file access modes of the Intel Paragon. For this purpose, we implemented the sparse matrix compression, only, that yields a reduced computation cost, easily surpassed by the I/O cost for $P \geq 2$. Our results for two different read sizes show a consistent good performance achieved with M_ASYNC. This file access mode also has the advantage over the M_RECORD mode, the second best performance-yielding one, of supporting MIMD-type implementations. Hence, the rest of our experiments is based on the M_ASYNC mode.

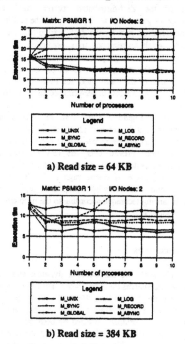

a) Read size = 64 KB

b) Read size = 384 KB

Figure 3. Sparse matrix compression using PFS file access modes available on an Intel Paragon

For the first case, in which we compare the Worker-Worker and Master-Worker approaches on a total-number-of-processor basis, we perform experiments based on PSMIGR 1 and extrapolate these results for prediction purposes to the Zipf-distributed matrix. In this case, our prediction that the switching from one approach to the other has to occur for P=7 coincides also with the

actual decision. The task of ensuring accurate prediction is eased by the fact that this is a course-grained decision-making process, as compared to a finer-grained situation depicted in [6]. The prediction of the optimum-performance parameter range is based on (7), by enforcing $\Delta T=0$. Obviously, a non-monotonic relative behavior of the performance of the two approaches complicates the decision-making process.

At the same time, we use the statistical metric for skewness to assess the relative performance of data scheduling schemes. Thus, for a comparison based on the number of compute nodes, we performed computations with both the benchmark matrix (PSMIGR 1) and the Zipf-distributed one. In Figure 5, we plot the speedup results for I/O, compression, and multiplication with 150 vectors, for both the static and dynamic allocations. Variations of processor workloads may occur, and the dynamic data scheduling is aimed at setting the basis for recovering the potential node delays. For PSMIGR 1, the dynamic (M-W) allocation yields a maximum speedup of 9.45 for 10 nodes, as compared to a speedup of 8.37 obtained with the static (W-W) allocation for the same number of processing elements.

In Figure 5, we also show speedup results for the Zipf-distributed matrix, based on the dynamic and static allocations. For the synthetic matrix, the speedup performance of the dynamic and static approaches is 8.88 and 7.39, respectively. The relative degradation of results for both the static and dynamic allocations in case

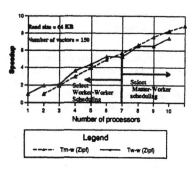

Figure 4. Optimum-performance decision for the W-W and M-W approaches when P and P-1 compute processors, respectively, are used.

of the Zipf-distributed matrix is due to the increased skewness and may be explained through equation (1). Thus, even if the total computation cost is larger than the total I/O cost, the computation cost may not always prevail over the I/O cost for every matrix data sub-set fetched from disk. Improvements can be obtained by increasing the size of each read operation from storage, thus obtaining an integration of the variations of non-zero elements per rows that would ultimately provide a more uniform workload distribution. However, the read-size increase is limited by computation and I/O overlapping considerations (see equations (1) and (2)). In Figure 5, we present a case in which the read-size is relatively small (1 stripe-unit or 64 KB). These results emphasize the robustness of the dynamic (m-w) data scheduling for data distributions with increased skewness.

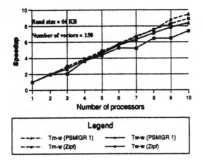

Legend

---·--- Tm-w (PSMIGR 1)	—— Tw-w (PSMIGR 1)
---·--- Tm-w (Zipf)	—•— Tw-w (Zipf)

Figure 5. Performance obtained with the Static and Dynamic data allocations

Based on experimental results obtained with matrix PSMIGR 1 and considering (18), the gain of performance of the dynamic allocation as compared to the static allocation for the Zipf-distributed matrix can be predicted and approximated, as previously shown, with the formula:

where $\Delta T = T_{static} - T_{dynamic}$. The simplified expression in

$$\Delta T_{Zipf_{P-P_0}} = \Delta T_{PSMIGR\ 1_{P-P_0}} \cdot \frac{NCOV_{Zipf}}{NCOV_{PSMIGR\ 1}} \quad (9)$$

(9) is due to the fact that the two matrices involved have the same order and density. Another way of employing the skewness metric is to analyze the obtained results and find out how they compare with the predicted values. For example, for P=10 nodes, we measured 12.9% performance gain generated by the dynamic allocation over the static allocation for matrix PSMIGR 1. In the case of the Zipf-distributed matrix, this same performance gain is 20.1%. By using (9), the prediction of performance gain for the Zipf-distributed matrix is 17.4%, which corresponds to a prediction error of 13.4%. However, larger prediction error is obtained for P=3 nodes, for example, due to the particularly poor performance of the static allocation in distributing the workload (a 44.5% loss of performance as compared to the dynamic allocation).

Results in Figure 5 show that, to achieve high performance, it is not enough to make the scalable operations, namely the sparse matrix compression and multiplication, prevail over the less scalable I/O performance. Additionally, developers of sparse matrix applications have to ensure high scalability of the computations themselves by appropriately selecting the load-balancing technique according to the problem and data requirements.

4. Conclusions

In this work, we introduced a dynamic allocation scheme that balances the overall workloads of processing and input/output operations. We showed that significant execution time improvements can be obtained with this technique on workloads affected by both random and persistent imbalances. Also, we studied the effect of overlapping scalable computations with I/O on alleviating the I/O bottleneck, and we showed that, for moderate and large sizes of scalable problems, the I/O latency can be effectively hidden using a combination of asynchronous calls and dynamic load balancing (as described in Figure 1). We showed that the dynamic (master-worker) allocation is more robust than the static (worker-worker) allocation for increased skewness of the data distributions. In our implementations, we used M_ASYNC, the most efficient PFS file access mode proven so by our experiments as well as other's, that enables flexible MIMD-type of parallel solutions.

Using NCOV as a metric for data distribution skewness, we introduced an approach for selecting the optimum-performance input/output scheduling. This methodology allows predicting the performance benefits as well as the associated overhead due to the use of a specific algorithm for load-balanced allocation and I/O scheduling. We experimented with sparse matrix computations and showed that, by using our approach, the best algorithm for a specific problem can be accurately selected.

References

[1] Arunachalan, M., Choudhary, A., Rullman, B., *Implementation and evaluation of prefetching in the Intel Paragon Parallel File System*, Proceedings of the 10th Int'l. Parallel Processing Symposium, Honolulu, Hawaii, April 15-19, 1996.

[2] Cheung, A. L., Reeves, A. P., *Sparse data representation*, Proceedings Scalable High Performance Computing Conference, 1992.

[3] Duff, I. S., *Sparse matrix test problems*, ACM Transactions on Mathematical Software, Vol. 15, No. 1, 1-14, 1989.

[4] Duff, I. S., Grimes, R.G., Lewis, J. G., *User's guide for the Harwell-Boeing Sparse Matrix Collection*, CERFACS Report TR/PA/92/86, 1992.

[5] Nastea, S. G., El-Ghazawi, T., Frieder, O., *Parallel input/output impact on sparse matrix compression*, Proceedings of the IEEE Data Compression Conference, Snowbird, Utah, 1996.

[6] Nastea, S., El-Ghazawi, T., Frieder, O., *A statistically-based multi-algorithmic approach for load-balancing sparse matrix computations*, Proceedings of the IEEE Frontiers of Massively Parallel Computations (Frontiers '96), Annapolis, Maryland, October 1996.

[7] Nastea, S., Frieder, O., El-Ghazawi, T., *"Load balancing in sparse matrix-vector multiplication"*, Proceedings of the IEEE Symposium of Parallel and Distributed Processing, New Orleans, Louisiana, October 1996.

[8] Paolini, G. V., Santangelo, P., *A graphic tool for the structure of large sparse matrices*, IBM Technical Report, ICE-0034 IBM ECSEC Rome (1989).

[9] Park, S. C., Draayer, J. P., Zheng, S. Q., *Fast sparse matrix multiplication*, Computer Physics Communications, Vol. 70, 1992.

[10] Peters, A., *Sparse matrix vector multiplication technique on the IBM 3090 VP*, Parallel Computing 17, 1991.

[11] Petiton, S., Saad, Y., Wu, K., Ferng, W., *Basic Sparse matrix computations on the CM-5*, International Journal of Modern Physics C vol. 4, No. 1, 63-83, 1993.

[12] Nitzberg, B., Fineberg, S. A., *Parallel I/O on Highly Parallel Systems*, Tutorial notes, Proceedings "Supercomputing '94", Washington D. C., 1994.

[13] Rothberg, E., Schreiber, R., *Improved load balancing in parallel sparse Choleski factorization*, "Supercomputing '94", Washington D.C., 1994.

Parallel Shared–Memory State–Space Exploration in Stochastic Modeling

Susann C. Allmaier, Graham Horton

Computer Science Department III, University of Erlangen–Nürnberg, Martensstr. 3, 91058 Erlangen, Germany. Email : { *snallmai* | *graham*} *@informatik.uni-erlangen.de.*

Abstract. Stochastic modeling forms the basis for analysis in many areas, including biological and economic systems, as well as the performance and reliability modeling of computers and communication networks. One common approach is the state–space–based technique, which, starting from a high–level model, uses depth–first search to generate both a description of every possible state of the model and the dynamics of the transitions between them. However, these state spaces, besides being very irregular in structure, are subject to a combinatorial explosion, and can thus become extremely large. In the interest therefore of utilizing both the large memory capacity and the greater computational performance of modern multiprocessors, we are interested in implementing parallel algorithms for the generation and solution of these problems. In this paper we describe the techniques we use to generate the state space of a stochastic Petri–net model using shared–memory multiprocessors. We describe some of the problems encountered and our solutions, in particular the use of modified B–trees as a data structure for the parallel search process. We present results obtained from experiments on two different shared–memory machines.

1   Introduction

Stochastic modeling is an important technique for the performance and reliability analysis of computer and communication systems. By performing an analysis of an appropriate abstract model, useful information can be gained about the behavior of the system under consideration. Particularly for the validation of a system concept at an early design stage, values for expected performance and reliability can be obtained. Typical quantities of interest in computer performance might be the average job throughput of a server or the probability of buffer overflow of a network node. In reliability analysis, probabilities for critical system states such as failures may be computed. As a result of such analyses, design parameters such as protocol algorithms, degrees of redundancy and component bandwidths may be optimized. Thus the ability to perform these analyses quickly and efficiently is of great importance [2].

One important class of techniques for stochastic modeling beside analytical and discrete–event simulation approaches is state–space analysis. Here, a high–level model such as a queuing network or stochastic Petri net is created, from which the entire state–space graph is generated, in which there is one node for each possible state which the model can assume. The states are linked by arcs which describe the timing characteristics for each state change. The state space is thus described by an annotated directed graph.

Using the simplest and most common assumptions on the transitions — that the time spent by the system in each state is exponentially distributed — the stochastic process described by the model is a Markov chain. In this case, the directed graph of the state space represents a matrix and the transient and steady–state analysis is performed by solving a corresponding system of ordinary differential equations and linear system of equations respectively.

Owing to the combinatorial nature of the problem, the state spaces arising in practical problems can be extremely large ($> 10^6$ nodes and arcs). The memory and computing requirements for the resulting systems of equations grow correspondingly. It is the size of the state space that is the major limiting factor in the application of these modeling techniques for practical problems. This motivates the investigation of parallel computing for this type of stochastic analysis.

One well–known technique for describing complex stochastic systems in a compact way are Generalized Stochastic Petri Nets (GSPNs) [7, 8]. Petri nets allow a graphical and easily understandable representation of the behavior of a complex system, including timing and conditional behavior as well as the forking and synchronization of processes.

We are interested in using shared–memory multiprocessors for the analysis of GSPN models. Such machines are becoming more widespread both as mainframe supercomputers and also as high–end workstations and servers. The shared–memory programming model is more general than the message–passing paradigm, allowing, for example, the concurrent access to shared data structures. On the other hand, the model contains other difficulties such as contention for this access, which requires advanced synchronization and locking methods. These will be the subject of this paper. We consider implementations on two different shared memory multiprocessors: the Convex Exemplar SPP1600 mainframe supercomputer using the proprietary CPS thread programming environment and a Sun Enterprise 4000 multiprocessor server using POSIX threads.

The significance of this work lies in the extension of the ability to model with GSPNs to shared–memory multiprocessors. To our knowledge, there has, until now, been no work published concerning parallel shared–memory state–space generation for stochastic models. An approach for distributed memory machines was published in [4]. The results of this work should provide faster state–space generation and, in conjunction with parallel numerical algorithms, overall acceleration of the analysis process. In particular, we will be able to better utilize the larger main memories of modern shared–memory multiprocessors. This will also enable the analysis of models whose size has prevented their computation on standard workstations.

In the following section we describe state–space generation for Petri nets. In Section 3 we describe the parallelization issues and the solution techniques we used. Section 4 contains results from the parallel programs and in Section 5 we give a short conclusion.

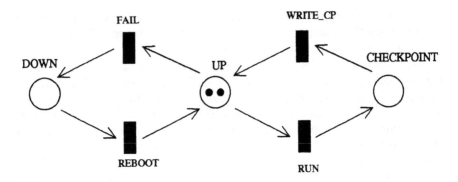

FAIL

WRITE_CP

DOWN

UP

CHECKPOINT

REBOOT

RUN

Fig. 1. Example GSPN Model of a Computer System with Failures and Checkpoints.

2 State–Space Generation for GSPNs

In this section we briefly describe stochastic Petri nets and the automatic generation of their underlying state spaces.

One of the most widely used high–level paradigms for stochastic modeling are *Generalized Stochastic Petri Nets (GSPNs)* [7, 8]. These are an extension to standard Petri nets to allow stochastic timed transitions between individual states. They have the advantages of being easy to understand and having a natural graphical representation, while at the same time possessing many useful modeling features, including sequence, fork, join and synchronization of processes. The state space, or *reachability graph*, of a GSPN is a semi–Markov process, from which states with zero time are eliminated to create a Markov chain. The Markov chain can be solved numerically, yielding probability values which can be combined to provide useful information about the net.

A GSPN is a directed bipartite graph with nodes called *places*, represented by circles, and *transitions*, represented by rectangles. Places may contain *tokens*, which are drawn as small black circles. In a GSPN, two types of transitions are defined, *immediate transitions*, and *timed transitions*. For simplicity, we will not consider immediate transitions any further in this paper. If there is a token in each place that is connected to a transition by an *input arc*, then the transition is said to be *enabled* and may *fire* after a certain delay, causing all these tokens to be destroyed, and creating one token in each place to which the transition is connected by an *output arc*. The state of the Petri net is described by its *marking*, an integer vector containing the number of tokens in each place. The first marking is commonly known as the *initial marking*. A timed transition that is enabled will fire after a random amount of time that is exponentially distributed with a certain rate.

Figure 1 shows a small example GSPN that models a group of computers, each of which may either be running (UP), writing a checkpoint (CHECK-POINT), or failed and rebooting from the last checkpoint (DOWN). The changes between these states are represented by timed transitions with appropriate exponentially distributed rates. In this case, we have modeled two computers (by

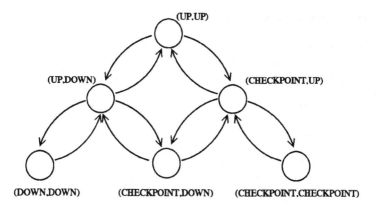

(UP,UP)

(UP,DOWN)

(CHECKPOINT,UP)

(DOWN,DOWN)   (CHECKPOINT,DOWN)   (CHECKPOINT,CHECKPOINT)

Fig. 2. State Space for Example GSPN.

inserting two tokens into the net, initially in place UP). Note that we could model any number of computers by simply adding tokens accordingly, assuming that the transition rates are independent of the states of the other machines.

Figure 2 shows the state space, or *reachability graph*, of this GSPN with two tokens. For readability, we have omitted the rates that are attached to the arcs and have used a textual description for the marking vector. Each state corresponds to one possible marking of the net, and each arc to the firing of a transition in that marking. Owing to the simplicity of this particular Petri net, the state space has a regular triangular structure. In general, however, the reachability graph is highly irregular.

In this example, the state–space graph has six nodes. It is easy to see that adding tokens to the net will lead to a rapid increase in the size of the graph. In the general case, the size grows as t^p, where t is the initial number of tokens and p the number of places in the Petri net. It is for this reason that Petri nets of even moderate complexity may have state spaces whose storage requirements exceed the capacities of all but the largest of computers. In addition, the computation times for the solution of the underlying equations grows accordingly. It is, of course, for these reasons that we are interested in parallel computation.

Figure 3 shows the sequential state–space generation algorithm in pseudo–code form. It utilizes a stack S and a data structure D, which is used to quickly determine whether or not a newly detected marking has previously been discovered. D is typically chosen to be either a hash table or a tree. One of the contributions of this work is the use of a modified B–tree to allow rapid search whilst at the same time minimizing access conflicts. The algorithm performs a depth–first search of the entire state space by popping a state from the stack, generating all possible successor states by firing each enabled transition in the Petri net and pushing each thus–created new marking back onto the stack, if it is one that has not already been generated. Replacing the stack by a FIFO memory would result in a breadth–first search strategy.

```
1 procedure generate_state_space
2 initial marking m₀
3 reachability graph R = ∅
4 search data structure D = ∅
5 stack S = ∅
6 begin
7       add state m₀ to R
8       insert m₀ into D
9       push m₀ onto S
10      while (S ≠ ∅)
11          mᵢ = pop(S)
12          for each successor marking mⱼ to mᵢ
13              if (mⱼ ∉ D)
14                  add state mⱼ to R
15                  insert mⱼ into D
16                  push mⱼ onto S
17              endif
18              add arc mᵢ → mⱼ to R
19          endfor
20      endwhile
21 end generate_state_space
```

Fig. 3. Sequential State–Space Generation Algorithm

3 Parallelization Issues

The state–space generation algorithm is similar to other state–space enumeration algorithms, such as the branch–and–bound methods used for the solution of combinatorial problems such as computer chess and the traveling salesman problem. Consequently it presents similar difficulties in parallelization, namely

- The size of the state space is unknown *a priori*. For many Petri nets it cannot be estimated even to within an order of magnitude in advance.
- All processors must be able to quickly determine whether a newly generated state has already been found — possibly by another processor — to guarantee uniqueness of the states. This implies either a mapping function of states to processors or an efficiently implemented, globally accessible search data structure.
- The state space grows dynamically in an unpredictable fashion, making the problem of load balancing especially difficult.

However, there are also two significant differences to a branch–and–bound algorithm:

- The result of the algorithm is not a single value, such as the minimum path length in the traveling salesman problem or a position evaluation in a game of strategy, but the entire state space itself.
- No cutoff is performed, i.e. the entire state space must be generated.

Our parallelization approach lets different parts of the reachability graph be generated simultaneously. This can be done by processing the main loop of algorithm generate_state_space (Lines 10–20) concurrently and independently on all threads, implying simultaneous read and write access to the global data structures R, D and S. Thus the main problem is maintaining data consistency on the three dynamically changing shared data structures in an efficient way. Our approach applies two different methods to solve this problem: S is partitioned onto the different threads employing a shared stack for load balancing reasons only, whereas the design of D and R limits accesses to a controlled part of the data structure which can be locked separately.

With respect to control flow there is not much to say: threads are spawned before entering the main loop of algorithm generate_state_space and termination is tested only rarely, namely when an empty shared stack is encountered[1]. Thus we can concentrate on the crucial and interesting part of the problem: the organization of the global data structures and the locking mechanisms that we have designed.

3.1 Synchronization

Synchronization is done by protecting portions of the global shared data with mutex variables providing mutual exclusive access.

Because arcs are linked to the data structures of their destination states (m_j in Figure 3), rather than their source states (m_i in Figure 3), synchronization for manipulating the reachability graph may be restricted to data structure D: marking m_j is locked implicitly when looking for it in D by locking the corresponding data in D and holding the lock until Line 18 has been processed. No barriers are needed in the course of the generation algorithm.

Synchronization within the Search Data Structure. We first considered designing the search data structure D as a hash table like some (sequential) GSPN tools do, because concurrent access to hash tables can be synchronized very easily, by locking each entry of the hash table before accessing it. But there are many unsolved problems in using hash tables in this context: As mentioned earlier, neither the size of the state space is known in advance — making it impossible to estimate the size of the hash table *a priori* — nor is its shape, which means that the hash function which maps search keys onto hash table entries would be totally heuristic.

For these reasons, we decided to use a balanced search tree for retrieving already generated states. This guarantees retrieval times that grow only logarithmically with the number of states generated.

The main synchronization problem in search trees is rebalancing: a non-balanced tree could be traversed by the threads concurrently by just locking the tree node they encounter, unlocking it when progressing to the next one, and,

---

[1] This can easily be done by setting and testing termination flags associated with threads under mutual exclusion.

if the search is unsuccessful, inserting a new state as a leaf without touching the upper part of the tree. Rebalancing — obviously obligatory for efficiency reasons with large state spaces — means that inserting a new state causes a global change to the tree structure. To allow concurrency, the portion of the tree that can be affected by the rebalance procedure must be anticipated and restricted to as small an area as possible, since this part has to be locked until rebalancing is complete. A state is looked up in D for each arc of R (Line 13 in Figure 3), which will generate a lot of contention if no special precautions are taken.

We found an efficient way to maintain the balance of the tree by allowing concurrent access through the use of B–trees ([5]): these are by definition automatically balanced, whereby the part of the tree that is affected by rebalancing is restricted in a suitable way.

Synchronization Schemes on B–trees. A B–tree node may contain more than one search key — which is a GSPN marking in this context. The B–tree is said to be of *order* σ if the maximum number of keys that are contained in one B–tree node is 2σ. A node containing 2σ keys is called *full*. The search keys of one node are ordered smallest key first. Each key can have a left child which is a B–tree node containing only smaller keys. The last (largest) key in a node may have a right–child node with larger keys.

Searching is performed in the usual manner: comparing the current key in the tree with the one that is being looked for and moving down the tree according to the results of the comparisons. New keys are always inserted into a leaf node. Insertion into a full node causes the node to split into two parts, promoting one key up to the parent node which may lead to the splitting of the parent node again and so on recursively. Splitting might therefore propagate up to the root. Note that the tree is automatically balanced, because the tree height can only increase when the root is split.

The entity that can be locked is the B–tree node containing several keys. Several methods which require one or more mutex variables per node are known [3]. We have observed that using more than one mutex variable causes an unjustifiable overhead, since the operations on each state consume little time. The easiest way to avoid data inconsistencies is to lock each node encountered on the way down the tree during the search. Since non–full nodes serve as barriers for the back propagation of splittings, all locks in the upper portion of the tree can be released when a non–full node is encountered [3]. However, using this approach, each thread may hold several locks simultaneously. Moreover, the locked nodes are often located in the upper part of the tree where they are most likely to cause a bottleneck. Therefore, and since we even do not know a priori if an insertion will actually take place, we have developed another method adapted from [6] that we call *splitting–in–advance*.

Our B–tree nodes are allowed to contain at most $2\sigma + 1$ keys. On the way down the B–tree each full node is split immediately, regardless of whether an insertion will take place or not. This way back propagation does not occur, since parent nodes can never be full. Therefore a thread holds at most one lock at

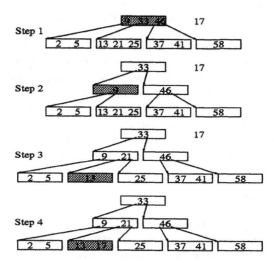

Fig. 4. Splitting–in–advance when Inserting Key 17 in a B–tree

a time. The lock moves down the tree as the search proceeds. For this reason, access conflicts between threads are kept to a small number, thus allowing high concurrency of the search tree processing. Figure 4 shows the insertion of key 17 in a B–tree of order $\sigma = 1$ using splitting–in–advance. Locked nodes are shown as shaded boxes. As the root node is full, it is split in advance in Step 1. The lock can be released immediately after the left child of key 33 has been locked (Step 2). The encountered leaf node is again split in advance, releasing the lock of its parent (Step 3). Key 17 can then be inserted appropriately in Step 4 without the danger of back propagation.

Using efficient storage methods, the organization of the data is similar to that of binary trees, whereby B–tree states consume at most one more byte per state than binary trees [5].

Synchronization on the Stack. The shared stack, which stores the as yet unprocessed markings, is the pool from which the threads get their work. In this sense the shared stack implicitly does the load balancing and therefore cannot be omitted. Since it has to be accessed in mutual exclusion, one shared stack would be a considerable bottleneck, forcing every new marking to be deposited there regardless if all the threads are provided with work anyway.

Therefore we additionally assign a private stack to each thread. Each thread uses mainly its private stack, only pushing a new marking onto the shared stack if the latter's depth drops below the number of threads N. A thread only pops markings from the shared stack if its private stack is empty. In this manner load imbalance is avoided: a thread whose private stack has run empty because it has generated no new successor markings locally, is likely to find a marking on the shared stack, provided the termination criterion has not been reached.

The shared stack has to be protected by a mutex variable. The variable containing the stack depth may be read and written with atomic operations, thus avoiding any locking when reading it. This is due to some considerations:

- The variable representing the depth of the shared stack can be stored in one byte because its value is always smaller than two times the number of threads: if the stack depth is $n - 1$ when all the threads are reading it every thread will push a marking there leading to a stack depth of $2N - 1 > N$ which causes the threads to use their private stacks again.
- The number of threads can be restricted to $N = 256 = 2^8$ so that $2N - 1$ can be represented in one byte without any loss of generality since the state–space generation of GSPNs is no massively parallel application.

3.2 Implementation Issues

Synchronization. Since we have to assign a mutex variable to each B–tree node in the growing search structure, our algorithm relies on the number of mutex variables being only limited by memory size.

Our algorithm can be adapted to the overhead that lock and unlock operations cause on a given machine by increasing or decreasing the order of the B–tree σ: an increase in σ saves mutex variables and locking operations. On the other hand a bigger σ increases both overall search time — since each B–tree node is organized as linear list — and search time within one node, which also increases the time one node stays locked. Measurements in Section 4 will show that the savings in the number of locking operations are limited and that small values for σ lead to a better performance for this reason.

Waiting at Mutex Variables. Measurements showed that for our algorithm — which locks small portions of data very frequently — it is very important that the threads perform a spin wait for locked mutex variables rather than getting idle and descheduled. The latter method may lead to time consuming system calls and voluntary context switches in some implementations of the thread libraries where threads can be regarded as light weight processes which are visual to the operating system (e.g. Sun Solaris POSIX threads). Unfortunately busy waits make a tool–based analysis of the waiting time for mutex variables difficult since idle time is converted to CPU time and gets transparent to the analysis process.

Memory Management. Our first implementation of the parallel state–space generation algorithm did not gain any speedups at all. This was due to the fact that dynamic memory allocation via the `malloc()` library function is implemented as a mutually exclusive procedure. Therefore two threads that each allocate one item of data supposedly in parallel always need more time that one thread would need to allocate both items.

Since our sparse storage techniques exploit dynamic allocation intensively, we had to implement our own memory management on top of the library functions: each thread reserves large chunks of private memory and uses special `allocate()` and `free()` functions for individual objects. In this way, only few memory allocation must be done under mutual exclusion.

4 Experimental Results

We implemented our algorithms using two different shared–memory multiprocessors:

- A Convex Exemplar SPP1600 multiprocessor with 4 Gbytes of main memory and 32 Hewlett/Packard HP PA-RISC 7200 processors. It is a UMA (uniform memory access) machine within each hypernode subcomplex consisting of 8 processors whereby memory is accessed via a crossbar switch. Larger configurations are of NUMA (non–uniform memory access) type, as memory accesses to other hypernodes go via the so–called CTI ring. We did our measurements on a configuration with one hypernode to be able to better compare with the second machine:
- A Sun Enterprise server 4000 with 2 Gbytes of main memory and 8 UltraSparc-1 processors which can be regarded as UMA since memory is always accessed via a crossbar switch and a bus system.

Our experiments use a representative GSPN adapted from [7]. It models a multiprocessor system with failures and repairs. The size of its state space can be scaled by initializing the GSPN with more or less tokens that represent the processors of the system. The state spaces we generated consist of 260,000 states and 2,300,000 arcs (Size S) and of 900,000 states and 3,200,000 arcs respectively (Size M).

Figure 5 shows the overall computation times needed for the reachability graph generation depending on the number of processors measured on the Convex SPP and on the Sun Enterprise for the GSPN of Size M. Figure 6 shows the corresponding speedup values and additionally the speedups for the smaller model (Size S). In the monoprocessor versions used for these measurements, all parallelization overhead was deleted and the B–tree order was optimized to $\sigma = 1$. The figures show the efficiency of our algorithms — especially of the applied synchronization strategies. For both architectures the speedup is linear. [1] shows that these speedups are nearly model independent.

Figure 7 shows the dependency between the computation times of the state–space generation and the B–tree order σ for Size M measured on the Sun Enterprise.

Table 1 gives the total number of locking operations and the number of used mutex variables for different B–tree orders σ for a parallel run with 8 processors. It can be seen that the number of locking operations decreases only by a factor of 3.3 whereby the total number of mutex variables — which is also the number of B–tree nodes — decreases by a factor of 34.7 when σ is increased from 1 to 32. This is due to the fact that for each arc in the state space at least one locking operation has to be performed and that the number of locking operations per arc depends only on how deep the search moves down the B–tree (see Section 3.1). Thus it becomes intelligible that $\sigma < 8$ leads to the best performance (compare Section 3.2).

The number of shared stack accesses turned out to be negligible when local stacks are used: 50 was the maximum number of markings ever popped from

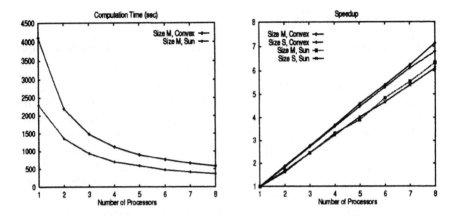

Fig. 5. Computation Times, Convex and Sun, Size M.

Fig. 6. Speedups, Convex and Sun, Size M and S.

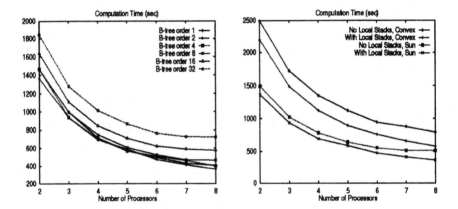

Fig. 7. Computation Times for Various Values of σ, Sun, Size M.

Fig. 8. Computation Times with and without Local Stacks, Convex and Sun, Size M.

B–tree order σ	Number of Locking Operations	Number of Mutex Variables
1	165,000,000	520,000
2	110,000,000	255,000
4	75,000,000	125,000
8	65,000,000	60,000
16	55,000,000	30,000
32	50,000,000	15,000

Table 1. Synchronization Statistics for Size M

the shared stack during all our experimental runs. Figure 8 compares the computation times with and without the use of local stacks on both multiprocessor architectures for Size M.

5 Conclusion and Further Work

We presented a parallel algorithm for generating the state space of GSPNs which is mainly based on the use of modified B–trees as a parallel search data structure. Measurements showed good linear speedups on different architectures.

One B–tree node could be organized as a balanced tree rather than a linear list. But measurements of the reduction in the number of locking operations when σ is increased (Table 1) let expect only moderate performance improvements this way.

On the other hand the maintenance of several B–trees rather than one seems to be a a promising improvement in the organization of the search data structure: conflicts at the root node could be avoided thus allowing a higher degree of parallelization.

Acknowledgments. We wish to thank Stefan Dalibor and Stefan Turowski at the University of Erlangen–Nürnberg for their helpful suggestions and for their assistance in the experimental work.

References

1. S. Allmaier, M. Kowarschik, and G. Horton. State space construction and steady-state solution of GSPNs on a shared–memory multiprocessor. In *Proc. IEEE Int. Workshop Petri Nets and Performance Models (PNPM '97)*, St. Malo, France, 1997. IEEE Comp. Soc. Press. To appear.
2. G. Balbo. On the success of stochastic Petri nets. In *Proc. IEEE Int. Workshop on Petri Nets and Performance Models (PNPM '95)*, pages 2–9, Durham, NC, 1995. IEEE Comp. Soc. Press.
3. R. Bayer and M. Schkolnick. Concurrency of operations on B-trees. *Acta Informatica*, 9:1–21, 1977.
4. G. Ciardo, J. Gluckman, and D. Nicol. Distributed state-space generation of discrete-state stochastic models. Technical Report 198233, ICASE, NASA Langley Research Center, Hampton, VA, 1995.
5. D. Comer. The ubiquitous B-tree. *Computing Surveys*, 11(2):121–137, 1979.
6. L.J. Guibas and R. Sedgewick. A dichromatic framework for balanced trees. In *Proc. 19th Symp. Foundations of Computer Science*, pages 8–21, 1978.
7. M. Ajmone Marsan, G. Balbo, and G. Conte. *Performance models of multiprocessor systems*. MIT Press, 1986.
8. M. Ajmone Marsan, G. Balbo, G. Conte, S. Donatelli, and G. Franceschinis. *Modelling with generalized stochastic Petri nets*. Wiley, Series in Parallel Computing, 1995.

Parallel Software Caches[*]

Arno Formella[1] and Jörg Keller[2]

[1] Universität des Saarlandes, FB 14 Informatik, 66041 Saarbrücken, Germany
[2] FernUniversität-GHS, FB Informatik, 58084 Hagen, Germany

Abstract. We investigate the construction and application of parallel software caches in shared memory multiprocessors. To re-use intermediate results in time-consuming parallel applications, all threads store them in, and try to retrieve them from, a common data structure called parallel software cache. This is especially advantageous in irregular applications where re-use by scheduling at compile time is not possible. A parallel software cache is based on a readers/writers lock, i. e., multiple threads may read simultaneously but only one thread can alter the cache after a miss. To increase utilization, the cache has a number of slots that can be updated separately. We analyze the potential performance gains of parallel software caches and present results from two example applications.

1 Introduction

In time consuming computations, intermediate results are often needed more than once. A convenient method to save these results for later use are software caches. When switching to parallel computations, the easiest method is to give each thread its own private cache. However, this is only useful if the computation shows some regularity. Then, the computation can be scheduled in such a way that a thread that wants to re-use an intermediate result knows which thread computed this result, and that this thread in fact did compute the result already. However, many challenging applications lack the required amount of regularity. Another disadvantage of private software caches in massively parallel computers is the fact that for n threads n times as much memory is needed for software caching as in the sequential case.

Irregular applications, when run on shared memory multiprocessors (SMM), can benefit from a shared parallel software cache. By this term we mean one software cache in the shared memory, where all threads place their intermediate results and all threads try to re-use intermediate results, no matter by which thread they were computed. To allow for concurrent read and ensure exclusive write access of the threads to the cache, a readers/writers lock is used.

Control of multiple accesses of different types to data structures, e.g. by using readers/writers locks, have been investigated in the areas file systems and

---

[*] The first author was supported by the German Science Foundation (DFG) under contract SFB 124, TP D4.

databases, see e.g. [6, 11]. While in the first area the focus was on providing functionality such as files being opened by several threads or processes, the focus in the latter area was on developing protocols so that these accesses can be made deadlock free. We will show that in parallel software caches, no deadlock can occur. Our goal is to investigate the potential performance benefits possible from re-using intermediate results, and the tradeoffs that one encounters while implementing such a parallel data structure. We use the SB-PRAM [1] as platform, but the concept should be portable to other shared memory architectures (see Sect. 4).

In Sect. 2, we define the notion of a cache and review the classical replacement strategies and possible cache organizations. The modifications for a parallel cache are explained in Sect. 3. The SB-PRAM as execution platform is briefly discussed in Sect. 4. Section 5 introduces the applications FViewpar and Rayo and presents the performance results we obtained with the parallel data structure on these applications. Section 6 concludes.

2 Sequential Caches

2.1 Definitions

The notion of a cache is primarily known in hardware design. There, the hardware cache is a well known means to speedup memory accesses in a computer system [8]. We adapted the concept of such an "intermediate memory" to the design of efficient shared memory data structures. Software caches can also be regarded as an implementation of the memorization concept in the field of programming languages, where a result for a function is stored rather than recalculated.

Let us introduce first some notations. An *entry* $e = (k, i)$ consists of a *key* k and associated *information* i. A *universe* U is a set of entries. Given key k the *address function* m returns the associated information i, i.e., $m(k) = i$, if $(k, i) \in U$. Usually, m is a computationally expensive function, let us assume a time $t_m(k)$ to compute $m(k)$. U can be large and is not necessarily given explicitly. We say that a universe is *ordered* if the keys of its entries can be ordered.

A *cache* C is a small finite subset of U together with a *hit function* h and an *update function* u. Given key k the hit function h returns information i associated with k if the entry $e = (k, i)$ is located in C, i.e., $h(k) = i$ iff. $(k, i) \in C$. The hit function h is a relatively simple function, let us assume a time $t_c(k)$ to compute $h(k)$. $t_c(k)$ should be much smaller than $t_m(k)$. For an entry $e = (k, i)$ the update function u inserts e in the cache C possibly deleting another entry in C. u usually implements some replacement strategy. Let us assume a time $t_u(k)$ to update C with an entry which has key k.

The cache C can be used to speedup addressing of U. Given key k, first try $h(k)$ which delivers the information i if $(k, i) \in C$. If an entry is found, we call it a cache hit, otherwise it is called a cache miss. In the latter case, use function $m(k)$ to calculate i. Now, the update function u can be invoked to insert the entry $e = (k, i)$ into C, such that a following request with same key k succeeds in calculating $h(k)$.

For j subsequent accesses to the cache C, i.e., computing $h(k_1), ..., h(k_j)$, the ratio $\alpha = s/j$ where s is the number of misses is called *miss rate*, analogously $\alpha' = (j - s)/j$ is called *hit rate*. For a sufficiently large number of accesses, we can assume an average access time $t_c = 1/j \cdot \sum_{l=1}^{j} t_c(k_l)$. Similarly, we assume an average access time t_m to access the universe, and an average update time t_u after a cache miss. The run time for a sequence of j accesses to U without a cache is $T_{no} = j \cdot t_m$ and with a cache it is

$$T_c(\alpha) = j \cdot (t_c + \alpha \cdot (t_m + t_u)).$$

Hence, in case of worst miss rate, i.e., $\alpha = 1$, the run time is increased by a factor $T_c(1)/T_{no} = 1 + (t_c + t_u)/t_m$, and in best case, i.e., $\alpha = 0$, the run time is decreased by a factor $T_c(0)/T_{no} = t_c/t_m$. Thus, the cache improves the run time of j consecutive accesses to U if $T_c(\alpha) < T_{no}$. Clearly, the improvement of the entire program depends on how large the portion of the overall run time is which is spent in accessing U.

2.2 Replacement Strategies and Cache Organization

For the update function u one has to decide how to organize the cache C such that subsequent accesses to the cache perform both fast and with a high hit rate. For a sequential cache the following update strategies are commonly used.

LRU, least recently used: The cache entries are organized in a queue. Every time a hit occurs the appropriate entry is moved to the head of the queue. The last entry in the queue is replaced in case of an update. Hence, an entry stays at least $|C|$ access cycles in the cache, although it might be used only once. For an unordered universe a linear search must be used by the hit function to examine the cache. Starting at the head of the queue ensures that the entry which was accessed last is found first.

FRQ, least frequently used: Here, the cache entries are equipped with counters. The counter is incremented with every access to the entry. The one with the smallest counter value is replaced in case of an update. Entries often used remain in the cache and the most frequently used are found first if a linear search is employed.

CWC, $|C|$-way cache: For a cache of fixed size the cache is simply implemented by a round robin procedure in an array. Thus, after $|C|$ updates an entry is deleted, independently of its usage count. The update of the cache is very fast, because the location in the cache is predetermined.

RND, random replacement cache: The cache entries are organized in an array as well. In case of an update, one entry is chosen randomly and replaced. For the first three strategies an adversary can always find some update patterns which exhibit poor cache performance. The probability that this happens to a random cache is usually low.

The organization of the cache partly depends on the structure of the universe. If the universe is not ordered, then the cache consists for LRU and FRQ in a linked list of entries. For a miss, the function h must search through the complete list. For CWC and RND the complete array must be searched, too. However, if the universe is ordered, then we can organize the cache such that the entries appear in sorted order. Given key k, function h must search until either (k, i) or an entry (k', i') with $k' > k$ is found. If the number of entries per cache gets larger, an alternative to speed up the search is to use a tree instead of a list.

3 Parallel Caches

We assume that our applications are formulated in a task oriented model. The work to be done can be split in a large number of similar tasks which can be executed in parallel. A number of concurrent threads $p_1, p_2, \ldots, p_i$ is used in the parallel program. Each thread computes a task and then picks up a new one until all tasks have been done. Here we mean real parallel threads that are running simultaneously on at least i processors. Hence, we assume that our program can be optimally parallelized, with the possible exception of conflicts in the case of concurrent accesses to a parallel software cache. Also, there will be a sequential part before the spawning of parallel threads. The spawning incurs some overhead.

We assume that the threads might access the universe U in parallel executing function m without restrictions, and that the access time t_m in the average does not depend on a specific thread nor the access.

3.1 Concurrent Accesses

If the program spends a large amount of time in accessing U and if many threads are accessing the cache, it happens more often that concurrent accesses to the cache become necessary. In the worst case all reads and writes to the cache are serialized. Often however, a more efficient solution is possible, because many SMMs efficiently handle concurrent read accesses, i.e. CC-NUMA architectures.

Updating the cache introduces some difficulties: i) one thread wants to delete an entry of the cache which is still or just in the same moment used by another one or ii) two threads might want to change the cache structure at the same time. To overcome the difficulties, a parallel data structure must be created which is protected by a so called readers/writers lock. A thread which wants to perform an update locks a semaphore; when all pending read accesses have finished, the writer gets exclusive access to the cache. During this time other readers and writers are blocked. After the update has been terminated, the writer releases the lock.

A thread p_i inspects first the cache as a reader. After a miss, the thread leaves the readers queue and calculates address function m. This gives other writers the chance to perform their updates. Once a new entry is found, p_i enters the writers queue. Because the writers are queued as well, p_i must check again whether the entry has been inserted already in the cache during its calculating and waiting time. By moving the execution of address function m outside the region protected by the readers/writers lock, we can guarantee that our protocol is deadlock free: while a thread is reader or writer, it only executes code that works on the cache. It cannot execute other functions that might again try to access the lock, and which could lead to a deadlock.

The readers/writers lock restricts the speedup to $1/\alpha$, because all misses are serialized. For an architecture that does not allow for concurrent reads, the speedup might be even less. To implement concurrent access to the lock data structure and to the reader and writer queues in a constant number of steps (i. e., without additional serialization), parallel prefix can be used. Thus, a machine with atomic parallel prefix and atomic concurrent access only serializes multiple writers (an example is the SB-PRAM, see Sect. 4).

3.2 Improvements

To overcome the speedup restrictions that the exclusive writer imposes, one can use several caches $C_0, \ldots, C_{j-1}$, if there is a reasonable mapping from the set of keys to $\{0, \ldots, j-1\}$. An equivalent notation is that the cache consists of j slots, each capable of holding the same number of entries, and each being locked independently. While this realizes the same functionality, it hides the structure from the user, with the exception of the mapping function. The distribution of the accesses to the different slots will have a significant impact on the performance.

A difference between sequential and parallel software caches is the question of how to provide the result. In a sequential software cache it is sufficient to return a pointer to the cached entry. As long as no entry of the cache is deleted, the pointer is valid. We assume that a thread will use the cached information, continue and access the cache again only at some time later on. Hence, the above condition is sufficient.

In a parallel cache, the cached entry a that one thread requested might be deleted immediately afterwards because another thread added an entry b to the cache and the replacement strategy chose to delete entry a to make room for entry b. Here, we have two possibilities. Either we prevent the replacement strategy from doing so by locking requested entries until they are not needed anymore. Or, we copy such entries and return the copy instead of a pointer to the original entry. If entries are locked while they are used, we have to think about possible deadlocks. However, as long as the application fulfills the above condition (each thread uses only one cached entry at a time) the protocol is deadlock free.

If all accesses to the cache use entries for about the same amount of time, then one can decide by example runs whether to copy or to lock entries. It depends on the application which one of the two methods lead to higher performance, i. e., how long a cache entry might be locked and how much overhead a copying would produce.

For an explicitly given universe, neither locking nor copying is necessary, because the cache contains only pointers to entries. In case of a hit, a pointer to the entry in the universe is returned. The update function safely can replace the pointer in the cache although another threads still makes use of the entry. Additionally, the second check before the cache is updated can be reduced to a simple pointer comparison.

3.3 Replacement Strategies

Another major difference between sequential and parallel software caches is the replacement strategy. The interactions between threads make it more difficult to decide which entry to remove from a slot. We adapt the classic replacement strategies from subsection 2.2 for parallel caches.

In the sequential version of LRU the entry found as a hit was moved to the beginning of the list. This does not work in the parallel version, because during a read no change of the data structure is possible. The reader would need writer permissions and this would serialize all accesses. In our parallel version of LRU, every reader updates the time stamp of the entry that was found as a hit. Previously to the update, a writer sorts all entries in the cache according to the time stamps. The least recently used is deleted. In order to improve the run time of a write access, the sorting can be skipped, but this might increase the subsequent search times for other threads.

Replacement strategy FRQ is implemented similarly. Instead of the time stamp, the reading thread updates an access counter of the entry that was found as a hit. A writer rearranges the list and deletes the entry e with lowest access frequency $f(e)$. The frequency is defined as $f(e) = a/n$ where a is the number of accesses to entry e since insertion and n is the total number of accesses to the cache. It appears a similar tradeoff between the sorting time and search time as for LRU.

Here arises the question whether the whole lifespan of a cached entry must be considered. For example, if an entry is in the cache for a large number of accesses and additionally it has a relatively high actual frequency, then the entry will remain in the cache for a significant amount of time, since its frequency is reduced very slowly. A possible solution to this problem is to use only the last x accesses to the cache to compute the actual frequency. Previous accesses can be just ignored or one might use some weight function which considers accumulatively blocks of x accesses while determining the frequency.

The replacement strategies CWC and RND can be implemented similarly to the sequential version.

3.4 Performance Prediction

We want to predict the performance of the parallel cache by profiling the performance of the software cache in a sequential program. To do this, we assume that the sequential program consists of a sequential part s and a part p that can be parallelized. We assume that all accesses to the universe (and, if present, to the software cache) occur within part p. We assume that the work in that part can be completely parallelized (see task model in Sect. 3). Thus, the time to execute the sequential program without a cache is $T_{seq}^{no} = T_s + T_p$, where T_s is the time to execute the sequential part and T_p is the time to execute the parallelizable part. The time $T_{no} = j \cdot t_m$ to access the universe is a fraction $1 - \beta$ of T_p, i.e. $T_{no} = (1-\beta) \cdot T_p$. This means that time $\beta \cdot T_p$ in part p is spent without accessing the universe. Then the time to execute the sequential program when a cache is present is $T_{seq}^{with} = T_s + \beta \cdot T_p + T_c(\alpha)$, where α is the miss rate of the cache as described in Sect. 2. By profiling both runs of the sequential program, we can derive T_s/T_{seq}^{no}, T_p/T_{seq}^{no}, β, α, j, t_u/t_m, t_c/t_m. With $T_{no} = j \cdot t_m = (1 - \beta) \cdot T_p$, the value of t_m/T_{seq}^{no} can be computed from the other parameters.

If we run our parallelized program without a cache on a parallel machine with n processors, then the runtime will be $T_{par}^{no}(n) = T_s + T_p/n + o$, where o is the overhead to create a set of parallel threads. It is assumed to be fixed and independent of n. We derive o/T_{seq}^{no} by running the sequential program on one processor, the parallel program on n processors of the parallel machine and solving $T_{par}^{no}(n)/T_{seq}^{no} = T_s/T_{seq}^{no} + T_p/(n \cdot T_{seq}^{no}) + o/T_{seq}^{no}$ (which is the inverse of the speedup) for the last term. Thus, we only need the runtimes of the programs and need not be able to profile on the parallel machine.

The runtime of the parallelized program with a parallel software cache will be $T_{par}^{with}(n) = T_s + \beta \cdot T_p/n + T_c(\alpha)/n + o + w(n)$. The term $w(n)$ represents the additional time due to the serialization of writers. In the best case, $w(n) = 0$. Now, we can compute a bound on the possible speedup. Here, we assume that the miss rate will be the same for the sequential program and each thread of the parallel program, which will be supported by our experiments.

4   Execution Platform

The results presented in Sect. 5.2 have been obtained on the SB-PRAM, a shared memory multiprocessor simulating a priority concurrent read concurrent write PRAM [1]. n physical processors are connected via a butterfly network to n memory modules. A physical processor simulates several virtual processors, thus the latency of the network is hidden and a uniform access time is achieved. Each virtual processor has its own register file and the context between two virtual processors is switched after every instruction in a pipelined manner. Thus, the user sees all virtual processors run in parallel. Accesses to memory are distributed with a universal hash function so memory congestion is avoided. The network is able to combine accesses on the way from the processors to the memory location. This avoids hot spots and is extended to employ parallel prefix operations which allow to implement very efficient parallel data structures without serialization, e.g. a readers/writers lock.

A first prototype with 128 virtual processors is operational [3]. Although most of the results have been obtained through simulations of the SB-PRAM on workstations, we have verified the actual run times on the real machine. Each virtual processor executes one thread. The predicted run times matched exactly with the run times obtained by simulation. We did no simulations with more virtual processors, because our workstations did not have enough memory to run such simulations.

Several other multiprocessors provide hardware support for parallel prefix operations: NYU Ultracomputer [7], IBM RP3 [10], Tera MTA [2], and Stanford Dash [9]. The presented concepts should be transferable to these machines. The DASH machine provides a cache-coherent virtual shared memory (CC-NUMA). Here, it would be useful for performance reasons to consider the mapping of the software cache to the hardware caches when designing size and data structures of the cache.

5 Experiments

5.1 Applications

A software cache is part of an application. Hence, we did not test its performance with standard cache benchmarks, but decided to use real applications.

Application FViewpar [5] realizes a fish-eye lens on a layouted graph, the focus is given by a polygon. Graph nodes inside and outside the polygon are treated differently. To determine whether a node is inside the polygon, we intersect the polygon with a horizontal scanline through the node. The parallelization is performed with a parallel queue over all nodes of the graph. Universe U is the set of all possible horizontal scanlines intersecting the polygon, thus it is not given explicitly. A key k is a scanline s, information i is a list of intersection points, and the address function m is the procedure which intersects a scanline with the polygon. To implement a cache with multiple slots application FViewpar maps a horizontal scanline s given as $y = c$ to slot $g(s) = c \bmod j$, where j is the number of slots.

Application Rayo [4] is a ray tracer. It is parallelized with a parallel queue over all pixels. The cache is used to exploit image coherency. In the case presented here, we reduce the number of shadow testing rays. Those rays are normally cast from an intersection point towards the light sources, so that possibly blocking objects are detected. An intersection point is only illuminated if no object is found in direction towards the light source. We use a separate cache for each light source which is a standard means to speedup ray tracing. If two light sources are located closely together, one might unify their caches.

Universe U is the set of all pairs (v, o) where v is a shadow volume generated by object o and the light source. Due to memory limitations U is not given explicitly. A key k is a shadow volume, information i is the blocking object o, and the address function m is simply the ray tracing procedure for ray r finding a possible shadow casting object. The hit function h examines for a new ray r, whether its origin is located in a shadow volume of an object in the cache associated with the light source. The cache makes use of the coherency typically found in scenes: if two intersection points are sufficiently close to each other then the same object casts a shadow on both points.

An alternative approach does not compute the shadow volume explicitly, because it might not have a simple geometrical shape. One verifies for a certain object in the cache whether the object really casts a shadow on the origin of the ray. Hence, an entry (k, i) can be replaced simply by the information (i), coding a previously shadow casting object. A cache hit returns a pointer to the object that casts the shadow. Now, the universe is explicitly given, because the objects are always available. Note, that all objects in the cache must be checked for an intersection with the ray, because no key is available to reduce the search time. For application Rayo, the mapping function g takes advantage of the tree structure while spawning reflected and transmitted rays. For each node in the tree a slot is created. Thus, the slots allow to exploit the coherency between ray trees for adjacent pixels.

5.2 Results

We tested several aspects of the concept of software caches: its scalability, the influence of the replacement strategies, whether copying or locking of the information is more effective, and the tradeoffs due to size of the cache and its organization.

First, we ran the sequential version of FViewpar without cache and found that the sequential part only comprises $T_s/T_{seq}^{no} = 0.004$ of the runtime. Thus with Amdahl's law the speedup can be 250 at most. The parallelizable part consumes the remaining $T_p/T_{seq}^{no} = 0.996$ of the runtime. In part p, a fraction $\beta = 0.121$ is spent without accessing the universe. The function m was executed $j = 6913$ times. When we used a software cache, we found that $t_u/t_m = 0.05$ and that $t_c/t_m = 0.071$. t_c is a bit larger than t_u because of the copying. The miss rate was $\alpha = .355$. By running the parallel program and the sequential program without a cache on the SB-PRAM, we found $o/T_{seq}^{no} = 0.00137$. Now we ran the program with a parallel software cache for $n = 2, 4, 8, \ldots, 128$. We computed $w(n)/T_{seq}^{no}$ from the program runtimes. For increasing n, the value approaches 0.0023 from above.

Then, we simulated application FViewpar for $n = 2^i$ processors, $i = 0, \ldots, 7$, with and without cache. For the cache, we used a fixed size of 16 slots, each capable of holding 4 entries. Accessed entries were copied from the cache to the memory space of the particular thread. Let $T_{par}^x(n)$ denote the runtime on n processors with replacement strategy x, where no indicates that no cache is used. Figure 1 depicts the speedups $s_x(n)$, where $s_x(n) = T_{seq}^{no}/T_{par}^x(n)$, for $x = no, lru, frq, rnd, cwc$. MAX denotes the maximum speedup which is possible by assuming a hit rate of 100% and $w(n) = 0$, while all other parameters are as before.

For $n = 1$, all replacement strategies give a runtime improvement by a factor of about 1.8. As n increases, the curves fall into two categories. RND and CWC strategies provide less improvement, until they make the application slower than without cache for $n = 128$. LRU and FRQ remain better than without a cache, with LRU being slightly faster than FRQ. Their curves slowly approach $s_{no}(n)$, but this might be partly caused by saturation, as the input graph used has only 3600 nodes to be moved, so with $n = 128$, each processor has to move just 28 nodes. Also, as $s_{no}(n)$ approaches the maximum possible speedup of 250 and

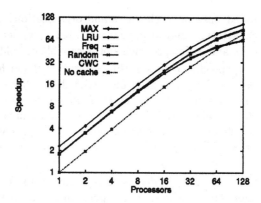

Fig. 1. Scalability of the software cache

hence $s_{max}(n)$, not much can be gained anymore from using a cache. Thus, at some larger number of processors the overhead in using a cache will be larger than the gain. This is supported by computing $s_{no}(256)$ and $s_{LRU}(256)$ with the formulas from Sect. 3.4. The program without a cache is predicted to be slightly faster than with a cache[3]. The miss rate for LRU was $\alpha = 0.355$ for $n \leq 32$ and sank to 0.350 for $n = 128$.

As LRU turns out to be the best of the replacement strategies, we used it to compare locking and copying of cached entries. Processor numbers and cache sizes were chosen as before. The size of the cached entries is 9 words. Locking is 15 to 25 percent faster than copying, so it is a definite advantage in this application.

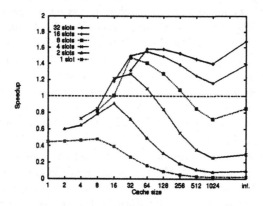

Fig. 2. Comparison of Cache Sizes

---

[3] This behavior has also been observed in a simulation with 256 processors, done by J. Träff, Max-Planck-Institute for Computer Science, Saarbrücken, Germany.

Last, we compared different cache sizes and organizations. Again, we used LRU as replacement strategy, and we fixed $n = 32$. Let $\widetilde{T_j}(k)$ denote the runtime with a cache of size k and j slots, so each slot is capable of holding k/j entries. Figure 2 depicts the speedup curves $s_j(k) = T^{no}_{par}(32)/\widetilde{T_j}(k)$ for $k = 2^i$, $i = 0,\ldots,10$, and for $k = \infty$, i.e., a cache of unrestricted size. Note that for a cache with j slots, $k \geq j$. For a fixed cache size k, $s_j(k)$ grows with j, if we do not consider the case $k = j$, where each cache slot can contain only one entry. This means, that for a cache of size k, one should choose $j = k/2$ slots, each capable of holding two entries. The only exception is $k = 16$. Here $j = 4$ is better than $j = 8$.

For fixed j, the performance improves up to a certain value of k, in our case $k = 4j$ or $k = 8j$. For larger cache sizes, the performance decreases again. Here, the searches through longer lists need more time than caching of more entries can save. If we give the cache an arbitrary size $k = \infty$, then the performance is increased again. The reason seems to be that from some k on, each entry is only computed once and never replaced. Note that the miss rate remains constantly close to 35.5 percent for $j = 32$ and $k \geq 64$.

If the cache size is chosen too small, the speedup is less than 1, i.e., the program is slower than without cache for $k \leq 8$. For $k \geq 16$, the gain when doubling the cache size gets smaller with growing k. In this spirit, the choice $k = 64$ and $j = 16$ for the comparison of speedups was not optimal but a good choice.

For application **Rayo** we decided to implement only the cache with LRU replacement strategy. The decision is based on the fact, that usually the object which was found last is a good candidate as blocking object for the next intersection point. As we will see in the sequel, the optimal cache size is quite small, so one can infer that at least for the presented scenes the update strategy has not a large impact on performance. The results are presented for a scene of 104 objects and four light sources. Image resolution was set to 128×128, 16384 primary rays and 41787 secondary rays are traced. Four light sources make 205464 shadow rays necessary, 85357 of them hit a blocking object. We measured the hit and miss rates in the cache respective to the actually hitting rays, because if the shadow does not hit any object we cannot expect to find a matching cache entry. The cache can only improve the run time for hitting shadow rays, thus it can improve at most 32 percent of the run time. We focus only on the inner loop of the ray tracer, where more than 95 percent of the run time is spent.

We simulated application **Rayo** for $n = 2^i$ processors, $i = 0,\ldots,7$. Figure 3 shows some relative speedups, where we varied the size and the number of slots. Let us denote with $T_x(n)$ the run of the inner loop running on n processors. x indicates the number of entries in the cache. s_1 is the relative speedup $T_0(p)/T_1(p)$, s_2 is the relative speedup $T_0(p)/T_2(p)$, and s_3 is the relative speedup $T_0(p)/T_4(p)$, respectively. s_4, the best one in Fig. 3, is obtained if we use one slot in the cache for every node in the ray tree. The size of the slot was set to only one entry. Increasing the slot size to two entries already led to a small loss of performance.

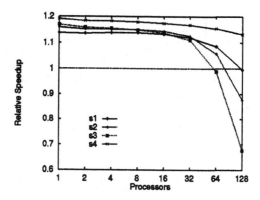

Fig. 3. Relative Speedups for Different Cache Sizes and Numbers of Slots

For small numbers of processors, a larger cache has some advantages, but with increasing number of processors the smaller cache becomes the better one. As the curve for s_4 implies, this is due to the conflicts during updating the cache. The processors are working at different levels in the ray tree and one single cache cannot provide the correct blocking object. As long as few processors are competing, the larger the cache the better the performance is. The search time in the larger cache together with the serialization during update has a negative impact on performance for a large number of processors. However, adapting the cache to the structure of the ray tree exhibits a large speedup s_4. Even for 128 processors a speedup of 13 percent has been achieved. Note that only 32 percent of the run time can be improved, thus, 40 percent of the run time during shadow determination has been saved.

For the run time of one single processor a slightly better update strategy was implemented, because we can afford an update of the cache during every access. After a cache miss, the least recently used object is removed from the cache if the update function u does not provide a blocking object. This performs better for a single processor because after a shadow boundary has been passed, it is quite unlikely that the previous object which cast the shadow will be useful again. Nevertheless, the run times in Fig. 3 demonstrate that the parallel cache even with the weaker replacement strategy outperforms the version with no cache.

Instead of sharing one data structure one might provide each processor with its own cache. This leads to n times the memory size occupied by the cache structure, such that for large numbers of processors memory limitations may become problematic. Figure 4 shows that the hit rate for the parallel cache is significantly larger than the average hit rate for the individual caches. The difference increases with larger numbers of processors. The difference for one processor in the figure is explained by the alternative implementation of the replacement strategy. If the cache is owned by a single processor we always deleted the least recently used object.

The large difference in the hit rates does not imply necessarily a large gain

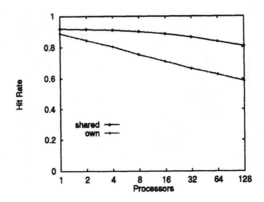

Fig. 4. Hit Rates for Individual and Parallel Cache

in run time, as it is illustrated in Fig. 5. The relative speedup between a version with individual caches and a version with a parallel cache is always close to one, but tends to be larger for 16 and 32 processors. Remembering that the cache improves at most the run time of 32 percent of the overall run time, in this portion of the program almost 5 percent are gained. The effect is due to the cache overhead and the serialization while updating. Nevertheless, the parallel cache saves memory and slightly improves the run time.

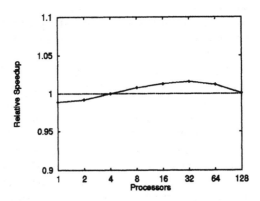

Fig. 5. Relative Speedup for Individual and Parallel Cache

6   Conclusion

We introduced the concept of a parallel cache and implemented the data structure on the SB-PRAM multiprocessor. In our applications, the software cache leads to performance improvements, but investigations on more diverse work-loads are necessary. Providing several slots in the cache which can be updated independently reduces serialization after cache misses. The modified LRU strategy together with locking of requested entries was found to be best in the presented applications. The other parameters have to be chosen carefully, too.

The concept of a parallel cache as a data structure might be useful for sequential programs consisting of several interacting threads as well. Here, there might exist data exchange between the threads which is not predictable statically in advance.

The SB-PRAM as simulation platform allows for a quantitative analysis, because as a UMA-architecture its performance is predictable and explainable. Once crucial parameters have been detected, the promising implementation should be portable to other shared memory architectures.

References

1. Abolhassan, F., Drefenstedt, R., Keller, J., Paul, W. J., Scheerer, D.: On the physical design of PRAMs. Comput. J. **36** (1993) 756–762
2. Alverson, R., Callahan, D., Cummings, D., Koblenz, B., Porterfield, A., Smith, B.: The Tera computer system. In Proc. Int.l Conf. on Supercomputing (1990) 1–6
3. Bach, P., Braun, M., Formella, A., Friedrich, J., Grün, T., Lichtenau, C.: Building the 4 Processor SB-PRAM Prototype. In Proc. Hawaii Int.l Symp. on System Sciences (1997) 14–23
4. Formella, A., Gill, C.: Ray Tracing: A Quantitative Analysis and a New Practical Algorithm. Visual Comput. **11** (1995) 465–476
5. Formella, A., Keller, J.: Generalized Fisheye Views of Graphs. Proc. Graph Drawing '95 (1995) 242–253
6. Fussell, D.S., Kedem, Z., Silberschatz, A.: A Theory of Correct Locking Protocols for Database Systems. Proc. Int.l Conf. on Very Large Database Systems (1981) 112–124
7. Gottlieb, A., Grishman, R., Kruskal, C. P., McAuliffe, K. P., Rudolph, L., Snir, M.: The NYU ultracomputer — designing an MIMD shared memory parallel computer. IEEE Trans. Comput. **C–32** (1983) 175–189.
8. Handy, J.: The Cache Memory Book. Academic Press, San Diego, CA (1993)
9. Lenoski, D., Laudon, J., Gharachorloo, K., Weber, W.-D., Gupta, A., Hennessy, J., Horowitz, M., Lam, M. S. The Stanford DASH multiprocessor. Comput. **25** (1992) 63–79
10. Pfister, G.F., Brantley, W.C., George, D.A., Harvey, S.L., Kleinfelder, W.J., McAuliffe, K.P., Melton, E.A., Norton, V.A., Weiss, J: The IBM research parallel processor prototype (RP3): Introduction and architecture. In Proc. Int.l Conf. on Parallel Processing (1985) 764–771
11. Silberschatz, A., Peterson, J. L., Galvin, P. B.: Operating System Concepts, 3rd Edition. Addison-Wesley, Reading, MA (1991)

Communication Efficient Parallel Searching *

Armin Bäumker and Friedhelm Meyer auf der Heide

Department of Mathematics and Computer Science and Heinz Nixdorf Institute,
University of Paderborn, Paderborn, Germany
email: {abk,fmadh}@uni-paderborn.de

Abstract. Searching is one of the most important algorithmic problems, used as a subroutine in many applications. Accordingly, designing search algorithms is in the center of research on data structures since decades. In this paper we aim to survey recent developments in designing parallel search algorithms where parallel machines are used to answer many search queries in parallel, so called multisearch algorithms. We briefly describe the current state of multisearch algorithms based on hashing and binary search, as they are developed for abstract parallel models like the PRAM. The main part of the paper describes deterministic and randomized multisearch algorithms that are very communication efficient. As a computation and cost model we employ Valiant's BSP model and its variant BSP* due to Bäumker et al.

1   Introduction

One of the most important algorithmic problems is that of searching in a given set of objects where each object can be identified by a search key. In order to illustrate the problem we give two examples.

Consider, for example, the set of student records of a university. The records have a field that contains the student's name. We will take the name as the search key. The other fields of the records contain additional information about the student. Given a name, the search problem may consist of retrieving the record of the student with that name, if there is such a student. Otherwise, a "not found" message is returned. Furthermore the set of students is not fixed. So student records may be added or deleted. Accordingly, we distinguish static search problems, where the set of objects is fixed, and dynamic search problems, where the set may be altered.

In sequential computing much work has been done on the subject of searching since the seventies. A deep treatment of this topic can be found in the books of Knuth [22], Mehlhorn [25] and Cormen et al. [11]. In this paper we present recent results on searching in the setting of parallel computing. Parallelism can be used in two different ways. First, parallelism can be employed in order to accelerate

---

* This research was supported by the EC Esprit Long Term Research Project 20244 (ALCOM-IT) and by DFG-Sonderforschungsbereich 376 "Massive Parallelität: Algorithmen, Entwurfsmethoden, Anwendungen".

the execution of a single search process. In [20], upper and lower bounds for this problem on the PRAM model can be found. Further work in this direction has been done by Tamassia et al. in [32]. Second, parallelism can be employed in order to execute many search processes at the same time. Such search problems are called *multisearch* problems.

In the technical part of this paper we will treat the following multisearch problem. Given a *universe* U and a partition of U in *segments* $S = \{s_1, \ldots, s_m\}$. The segments are ordered in the sense that, for each $q \in U$ and segment s_i, it can be determined with unit cost whether $q \in s_i, q \in \{s_1 \cup \ldots \cup s_{i-1}\}$, or $q \in \{s_{i+1} \cup \ldots \cup s_m\}$. The *Multisearch Problem* is: Given a set of queries $Q = \{q_1, \ldots, q_n\} \subseteq U$ and a set of segments $S = \{s_1, \ldots, s_m\}$, find, for each q_i, the segment it belongs to (denoted $s(q_i)$). Sequentially, this needs time $O(n \log m)$ in the worst case.

An important example is: A strip in 2D is partitioned into segments, and queries are points in the strip, see Figure 1. The task is to determine for each query point which segment it belongs to. Note that sorting the points and merging them with the segments would not solve the problem, as our example shows.

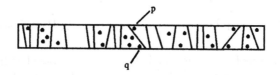

Fig. 1. Strip with segments and query points. Note, that p lies left to q but $s(p)$ is right to $s(q)$.

The computational model we use is the bulk synchronous model of parallel computing (BSP model, [33]). In this model a parallel computation proceeds in alternating computation and communication phases, separated from each other by barrier synchronizations. We focus on work-optimal algorithms and on reducing the number and complexity of communication phases. In order to measure the complexity of a communication phase we employ the BSP* cost model [5], which extends the original BSP cost model by accounting for the number of words of data sent or received by each processor (as in the BSP model) and for the number of messages sent or received by each processor (which is in addition to the BSP model). So the BSP* cost model is more accurate than the BSP model with respect to many parallel systems that suffer from high start-up costs involved in each message transmission.

In order to solve the above multisearch problem on the BSP* model we employ the following approach. First, a search tree of size m is generated from the sorted sequence of segments, and then the nodes of the search tree are mapped to the processors. Second, the n search processes are performed in parallel by using the distributed representation of the tree. We present randomized and deterministic mapping schemes and a multisearch algorithm that efficiently works on each kind of mapping. The technical part of this paper is intended to present a unified survey of deterministic and randomized multisearch algorithms presented in [5], [6] and [7]. The descriptions of the mapping schemes and the multisearch algorithm presented here are informal. For detailed descriptions and proofs see [5], [6] and [7].

1.1 Organization of the paper

In the next section we briefly survey results on sequential and parallel searching that can be found in the computer science literature. Here we survey methods based on (universal) hashing and on binary search. In Section 3 we present the BSP and the BSP* model and give the runtimes of some basic algorithms that are used as subroutines in the multisearch algorithm. The multisearch algorithm given in Section 4 works on each kind of mapping. In Section 5 and 6 we present the randomized and the deterministic mapping scheme, respectively, together with appropriate interfaces that are used by the multisearch algorithm in order to get access to the tree.

2 Sequential and Parallel Searching: A Brief Survey

In this section we briefly describe search data structures based on hashing and on binary searching, and their parallel variants, mainly for abstract parallel models like the PRAM. Further we describe basic problems arising when the parallel search algorithms are supposed to use very few, cheap communication rounds.

2.1 Hashing

Consider the following type of search problems. Let a set of objects be given where a unique key from a universe U is associated to each object. The task is to process search queries of the following type: For a key $k \in U$, return a pointer to the object associated to k, if such an object is in the set. Otherwise, return a pointer to NIL.

This kind of search problem can be solved by using hashing, a technique that is used to store sets of objects such that they can be accessed very fast. The objects are stored in a so-called hash table by using a hash function h that maps the key values into the hash table. Thus, a pointer to the object with key k is stored into position $h(k)$ of the hash table. Searching for an object with key k can then be done by looking up the objects in position $h(k)$ of the hash table.

A problem that arises is that of collisions. Collisions occur if the hash function maps more then one object into the same hash table position. A method that resolves collisions is chaining, i.e. objects that are mapped to the same position are stored in a secondary data structure associated to the position. Another method of resolving collisions is that of open addressing, i.e. if a collision occurs, the next hash function in a predefined sequence of hash functions is taken.

In general, the goal of hashing is to reduce collisions. For this purpose, Carter and Wegman introduced the notion of universal hashing [10]. Extensions are described by Dietzfelbinger et al. in [14]. In this setting the hash functions used are randomly drawn from a universal class H of hash functions. Such classes H have the properties that its members are easy to evaluate and need only little space to be stored. On the other hand, random members of H have properties similar to random functions. Universal hashing as described e.g. in the textbooks [22], [25] and [11] refers to hashing techniques based on universal classes of hash functions. Universal hashing guarantees constant *expected* access time to the objects stored in the hash table. Constant *worst case* access time is guaranteed by perfect hashing schemes. Fredman et al. [16] describe a perfect hashing scheme in which n elements are mapped into a hash table of size $O(n)$.

Hashing is also used for storing dynamic sets such that insertions, deletions and searching of objects can be done very fast. In [14], Dietzfelbinger et al. present a dynamic perfect hashing scheme that extends the static perfect hashing scheme from [16]. With the dynamic perfect hashing scheme the objects can be accessed in constant time, while insertions and deletions take constant amortized time. A real time variant where also insertions and deletions take worst case constant time is presented in [15]. These schemes require space proportional to the current size of the set. The algorithms are randomized and the time bounds hold with high probability.

The technique of hashing can easily be extended to parallel computing if the PRAM model is used. This is an abstract model of a parallel computer where processors are considered to have fast access to a large shared memory. For the CREW-PRAM, a PRAM variant in which the processors may concurrently read from the same shared memory location in one time unit, the parallel solution is trivial. For all queries the hash functions can be evaluated in parallel. Afterwards, the accesses to the hash table can be performed in parallel. Note that concurrent reads to the same hash table position may occur in the case of collisions.

Many more problems arise if hashing is to be used in a distributed memory environment, where memory is distributed among a number of modules and accesses to the same module are served sequentially. Assume that there are p processor/memory modules and n search queries. Even if we assume that the hash table is distributed evenly among the modules it can happen that many search queries are directed to hash table positions that reside on the same module, thereby inducing high congestion. By using universal hashing to distribute the objects among the modules, one can ensure that the hash table positions that are accessed by the n queries are evenly distributed among the processors, provided that $n = \Omega(p \cdot \log p)$. So, in case of $\log p$ queries per processor, this approach can be used for parallel hashing on distributed memory models like the BSP model.

If $n = o(p \cdot \log p)$ the approach from above does not work efficiently, because, with high probability, the accessed hash table position will not be distributed evenly among the processors. Instead, one module will hold $\Theta(\log p / \log \log p + n/p)$ accessed hash table positions, with high probability. In order to allow for smaller values of n/p, namely $n/p = 1$, one has to introduce redundancy, i.e. the objects in the set are now stored in more than one module. Meyer auf der Heide et al. [26] present a scheme in which r randomly chosen hash functions are used in order to distribute r copies of each object among the processors. If the set has size m, space $m \cdot r/p$ is used per module, with high probability. The model they use is the DMM (Distributed Memory Machine). In contrast to the BSP model this model has the nice property that accesses to the same memory module are not served sequentially. Instead, a DMM module may choose one of the accesses directed to it and forget about the others (arbitrary DMM). In another variant the DMM module may answer all accesses directed to it, provided that these are not more than c many, otherwise the accessing processors will be notified that there are more than c accesses to the same module and no access goes through (c-collision DMM). In [26] it is shown that n queries can be answered by using redundant hashing with $r = 2$, in time $O(\log \log n)$ on a n-processor c-collision DMM. If we allow for $r = \log p$ copies, time $O(\log^* n)$ can be reached. Results from [27] imply a hashing scheme on an arbitrary DMM that only needs time $O(\log \log \log n \cdot \log^* n)$, with high probability, for $r = 2$.

2.2 Binary searching

A drawback of hashing is that it cannot be used for certain types of search problems. Assume that we have a static set S of objects with search keys from a totally ordered universe $(U, <)$. Consider the following search problem: For a key k, find the smallest key in S that is larger than or equal k. Hashing destroys the ordering of the keys. So it is not applicable in this setting. Instead the set S can be sorted and stored in an array. Then for a key k the smallest key in S larger than or equal k can be found by the well known binary search algorithm.

The idea of binary search can also be translated to dynamic sets. The set can be represented by a binary search tree, see [25]. The objects (or pointers to the objects) are represented by the nodes. A query for a key k can easily be answered by travelling down the tree until the node with key k or a leave node is reached. Insertions and deletions can be done by adding and deleting nodes, without destroying the search tree property. The worst case time for searching is then proportional to the depth of the tree. So it is important to keep the tree balanced while performing insertions and deletions. This is done be balancing operations. The search requests can than be performed in $O(\log m)$ time, when m is the current size of the set. Adel'son-Velskii and Landis [1] introduced a scheme in which the search keeps balanced under insertions and deletions. Their solution is known as the AVL-tree. Insertions, deletions and searching takes time $O(\log m)$. Many other variants of dynamic search trees are known, like the (2,3)-tree and the BB[α]-tree (see [25]). The time bounds are roughly the same for the most kinds of these trees. They mainly differ in the number of balancing operations needed while performing a number of insertions and deletions.

Binary searching can easily be extended to parallel computing if the CREW PRAM model is considered. Time $O(\log m)$ is needed in order to perform n binary searches on an array of size m, provided that n processors are available.

Performing n binary searches in parallel on the EREW-PRAM, is already a complicated problem. Reif and Sen [31] developed a randomized EREW-PRAM algorithm that performs n binary searches on an array of size m in time $O(\log m)$ with high probability, provided that n processors are available and $m = O(n)$. It is not obvious how the algorithm can be extended such that it works optimally in the case $m = \omega(n)$.

Paul et al. ([28]) developed a parallel version of the (2,3)-tree on the EREW PRAM. If the search tree has size m, n deletions, insertions or searches can be performed in time $O(\log m + \log n)$.

In the PRAM model, communication among the processors is via the shared memory and thereby as cheap as local computations. There is no incentive to reduce communication. Accordingly, the above PRAM algorithms are very communicaton intensive. Further, the constant runtime factors hidden in the asymptotical analysis are rather high. Thus, translating the above PRAM algorithms to more realistic models that accurately charge for communication costs would result in a bad performance.

Another class of models that have been investigated in theory are the network models. Here processors cannot communicate via a shared memory, rather a interconnection topology is given and communication is only possible between neighbouring processors. The EREW-PRAM algorithm for binary searching from Reif and Sen mentioned above can be extended such that it also works on a butterfly network with n processors, i.e. n binary searches on an array of size n can be performed in time $O(\log n)$ with high probability on a butterfly network of size n. Again, it is not clear how to extend the algorithm such that it works efficiently on larger arrays. Atallah and Fabri [4] achieved (non-optimal, deterministic) time $O(\log n(\log \log n)^3)$ on a n-processor hypercube. A $O(\sqrt{n})$ time algorithm on a $\sqrt{n} \times \sqrt{n}$ mesh network is from Atallah et al. [3].

There is yet another class of models of parallel computation that emerged in recent years. These models try to charge accurately for communication cost. The cost models capture critical aspects of communication in real parallel machines, like limited communication bandwidth and high latency, by introducing appropriate parameters. So in these models there is an incentive to focus on reducing communication. Models of this kind are the Bulk Synchronous Parallel (BSP) model due to Valiant [33], the LogP model due to Culler et al. [12], the BPRAM of Aggarwal et al. [2], and the CGM due to Dehne et al. [13], to name a few.

In [13], Dehne et al. developed an algorithm that performs n binary searches on a set of n elements on the CGM model. The focus there was on reducing the number of communication rounds while allowing a total exchange of local memory contents in each round. The algorithm achieves runtime $O((n/p) \cdot \log n)$ and a constant number of communication rounds on a p-processor CGM, provided that $n > p^2$. So the input size needed to be optimally is rather high in this algorithm. Further, if the set of elements to be searched is larger than n, the algorithm is not longer efficient.

Goodrich [18] presents an algorithm for the same problem on the BSP model that needs an optimal number of communication rounds. Again, the algorithm is not suitable if the set of elements to be searched is larger than n.

In [8], Bäumker et al. extend the results on communication efficient parallel searching presented in this paper to the dynamic case. In particular, they give BSP* algorithms for maintaining (2,3)-trees under insertions and deletions. More results on parallel searching on the BSP* model are presented in [9], where priority queues are investigated.

2.3 Communication efficient searching: An overview

In this subsection we shed some light on the problems that one encounters while trying to do communication efficient multisearch. We want to solve the multi-search problem with m segments and n queries as defined in the introduction, see Figure 1.

In the following we assume that the set of segments is given in the form of a d-ary search tree T with m leaves. Each leave represents a segment. For each query $q \in U$ there is a search path from the root to the leave that represents the segment that contains q. So for each query q, the correct segment can be found by travelling down the tree, along q's search path, from the root to a leave node. The search paths are not known in advance, instead they have to be constructed step by step. We assume that each node holds an array that stores d segments, the *next-array*, where the i-th array entry corresponds to the i-th child of the node. If a query q has a node v on its search path, then the node to be visited next can be determined by performing a binary search on the next-array stored in v.

The parallel solution for the multisearch problem presented in this paper consist of two parts:

Mapping: In a preprocessing step, the nodes of the search tree T are mapped to the processors of the parallel computer.

Multisearch: After having mapped T to the processors of the parallel computer, the search processes related to the n queries are performed in parallel by using the distributed representation of the tree generated in the preprocessing.

From a high level point of view, our solution for the multisearch problem is very simple. As in the sequential algorithm, the queries are routed along their search paths from the root to the leaves. The search algorithm proceeds in rounds. In the ℓ-th round the queries reach level ℓ and each query is brought together with the node that it visits on level ℓ of the search tree, i.e. each query meets the node that it wants to visit on some processor. Then, for each query the $(\ell+1)$-st node on its search path can be determined in time $\log d$, by binary searching the next-list of the node. This approach entails two main problems: When trying to bring together each query with the node it visits on the current level, two kinds of congestion occur that we call *node congestion* and the *processor congestion*.

Definition 1. Let T be a search tree mapped to the processors of the parallel computer and let Q be a set of search queries for T. For a node v of the tree the *node congestion* is defined as the number of queries in Q that have v on their search path. For a processor P the *processor congestion* is defined as the number of nodes mapped to processor P that are accessed while performing the search processes related to queries in Q.

The node congestion differs very much for different nodes. For example, the node congestion of the root is always n, while other nodes may be visited by only one query. For the multisearch algorithm this means that nodes with low node congestion have to be brought together with few queries only, while nodes with high node congestion need to be brought together with many queries. The multisearch algorithm copes with this problem by employing different strategies for nodes with high node congestion and for nodes with low node congestion: If a node v has high congestion, then it will be broadcasted to the processors that hold the queries that want to visit node v. If a node v has low congestion, then the queries that want to visit v are moved to the processor on which v resides.

Processor congestion becomes a problem if the tree is much larger than the number of processors. In order to make things clearer consider the following situation: Assume that we use some deterministic mapping scheme where each node is mapped to one processor. Consider a level ℓ of the tree where more than n keys are mapped to the same processor. Now we can choose n input queries such that they visit n nodes from level ℓ that are all mapped to processor P. So the processor congestion on level ℓ for processor P is n. This situation is fatal for the runtime of the multisearch algorithm, because, in order to bring queries together with visited nodes, processor P would have to send n nodes or receive n queries, or a mix of both. In any case P would have to communicate $\Omega(n)$ words of data. To circumvent such high processor congestion, we describe two mapping schemes, a randomized and a deterministic mapping scheme. While the randomized mapping scheme only maps one copy of each node to the processors, the deterministic mapping scheme is redundant, i.e. several copies of a node may be mapped to different processors.

The multisearch algorithm given in Section 4 works on each kind of mapping via an interface that provides access to the nodes of the tree. For each kind of mapping a different interface is given. While the interface is simple for non-redundant mapping schemes, it is more complicated in the case of redundant mapping schemes. This is because an access to nodes in the tree involves selecting copies such that the processor congestion is low.

3 The Bulk Synchronous Model of Parallel Computing

In this section we describe Valiant's BSP model and its variant BSP*; we further present results on basic communication primitives for the BSP* model, as we will need them later.

3.1 BSP and BSP*

In the BSP (Bulk Synchronous Parallel) model of parallel computing [33] a parallel computer consists of a number of processors, each equipped with a large local memory. The processors are connected by a router (or communication network) that transports messages between processors. Furthermore, the BSP model presupposes that the parallel computer is equipped with a mechanism for synchronizing the whole system in barrier style.

A computation in the BSP model proceeds in a succession of *supersteps*. A superstep consists of a computation and a communication phase followed by a barrier synchronization. In the computation phase each processor independently performs operations on data that resides in its local memory at the start of the superstep and generates messages for other processors. In the communication phase the messages are transported to their destinations by the router.

BSP* cost model: Many routers of real parallel machines support the exchange of large messages and achieve much higher throughput for large messages compared to small ones. In order to make things clear consider the following experiments: We let the router realize a communication pattern in which each processor sends and receives at most h messages of size one. We measure the throughput in terms of bytes delivered per second. Then, we perform the same experiment several times, each time increasing the size of the messages. A common observation is that the throughput increases with increasing message size until a certain message size, the "optimal" message size, is reached. The main reason for this is that in real parallel systems high start-up costs are involved in the transmission of each message (see [12]). So the optimal message size may be considerably high. Experiments suggest that on the Parsytec GCel transputer system the messages should be of size at least 1 KBytes. On the Intel Paragon under the OSF/1 operating system and the NX message passing library the messages should have the size of 10 KBytes in order to get half of the possible communication throughput (see [21]). What we can learn from this is that parallel algorithms should be designed such that the communicated data can be combined into few large messages. This is what we call blockwise communication. To incorporate this aspect of real parallel systems in our model we extend the BSP model to the BSP* model [5] by adding the parameter B. In the BSP* cost model a parallel system is characterized by the following parameters:

- As in the BSP model the parameter p is the number of processor/memory components.
- As in the BSP model L is the minimum time between successive synchronization operations. Thus L is the minimum time for a superstep.
- The parameter B is the minimum size the messages must have in order to fully exploit the bandwidth of the router.
- g^* is the reciprocal of the throughput of the router in terms of words delivered per time unit when the messages have size B.

Consider a superstep where in the computation phase the processors perform a maximum number w of local operations and where in the communication phase the router realizes an (h, s)-relation, i.e. a communication pattern in which each processor sends or receives at most s words of data that are combined into at most h messages. The BSP* model charges runtime $w + g^* \cdot (s + h \cdot B) + L$ for this superstep. Now consider a computation that consists of $\mathcal{T}$ supersteps where in the i-th superstep an (h_i, s_i)-relation is realized and a maximum number of w_i local operations are performed by each processor. Let $\mathcal{S} = \sum_{i=1}^{\mathcal{T}} s_i$, $\mathcal{H} = \sum_{i=1}^{\mathcal{T}} h_i$ and $\mathcal{W} = \sum_{i=1}^{\mathcal{T}} w_i$. Then the total runtime for the computation is

$$\mathcal{W} + g^* \cdot (\mathcal{S} + B \cdot \mathcal{H}) + \mathcal{T} \cdot L.$$

Thus, designing algorithms in the BSP* model aims at minimizing $\mathcal{W}$, $\mathcal{S}$, $\mathcal{H}$ and $\mathcal{T}$. However, $\mathcal{W}$, $\mathcal{S}$, $\mathcal{H}$ and $\mathcal{T}$ can not be minimized independently. For instance, a constant value for $\mathcal{T}$ can always be reached, if one pays for it with high values for $\mathcal{S}$ and $\mathcal{W}$. So the decision of which is the best algorithm for a given machine must base on the actual parameter values p, g^*, B and L. Note that the messages should be of size B in the BSP* cost model. Then the runtime becomes $\mathcal{W} + g^* \cdot \mathcal{S} + \mathcal{T} \cdot L$. Otherwise, the runtime would be higher. Let $T_{\text{seq}}(n)$ be the sequential runtime for some problem. We say that a parallel algorithm for this problem is *optimal*, if it has runtime $O(T_{\text{seq}}(n)/p)$. It is *1-optimal*, if it has runtime $(1 + o(1)) \cdot T_{\text{seq}}(n)/p$. We aim at algorithms that are optimal or even 1-optimal for a broad range of BSP* parameters.

3.2 Basic communication primitives

In this section we consider BSP* algorithms for some basic problems. These algorithms are later used as subroutines in the multisearch algorithms. We only state the results. Details can be found in [5] and [6].

Broadcasting: The first problem we consider is the problem of broadcasting. Consider a vector of size ℓ stored in the first processor. After the broadcast, each processor has to store a copy of the vector. The following result can be achieved by organizing the processors as a balanced d-ary tree and by routing the vector in a pipelined fashion along all possible paths from the root to the leaves.

Result 1 *Let d be an arbitrary integer with $2 \le d \le p$. Broadcasting a vector of size ℓ in the BSP* model takes runtime $\mathcal{W} + g^* \cdot (\mathcal{S} + B \cdot \mathcal{H}) + L \cdot \mathcal{T}$, with $\mathcal{W} = O(d \cdot (\ell + \log_d p))$, $\mathcal{S} = O(d \cdot (\ell + \log_d p))$, $\mathcal{H} = O(d \cdot \log_d p)$ $\mathcal{T} = O(\log_d p)$. Further, space $O(\ell)$ per processor is needed.*

Parallel Scanning: In the multisearch algorithms we often need to scan a list of items distributed among the processors. In particular we need to solve the following problem: Consider an array A of size n that consists of k subarrays. Let A be partitioned into p blocks of size $\frac{n}{p}$ such that the i-th block is held by the i-th processor. The task is to label each element of A with the size of the subarray to which the element belongs. Further each element is labeled with the first and the last processor that holds an element of the same subarray and with its position within the subarray.

Result 2 *Parallel scanning a list of items as described above on the BSP* model takes runtime $W + g^* \cdot (S + B \cdot \mathcal{H}) + L \cdot \mathcal{T}$, with $W = O((n/p) + d \cdot \log_d p)$, $S = O(d \cdot \log_d p)$, $\mathcal{H} = O(d \cdot \log_d p)$ and $\mathcal{T} = O(\log_d p)$. Further, space $O(n/p)$ per processor is needed.*

Sorting: Another basic problem is that of sorting. We consider the integer sorting problem, the problem of sorting a number of integer keys from a certain range. We define the s-sorting problem as follows: Let a set of integer keys be given distributed among the processors such that each processor holds at most s keys. Let n be the total number of keys. In order to s-sort the keys they have to be redistributed such that afterwards the following holds:

- The first $\lfloor n/s \rfloor$ processors hold s keys and the $(\lfloor n/s \rfloor + 1)$-st processor holds the rest of the keys.
- On each processor the keys are sorted.
- For $1 \leq i < p$ the largest key on the i-th processor has at most the value of the smallest key on the $(i+1)$-st processor.

Result 3 *Fix arbitrary constants $c_1, c_2 \geq 1$. Then for $s = \Omega(\log^{c_1} p)$ the s-sorting problem for integer keys from the range $[0, \ldots, \ell]$ with $\ell = O(s^{c_2/c_1})$ can be solved on the BSP* model in runtime $c_1 \cdot c_2 \cdot (W + g^* \cdot (S + B \cdot \mathcal{H}) + L \cdot \mathcal{T})$, with $W = O(s)$, $S = O(s)$, $\mathcal{H} = O(s^{1/c_1})$ and $\mathcal{T} = O(\log p)$. Further, space $O(s)$ per processor is needed.*

Fix an arbitrary constant $\epsilon > 0$. If $s = \Omega(p^{c_1 \cdot \epsilon})$, with $\epsilon > 0$ a constant, then the problem can be solved with $W, S, \mathcal{H}$ as above, and with $\mathcal{T} = O(1)$. Again, space $O(s)$ per processor is needed.

4 The Multisearch Algorithm

In this section we present the multisearch algorithm. It is used in similar form in [5], [6] and [7]. It works on each kind of mapping, provided that there is an interface that gives access to the nodes of the tree. As mentioned earlier, the main task of the multisearch algorithm is to solve the problem of node congestion, while processor congestion is tackled by the mapping and the interface between mapping and search algorithm, see Definition 1.

In order to focus on how the multisearch algorithm resolves node congestion, we present the multisearch algorithm as it works on trees of size $O(p)$. In that case we can assume that each processor holds only a constant number of nodes and the problem of high processor congestion does not occur. Later, when we employ other kinds of mapping for larger trees, the multisearch algorithm can easily be reanalyzed.

So, let T be a d-ary search tree of depth h that has at most p leaves. For simplicity, we assume that $p = d^h$. We simply map the tree such that each processor holds at most 1 node of the same level.

As in the sequential algorithm, we route the queries along their search paths level by level until they reach the leaves. The algorithm proceeds in rounds. In the

ℓ-th round the queries reach level ℓ. Then the queries are brought together with the nodes that the queries visit on level ℓ. After that the queries are executed, i.e. for each query a binary search on the next-list of the node that it currently visits is performed in order to find the node it has to visit on the next level.

So we have to deal with the problem of node congestion which arises from the fact that some nodes may be visited by only few queries and other nodes may be visited by many queries.

Definition 2. Let v be a node of the search tree T. The queries that visit v while travelling through the search tree form the *job at node v*. We denote the job at node v by $J(v)$.

We distinguish small jobs and large jobs. For this purpose we introduce the parameter t. Small jobs are jobs of size smaller than t and large jobs are jobs of size at least t. The analysis in [5] shows that it is advantageous to choose $t = (\frac{n}{p})^\alpha$, for some α with $0 < \alpha < 1$. In order to bring together the queries with the nodes they have to visit we pursue different strategies. One for large jobs and one for small jobs. The two strategies work as follows:

- **Small Jobs:** For a node v of T, let $J(v)$ be a small job. In order to bring together the queries of $J(v)$ with the node v, we move $J(v)$ to the processor that holds the node. We call this the **moving strategy**.
- **Large Jobs:** For a node v of T let $J(v)$ be a large job. In order to bring together the queries of $J(v)$ with the node v, we cannot move $J(v)$ to the processor that holds the node, because that would imply severe congestion if $J(v)$ is very large. Instead we follow another strategy: First, the queries of $J(v)$ are collected on a group of consecutive processors such that each processor of the group holds at most n/p queries of $J(v)$. Then the first processor of the group fetches the node v from the processor that holds the node and broadcasts it to the other processors of the group. We call this the **fetching strategy**.

In this survey paper we only give a informal description of the algorithm. Details and proofs can be found in [5]. The multisearch algorithm can be understood best, if one considers the cases of small and large jobs separately. So assume first that we have only small jobs. At the beginning of round ℓ each query resides on the processor that holds the node that it visits on level ℓ. Hence, by binary searching each query can find out which node it has to visit on the next level. Afterwards, each query is moved to the processor that holds the node that the query visits on level $\ell + 1$, and round $\ell + 1$ may begin.

Now assume that we have only large jobs. At the beginning of round ℓ the queries are (n/p)-sorted according to the numbers of the nodes that the queries visit on level ℓ. This can be done by using the sorting algorithm from Subsection 3.2. Note, that it is sufficient to sort the queries that visit different nodes on the previous level separately from each other. So we have to solve a number of sorting problems with keys from a range of size d only. In order to identify the different processor groups that are responsible for the different sortings, the parallel scanning algorithm presented in Subsection 3.2 can be used.

The effect of the sorting is that queries from the same job are consecutive in the ordering. Then, for each job, the processor that holds the first query of the job, fetches the node that the queries of the job visit on level ℓ and broadcasts it to the other processors that hold queries of the job. After that, by using binary search, for each query the next node to be visited is determined, and the next round may begin.

Note, that we need both strategies. Employing the fetching strategy for small jobs, is not a good solution, because, if the jobs are smaller than the nodes, one would move the smaller objects to the larger objects. In particular, if a processor P holds $\frac{n}{p}$ small jobs, each of size one, employing the fetching strategy would cause P to fetch $\frac{n}{p} \cdot d$ words of data, which might be too much if d is large.

A proof of the following result can be found in [5] or [6]. Let $T_{\text{bin}}(x, y)$ be the sequential time for x binary searches on an array of size y. Note that $T_{\text{bin}}(x, y) = x \cdot \log y$, and that the sequential time for performing n search processes on a d-ary search tree of depth h is $h \cdot T_{\text{bin}}(\frac{n}{p}, d)$.

Result 4 *Let $c \geq 1$ be an arbitrary constant and $d = (n/p)^{1/c}$, then the algorithm solves the multisearch problem with n queries on d-ary trees with depth $h = \log_d p$ in runtime $h \cdot T_{\text{bin}}(n/p, d) \cdot (1 + o(1))$, i.e. the algorithm is 1-optimal, for the following parameter constellations:*

- *$n/p = \Omega(\log^c p)$, $g^* = o(\log(n/p))$, $B = o((n/p)^{1-(1/c)})$ and $L = o((n \cdot \log(n/p))/(p \cdot \log p))$.*
- *$n/p = \Omega(p^\epsilon)$ for an arbitrary constant $\epsilon > 0$, g^* and B as above, and $L = (n/p) \cdot \log(n/p)$.*

5 Randomized Mapping

In this section we resolve the problem of processor congestion by using a randomized mapping scheme. It is a simplified variant of the scheme presented in [5] and [6]. In order to make the problem clear, assume that the tree T is very large such that there is a level with $p \cdot n$ nodes. Then we need to map n nodes to each processor. In this case high processor congestion would be caused if the n queries happen to visit different nodes mapped to the same processor. In order to make high processor congestion unlikely one could map the nodes of T to random processors. But then there is a problem concerning blockwise communication. The critical part is the treatment of small jobs in the multisearch algorithm. For clarity assume that we are on some level ℓ of the tree where we have n small jobs of size one. Then, the multisearch algorithm moves each query from the processor that holds the node it visits on level $\ell - 1$ to the processor that holds the node it visits on level ℓ. If the nodes are mapped randomly to the processors, this corresponds to moving the queries from random source processors to random target processors. High processor congestion is unlikely in this scenario, i.e. each processor is source and target of $O((n/p) + \log p)$ queries with high probability. But, if n/p is much smaller than p, the random mapping ensures with high probability that queries from the same source processor have many different target processors. Hence, each processor may send and receive $\Omega((n/p) + \log p)$ messages of size one, which is not according to the philosophy of BSP*.

In order to achieve blockwise communication while keeping the benefits of the random mapping, we introduce the z-mapping, which is a two-step randomized process, described in the next subsection.

5.1 The z-Mapping

For a d-ary search tree T of depth h the z-mapping performs h iterations. In the ℓ-th iteration, $0 \leq \ell \leq h - 1$, the nodes of level ℓ are mapped as follows:

- If $\ell \leq \log_d p$, we have at most p nodes on the current level. These nodes are mapped to different processors.
- If $\ell > \log_d p$, level ℓ has more then p nodes. For each processor P let $R(P)$ be the subset of nodes on level $\ell - 1$ of T that has been mapped to processor P in the previous iteration. Each processor P chooses a random set of z, $z \leq p$, processors (successor processors of P on level ℓ) and distributes the children of nodes of $R(P)$ randomly among these successor processors.

The next lemma captures properties of the z-mapping that are crucial for the use of blockwise communication.

Lemma 3. *For $z = \Omega(\log p)$ and an arbitrary constant $c > 0$ the following holds: Fix an arbitrary processor P and an arbitrary level ℓ, $1 \leq \ell < h$. Let R be the set of nodes on level ℓ which the z-mapping maps to processor P. Then the parents of the nodes in R are mapped to $O(z)$ different processors by the z-mapping with probability at least $1 - n^{-c}$.*

This lemma guarantees that in the case of small jobs each processor sends and receives $O(z)$ messages. Note that z can be chosen small such that we can achieve blockwise communication, even if n/p is small. The next lemma bounds the processor congestion.

Lemma 4. *Consider a tree T with degree d. Let ℓ be the number of an arbitrary level. Fix arbitrary nodes on level ℓ and mark them with a weight such that the total weight is n. Let t be the maximum weight of a marked node. Fix an arbitrary processor P and apply the z-mapping to the tree T, with $z = \omega(\log n \cdot d)$. Afterwards, processor P holds marked nodes of weight at most*

$$(1 + o(1)) \cdot (\frac{n}{p} + r)$$

with probability at least $(1 - n^{-c})^\ell$, for an arbitrary constant $c > 0$, provided that $r = \omega(c \cdot t \cdot \log n)$.

Now we observe how the multisearch algorithm from Section 4 behaves in the case of small jobs (size smaller than t), if the z-mapping is employed. From Lemma 4 we can conclude that with high probability before and after Step 4 of round ℓ each processor holds at most $(1 + o(1)) \cdot \frac{n}{p} + t \cdot \omega(c \cdot \log n)$ queries of small jobs. Further, we know that each processor can combine its queries into at most z messages before sending them. From Lemma 3 we know that each processor receives only $O(z)$ messages. So altogether, a $((1 + o(1)) \cdot \frac{n}{p} + t \cdot \omega(c \cdot \log n), z)$-relation is realized, i.e. for $\frac{n}{p} = \omega(\log p)$ we can choose t such that a $(\frac{n}{p}, z)$-relation is realized. The multisearch algorithm can be reanalyzed such that the following theorem is obtained. For details see [5] or [6].

Theorem 5. *Let $c, c' \geq 1$ be arbitrary constants and choose $d = (n/p)^{1/c}$. Let T be a d-ary tree with depth h that has been mapped to the processors by using the z-mapping with $z = \omega(d \cdot \log p)$.*

Then, with probability at least $1 - (h/p^{c'-1})$, we get a 1-optimal for the multisearch problem with n queries on T, i.e. runtime $h \cdot T_{\text{bin}}(n/p, d) \cdot (1 + o(1))$, for the following parameter constellations:

- *$n/p = \omega(\log^c p)$, $g^* = o(\log(n/p))$, $B = o((n/p)^{1-(1/c)})$ and $L = o((n \cdot \log(n/p))/(p \cdot \log p))$*
- *$n/p = \Omega(p^\epsilon)$ for an arbitrary constant $\epsilon > 0$, $g^* = o(\log(n/p))$, $B = o((n/p)^{1-(1/c)})$ and $L = (n/p) \cdot \log(n/p)$*

6 Deterministic Mapping

Let again T be a d-ary search tree of depth h. As in the previous section we consider large trees, i.e. we allow that the number of nodes in T is much higher than the number of processors. In contrast to the previous section, where we used a randomized mapping, we now want to use a deterministic mapping scheme in order to map the nodes of T to the processors. For the multisearch part we use, as in the previous sections, the multisearch algorithm from Section 4.

The problem with a deterministic mapping scheme is that high processor congestion may occur if the tree is large, as described at the beginning of the previous section. The solution is to use redundancy, i.e., for each node, several copies are mapped to different processors. Whenever the multisearch algorithm needs to access a set of nodes S, copies of nodes of S have to be selected carefully in order to avoid high processor congestion. Of course, this is only possible if the copies have been mapped in a clever way.

In the following we first present a redundant mapping scheme. Then we sketch a deterministic and randomized copy selection algorithm that provides access to the nodes such that processor congestion is low. The copy selection algorithm can then be used by the multisearch algorithm as an interface that provides access to the tree. Detailed descriptions and proofs can be found in [7].

6.1 A deterministic redundant mapping scheme

Let T be a d-ary tree of depth h. Further, let $\mathcal{V}$ be the set of nodes of T and $\mathcal{P}$ be the set of processors. Let $|\mathcal{V}| = m$ and $|\mathcal{P}| = p$. For each $v \in \mathcal{V}$, the redundant mapping scheme generates r copies and maps them to different processors. The mapping is controlled by a mapping function $\Gamma : \mathcal{V} \to 2^{\mathcal{P}}$. For $v \in \mathcal{V}$, $\Gamma(v)$ is a set of r processors. The r copies of v are mapped to the processors in $\Gamma(v)$. Then, the triple $(\mathcal{V}, \mathcal{P}, \Gamma)$ defines a bipartite graph $G = G(\mathcal{V}, \mathcal{P}, \Gamma)$. $\mathcal{V} \cup \mathcal{P}$ is the set of nodes of G, and $(v, P) \in \mathcal{V} \times \mathcal{P}$ is an edge of G, if $P \in \Gamma(v)$. We choose Γ such that the resulting bipartite graph G is r-regular. That means, every vertex from $\mathcal{V}$ has degree r and every vertex from $\mathcal{P}$ has degree $m \cdot r/p$. Further, Γ is chosen such that G has high *expansion*. The notion of expansion is explained below.

Definition 6. (a) For $S \subseteq \mathcal{V}$, a *k-bundle for S* is a set E of edges in $G(\mathcal{V}, \mathcal{P}, \Gamma)$ such that for each $v \in S$, E contains k edges that are adjacent to v. Note, that $|E| = k \cdot |S|$. If $k = 1$, we call E a *target set for S*. E has congestion c if, for every processor $P \in \mathcal{P}$, P is adjacent to a maximum number of c edges from E.

(b) For a set of edges E, $\Gamma^E(S)$ denotes the set of processors in $\Gamma(S)$ that are adjacent to an edge in E.

Definition 7. $G(\mathcal{V}, \mathcal{P}, \Gamma)$ has (ϵ, δ)-*expansion*, for some $\epsilon > 0$ and $0 < \delta < 1$, if for any subset $S \subseteq \mathcal{V}$, with $|S| \leq p/r$, and any $\lceil r/2 \rceil$-bundle E for S,

$$|\Gamma^E(S)| > \epsilon r |S|^{1-\delta}.$$

The existence of graphs with suitable expansion is established by the following lemma, which is shown by Pietracaprina and Pucci in [29].

Lemma 8. *Let $\Gamma : \mathcal{V} \to 2^{\mathcal{P}}$ be a random function such that for each $v \in \mathcal{V}$, $\Gamma(v)$ is a random subset of $\mathcal{P}$ of size r. Then there is a constant $c = \Theta(1)$ such that for every $r > c \cdot (\log m / \log p) \cdot \log(\log m / \log p)$, $G(\mathcal{V}, \mathcal{P}, \Gamma)$ has (ϵ, δ)-expansion, with $\delta = (\log(m/p) / \log p) / \lceil r/2 \rceil$ and $\epsilon = \Theta(1)$, with high probability.*

An important case is that of $r > c \cdot 2 \cdot \log(m/p)$. Then, according to Lemma 8, there is a mapping Γ such that $G(\mathcal{V}, \mathcal{P}, \Gamma)$ has (ϵ, δ)-expansion, with $\epsilon = \Theta(1)$ and $\delta = 1/\log p$. Then, for every $S \subseteq \mathcal{V}$ with $|S| = p/r$ and every $\lceil r/2 \rceil$-bundle E for S we have $|\Gamma^E(S)| > (\epsilon/2) \cdot p$.

In the following we assume that the tree T is mapped to the processors by using a mapping function Γ such that $G(\mathcal{V}, \mathcal{P}, \Gamma)$ is r-regular and has (ϵ, δ)-expansion, with ϵ and δ chosen as specified in Lemma 8.

As will be apparent later, the performance of the multisearch algorithm heavily relies on the ability to select a target set of low congestion for any arbitrary subset $S \subseteq \mathcal{V}$.

6.2 Deterministic Copy Selection

The multisearch algorithm needs to access the nodes of T. Let $S \subseteq \mathcal{V}$ be the set of nodes that the algorithm wants to access at some point. For each $v \in S$ there are r copies of v. Thus, the access of nodes in S involves a copy selection for each $v \in S$. In order to reduce processor congestion we need to select the copies such that the selected copies are evenly distributed among the processors. This corresponds to the problem of constructing a target set of low congestion for S (see Definition 6(c)). In the following we sketch algorithms for the target set selection problem. The first one is suitable for $|S| < p/r$. The second one is suitable for S of arbitrary size, it uses the first algorithm as subroutine.

Let Γ be chosen such that $G = G(\mathcal{V}, \mathcal{P}, \Gamma)$ is an r-regular graph with (ϵ, δ)-expansion, with ϵ and δ chosen as specified in Lemma 8. Let S be a subset of $\mathcal{V}$, with $|S| \leq p/r$ and let E be an arbitrary $\lceil r/2 \rceil$-bundle for S. A target set for S of congestion at most $k = (2/\epsilon)|S|^\delta$ can be constructed from E using Procedure 1 below. As for the notation, for a set of edges F, for a set of processors $Q \subseteq \mathcal{P}$ and for a set of nodes $V \subseteq \mathcal{V}$, let $F(Q)$ and $F(V)$ denote the set of edges in F that are adjacent to elements of Q and V, respectively.

Procedure 1:

1. $R := S;\ F := E;\ T := \emptyset;\ k = (2/c)|S|^\delta$
2. while $R \neq \emptyset$ do
   (a) Identify the set $Q \subseteq \mathcal{P}$ of processors that are adjacent to at least k edges from F.
   (b) Set $F' = F - F(Q)$.
   (c) Let R' be the set of nodes from R that have an adjacent edge in F'. For each $v \in R'$ select an arbitrary adjacent edge from F'. Add the selected edge to the target set T.
   (d) $F := F - F(R');\ R := R - R'$

A proof for the following lemma can be found in [7].

Lemma 9. *a) For $i > 0$ the following holds: At the beginning of the i-th iteration of Step 2, $|R| \leq |S|/2^{i-1}$.*
b) Procedure 1 produces a target set for S of congestion at most $(2/\epsilon)|S|^\delta$.

Notice, that when $r = 2 \cdot \log(m/p)$ and $\delta = 1/\log p$ (in this case we say that G has high expansion), the target set determined by Procedure 1 has congestion $O(1)$, which is asymptotically optimal, if $|S| \leq p/r$.

Based on Procedure 1, we now show how to construct target sets of low congestion for subsets of $\mathcal{V}$ of larger size. Consider a set $S \subseteq \mathcal{V}$ of size $|S| \geq p/r$. A target set for S from an $\lceil r/2 \rceil$-bundle E for S, can be constructed by using the following procedure.

Procedure 2:

1. Let $\tau = \lceil |S|/(p/r) \rceil$.
2. Partition S into $S(1), S(2), \ldots, S(\tau)$, with $|S(i)| \leq p/r$.
3. $T := \emptyset;\ Q := \mathcal{P};\ F := E;\ k = (|S|/p)(p/r)^\delta (1/c)$
4. For $i := 1$ to τ do

(a) Let $S'(i)$ be the set of vertices from $S(i)$ that are adjacent to at least $r/2$ edges in F.

(b) $F := F(S'(i))$

(c) Run Procedure 1 to select a target set $T_i \subseteq F$ for $S'(i)$

(d) $T := T \cup T_i$

(e) Let Q be the set of processors that are adjacent to more than k edges in T.

(f) $F := F(Q)$

5. Let $\tilde{S} = S - \bigcup_{i=1}^{r} S'(i)$.

6. Run Procedure 1 to select a target set $\tilde{T} \subset E(\tilde{S})$ for $\tilde{S}$.

7. $T := T \cup \tilde{T}$

Lemma 10. *A target set for S of congestion at most $((|S|/p) + 4) \cdot (p/r)^\delta/\epsilon$ is produced by Procedure 2.*

Notice, that when $r = 2 \cdot \log(m/p)$ and $\delta = 1/\log p$, the target set determined by Procedure 1 has congestion $O(|S|/p)$, which is asymptotically optimal.

Actually, for the multisearch algorithm we need a procedure that produces a target set with low congestion for the case that the elements of S are weighted. Procedure 2 can easily be extended to this case. So, we do not further discuss the case of weighted sets.

Procedures 1 and 2 can be implemented on the BSP* model using the primitives from Section 3.2. Here we only state the runtime on the BSP* model for Procedure 2.

Lemma 11. *Let $c > 1$ be an arbitrary positive constants and let $|S| = n$. Then, if $n/p = \Omega(p^\eta)$ for a constant $\eta > 0$, Procedure 2 can be implemented on the BSP* model such that $W + g^* \cdot (S + B \cdot \mathcal{H}) + L \cdot \mathcal{T}$ time steps are needed, with $W = O(r \cdot n/p)$, $S = O(r \cdot n/p)$, $\mathcal{H} = O(r \cdot (n/p)^{\frac{1}{c}})$ and $\mathcal{T} = O(r \cdot \log p)$.*

If $\frac{n}{p} = \Omega(p^{1+\eta})$, we have $W = O(r \cdot n/p)$, $S = O(n/p)$, $\mathcal{H} = O((n/p)^{\frac{1}{c}})$ and $\mathcal{T} = O(1)$.

So, in the case of $\frac{n}{p} = \Omega(p^\epsilon)$ the performance is not very good because the redundancy r is a runtime factor in the computation time as well as in the communication time. This can be avoided by the use of randomization, as we show in the next subsection.

6.3 Randomized Copy Selection

The deterministic copy selection algorithm is not very efficient for high redundancy r. On the other hand, with low redundancy the copy selection causes high congestion. In this subsection we present a randomized copy selection algorithm that is efficient even for high redundancy r.

Let Γ be chosen such that $G(V, \mathcal{P}, \Gamma)$ is an r-regular graph with (ϵ, δ)-expansion as specified above. For $S \subseteq V$ let E be a set of edges in G which contains an $\lceil r/2 \rceil$-bundle for S. A target set for S with congestion $k = (n/p) \cdot (p/r)^\delta \cdot (1/\epsilon) + \log p$ can be constructed from E using the following algorithm. We give a high level description. As for the notation, for a set of edges E and a set of vertices V of G we denote the edges in E adjacent to vertices in V be $E(V)$.

Procedure 3:
1. $R := S$; $F := E$; $T := \emptyset$; $i := 0$;
2. while $|R| \neq \emptyset$ do
   (a) For each $v \in R$ randomly choose an edge from $F(v)$. Let F' be the set of edges randomly chosen in this step.
   (b) Identify the set Q of processors that are adjacent to edges in F' with a total weight of at least $k/2^i$.
   (c) $T := T \cup (F' - F'(Q))$ Let R' be the set of elements in R that are adjacent to an edge in $F' - F'(Q)$. Set $F := F - F(R')$; $R := R - R'$; $i := i + 1$.

Lemma 12. *(a) In round i of the algorithm $|R| = |S|/2^i$ with high probability. So the algorithm terminates after $\log |S|$ rounds with high probability.*
(b) The algorithm produces a target set for S of congestion $\frac{n}{p} \cdot (\frac{p}{r})^\delta \cdot \frac{1}{c} + \log p$

A BSP* implementation of the randomized copy selection can easily be done by using the basic primitives from Section 3.2. We get the following result.

Lemma 13. *Let $c_1, c_2 > 1$ be arbitrary positive constants and let $|S| = n$. Then, if $n/p = \Omega(p^\eta)$ for a constant $\eta \geq 0$, the randomized copy selection can be implemented on the BSP* model such that $W + g^* \cdot (S + B \cdot \mathcal{H}) + L \cdot T$ time steps are needed with probability at least $1 - p^{-c_2}$, where $W = O(n/p)$, $S = O(n/p)$, $\mathcal{H} = O((n/p)^{\frac{1}{c_2}})$ and $T = O(\log p)$.*
If $\frac{n}{p} = \Omega(p^{1+\eta})$, then we have W, S and $\mathcal{H}$ as above, and $T = O(1)$.

Remark: Using a special class of expander graphs and redundancy $r = \omega(\log \frac{m}{p})$ the algorithm achieves congestion $(\frac{n}{p} + \log p) \cdot (1 + o(1))$ within the same time and probability bounds as above.

6.4 Reanalyzing the multisearch algorithm for the deterministic mapping

Let T be a d-ary tree of depth h that has been mapped to the processors with redundancy r such that the resulting structure has (ϵ, δ)-expansion, as specified in Lemma 8. For the multisearch we employ the multisearch algorithm from Section 4 and use the deterministic or the randomized copy selection algorithm from above as an interface that provides access to the tree. Here we only present the results for high redundancy. For small redundancy one can proof that high congestion will occur and that the runtime will be far away from optimality. Corresponding upper and lower bounds can be found in [7]. Let again $T_{\text{bin}}(x, y)$ be the sequential time for performing x binary searches on an array of size y. Then, the following theorems can be proven.

Theorem 14. *Let T be a d-ary tree of depth h such that the number of nodes is polynomial in p. Let further $c \geq 1, \eta > 0$ be arbitrary constants and let $d = (n/p)^{1/c}$.*

Then the tree T can be mapped to the p processors with redundancy $r = \log p$ such that the multisearch problem with n queries can be solved, by using the deterministic copy selection scheme, in time $O(h \cdot T_{bin}(n/p, d))$, i.e. the algorithm is optimal, for the following parameter constellations:

- $n/p = \Omega(p^\eta)$, $g^* = O(1)$, $B = O((n/p)^{1-(1/c)})$ and $L = O(n/(p \cdot \log^2 p))$
- $n/p = \Omega(p^{1+\eta})$, $g^* = O(\log p)$, $B = O((n/p)^{1-(1/c)})$ and $L = O(n/(p \cdot \log p))$

Theorem 15. *Let T be a d-ary tree of depth h such that the number of nodes is polynomial in p. Let further $c \geq 1, c' > 0, \eta > 0$ be arbitrary constants and let $d = (n/p)^{1/c}$.*

Then the tree T can be mapped to the p processors with redundancy $r = \log p$ such that the multisearch problem with n queries can be solved with probability at least $1 - (1/p^{c'})$, by using the randomized copy selection scheme, in time $O(h \cdot T_{bin}(n/p, d))$, i.e. the algorithm is optimal, for the following parameter constellation: $n/p = \Omega(p^\eta)$, $g^* = O(\log p)$, $B = O((n/p)^{1-(1/c)})$ and $L = O(n/(p \cdot \log p))$

If $r = \omega(\log p)$, then multisearch can be done in time $h \cdot T_{bin}(n/p, d) \cdot (1 + o(1))$, i.e. the algorithm is 1-optimal, for the same parameter constellation.

7 Discussion and Experimental Results

Comparing the multisearch algorithms: We have presented three multisearch algorithms. The first algorithm (Theorem 5) uses a randomized mapping and deterministic searching algorithms. This is by far the simplest and most efficient one. It is 1-optimal for a very wide range of BSP* parameters, and very small input sizes, and uses (optimal) space $O(m/p)$ per processor.

The second and third algorithm (Theorems 14, 15) use a deterministic mapping. In order to reduce processor congestion, it is now necessary to use $\log p$ many copies of each object, i.e. we now need space $O(\log p \cdot m/p)$ per processor. Further the randomized (Theorem 15) and even more the deterministic search algorithm (Theorem 14) are very complicated. In case of deterministic searching, the algorithm is optimal only for large inputs but cannot be made 1-optimal. The randomized searching (Theorem 15) can be made 1-optimal, but requires much larger input sizes then the multisearch algorithm based on randomized mapping (Theorem 5).

Experiments: We have implemented the multisearch algorithm on the GCel from Parsytec. The GCel is a network of T800 transputers as processors, a 2-dimensional mesh as router and Parix as its operation system. In order to implement our algorithm in BSP style we have realized a library on top of Parix containing the basic routines mentioned above. We measured the BSP* parameters: In order to do this we used random relations as communication patterns. The router reaches maximal throughput in terms of Bytes/sec when we have messages of size 1 KByte. Therefore the value for parameter B is 1 KByte. The value for g is around 10000 ops, where ops is the time for an integer arithmetic operation on the T800 Transputer. The value for L is 17000 ops. We made experiments for the case $n \geq m$. The related search tree is of size p, such that a injective mapping is appropriate. The experiments show, for $n/p = 6144$ and $m/p = 2048$, a speed-up of 31 with 64 processors, and for $n/p = 16384$ and $m/p = 8192$, a speed-up of 64 with 256 processors. These results are quite good if one takes into account that the sequential binary search algorithm, with which we compare our algorithm, can be implemented very efficiently.

References

1. G.M. Adel'son-Velskii and Y.M. Landis, An Algorithm for the organization of information, *Soviet Math. Dokl.*, 3 (1962) 1259–1262.
2. A. Aggarwal, A.K. Chandra and M. Snir, On communication latency in PRAM computations, in: *Proc. ACM Symp. on Parallel Algorithms and Architectures* (1989) 11–21.
3. M.J. Atallah, F. Dehne, R. Miller, A. Rau-Chaplin and J.-J. Tsay, Multisearch techniques for implementing data structures on a mesh-connected computer, in: *Proc. ACM Symp. on Parallel Algorithms and Architectures* (1991) 204–214.
4. M.J. Atallah and A. Fabri, On the multisearching problem for hypercubes, in: *Parallel Architectures and Languages Europe* (1994).
5. A. Bäumker, W. Dittrich, and F. Meyer auf der Heide, Truly efficient parallel algorithms: c-optimal multisearch for an extension of the BSP model, in: *Proc. European Symposium on Algorithms* (1995) 17–30.
6. A. Bäumker, W. Dittrich, and F. Meyer auf der Heide, Truly efficient parallel algorithms: 1-optimal multisearch for an extension of the BSP model, Technical Report No. tr-rsfb-96-008, Universität-Gesamthochschule Paderborn (1996), accepted for *Theoretical Computer Science*.
7. A. Bäumker, W. Dittrich, A. Pietracaprina, The deterministic complexity of parallel multisearch, in: *Proc. 5th SWAT* (1996).
8. A. Bäumker and W. Dittrich, Fully dynamic search trees for an extension of the BSP model, in: *Proc. 8th SPAA* (1996) 233–242.
9. A. Bäumker, W. Dittrich, F. Meyer auf der Heide and I. Rieping, Realistic parallel algorithms: Priority queue operations and selection for the BSP* model, in: *Proc. of 2nd Euro-Par* (1996).
10. J.L. Carter and M.N. Wegman, Universal classes of hash functions, *J. Comput. Syst. Sci.*, 18 (1979) 143–154.
11. T.H. Cormen, C.E. Leiserson and R.L. Rivest, *Introduction to Algorithms* (MIT Press, Massachusetts, 1990).
12. D. Culler, R. Karp, D. Patterson, A. Sahay, K.E. Schauser, E. Santos, R. Subramonian and T. von Eicken, LogP: Towards a realistic model of parallel computation, in: *Proc. ACM SIGPLAN Symposium on Principles and Practice of Parallel Programming* (1993).
13. F. Dehne, A. Fabri and A. Rau-Chaplin, Scalable parallel computational geometry for coarse grained multicomputers, in: *Proc. ACM Conference on Comp. Geometry* (1993).

14. M. Dietzfelbinger, A. Karlin, K. Mehlhorn, F. Meyer auf der Heide, H. Rohnert and R.E. Tarjan, Dynamic Perfect Hashing: Upper and Lower Bounds, *SIAM J. Comput*, Vol. 23, No. 4 (1994) 738–761.

15. M. Dietzfelbinger and F. Meyer auf der Heide, A new universal class of hash functions and dynamic hashing in real time, in: *Proc. 17th ICALP* (1990) 6–19. Final version in: *J. Buchmann, H. Ganzinger and W.J. Paul, eds., Informatik · Festschrift zum 60. Geburtstag von Günter Hotz* (Teubner, Stuttgart, 1992) 95–119.

16. M.L. Fredman, J. Komlós, E. Szemeredi, Storing a sparse table with O(1) worst case time, in: *Proc. 23nd. FOCS*, (1982), 165–169.

17. A.V. Gerbessiotis and L. Valiant, Direct Bulk-Synchronous Parallel Algorithms, *Journal of Parallel and Distributed Computing* (1994).

18. M.T. Goodrich, Randomized fully-scalable BSP techniques for multi-searching and convex hull construction, in: *8th SODA* (1997).

19. W. Hoeffding, Probability inequalities for sums of bounded random variables, *American Statistical Association Journal* (1963) 13–30.

20. J. JáJá, *An Introduction to Parallel Algorithms* (Addison-Wesley, 1992).

21. B.H.H. Juurlink, P.S. Rao and J.F. Sibeyn, Worm-hole gossiping on meshes, in: *Proc. Euro-Par'96*, Lecture Notes in Computer Science, Vol. I (Springer, Berlin, 1996) 361–369.

22. D.E. Knuth, *The Art of computer Programming, Vol. 1: Fundamental Algorithms* (Addison-Wesley, 1968).

23. C.P. Kruskal, L. Rudolph and M. Snir, A complexity theory of efficient parallel algorithms, in: *Proc. 15th Int. Coll. on Automata, Languages, and Programming* (1988) 333–346.

24. W.F. McColl, The BSP approach to architecture independent parallel programming, Technical Report, Oxford University Computing Laboratory, 1994.

25. K. Mehlhorn, *Data Structures and Algorithms 1: Sorting and Searching* (Springer, Berlin, 1984).

26. F. Meyer auf der Heide, C. Scheideler and V. Stemann, Exploiting storage redundancy to speed up randomized shared memory simulations, *Theoretical Computer Science* 162 (1996) 245–281.

27. A. Czumaj, F. Meyer auf der Heide and V. Stemann, Shared Memory Simulations with Triple-Logarithmic Delay, in: in: *Proc. European Symposium on Algorithms* (1995).

28. W. Paul, U. Vishkin and H. Wagener, Parallel dictionaries on 2-3 trees, in: *Proc. of the 10th ICALP* (1983) 597–609.

29. A. Pietracaprina and G. Pucci, Tight bounds on deterministic PRAM emulations with constant redundancy, in: *Proc. 2nd European Symposium on algorithms* (1994) 319–400.

30. Abhiram Ranade, Maintaining dynamic ordered sets on processor networks, in: *Proc. 4th ACM Symp. on Parallel Algorithms and Architectures* (1992) 127–137.

31. J.H. Reif and S. Sen, Randomized algorithms for binary search and load balancing on fixed connection networks with geometric applications, *SIAM J. Comput.*, Vol. 23, No. 3 (1994) 633–651.

32. R. Tamassia and J.S. Vitter, Optimal cooperative search in fractional cascaded data structures, in: *Proc. ACM Symp. Computational Geometry* (1990) 307–315.

33. L. Valiant, A Bridging Model for parallel Computation, *Communications of the ACM*, Vol. 33, No. 8 (1994).

Parallel Sparse Cholesky Factorization

Burkhard Monien and Jürgen Schulze

University of Paderborn, Department of Computer Science
Fürstenallee 11, 33102 Paderborn, Germany

Abstract. In this paper we describe algorithms for the ordering and numerical factorization step in parallel sparse Cholesky factorization. Direct methods for solving sparse positive definite systems play an important role in many scientific applications such as linear programming and structual engineering. The importance of direct methods is mainly due to their generality and robustness. The paper describes minimum degree and nested dissection based ordering methods and presents a scalable parallel algorithm for the factorization of sparse matrices. The interested reader will find many references to the relevant literature.

1 Introduction

In contrast to dense matrix algorithms that can be implemented efficiently on distributed-memory parallel computers (see, e.g. [16, 17, 47]) complex algorithms and sophisticated data structures must be developed to achieve comparable performances when the underlying matrices are sparse. In this paper we will focus our attention on the solution of sparse symmetric positive definite linear systems by Cholesky factorization. Cholesky's method is a special variant of Gaussian elimination tailored to symmetric positive definite matrices $A \in M(n, \mathbb{R})$, $A = (a_{ij})$. Applying the method to A yields a lower triangular factor L such that $A = L \cdot L^t$ (the superscript t denotes the transpose operation). Let

$$A = \begin{pmatrix} d & v^t \\ v & B \end{pmatrix},$$

where d is a positive scalar, v a $(n-1)$-vector, and $B \in M(n-1, \mathbb{R})$. The Cholesky factor L is computed column by column using Eq. 1.

$$A = \begin{pmatrix} \sqrt{d} & 0 \\ v/\sqrt{d} & I \end{pmatrix} \begin{pmatrix} 1 & 0 \\ 0 & B - vv^t/d \end{pmatrix} \begin{pmatrix} \sqrt{d} & v^t/\sqrt{d} \\ 0 & I \end{pmatrix} \tag{1}$$

Vector $(\sqrt{d}, v/\sqrt{d})^t$ represents the first column of L. The remaining columns are obtained by recursively applying Eq. 1 to the submatrix $B - vv^t/d$. The linear system $A \cdot x = b$ can now be written as $LL^t \cdot x = b$. Solving $Ly = b$ and $L^t x = y$ yields the solution vector x. Eq. 1 also shows that the factorization process can introduce some fill into L, i.e., an element $a_{ij} = 0$ may become nonzero in L. To demonstrate this, let $v = (v_1 \ldots v_i \ldots v_j \ldots v_{n-1})$ with $v_i, v_j \neq 0$. Then, $v_{ij} \neq 0$ in vv^t and, thus, the corresponding element in $B - vv^t/d$ is nonzero even if $a_{ij} = 0$.

In general, L contains much more nonzeros than A and this fill heavily influences the performance of the overall factorization process. It is well known that the amount of fill can be reduced by reordering the columns and rows of A prior to factorization. More formally, we are searching a permutation matrix P so that the Cholesky factor $\tilde{L}$ of PAP^t has less fill than L. It should be noticed that for general indefinite matrices row and column interchanges are necessary to ensure numerical stability. But Cholesky's method applied to positive definite matrices does not require this pivoting. Therefore, we can choose any permutation matrix P to reorder A without affecting the numerical stability of the factorization process. Since P can be determined before the actual numerical factorization begins, the location of all fill entries and, thus, the data structure used to store L can be determined in advance. As a consequence, the overall factorization process can be divided in four independent steps: (1) determine a permutation matrix P so that the Colesky factor $\tilde{L}$ of PAP^t suffers little fill, (2) set up an appropriate data structure to store the nonzero elements of $\tilde{L}$ (symbolical factorization), (3) compute the nonzero elements of $\tilde{L}$ (numerical factorization), and (4) solve the triangular systems $\tilde{L}y = Pb$, $\tilde{L}^t z = y$, and set $x = P^t z$.

This paper presents algorithms for steps (1) and (3). A comprehensive treatment of the remaining steps can be found in [1, 21, 22, 47]. The paper is organized as follows: Section 2 describes minimum degree and nested dissection based ordering algorithms that determine a fill-reducing permutation matrix P. Compared to the numerical factorization step the ordering algorithm contributes only a small fraction to the total execution time of the entire factorization algorithm. Therefore, all ordering algorithms presented in Section 2 are sequential algorithms. The ordering can be considered as a sequential preprocessing step of the parallel numerical factorization. In Section 3 we introduce a parallel algorithm for sparse matrix factorization based on the multifrontal method [12, 57]. The multifrontal method reduces the factorization of a sparse input matrix A to the factorization of several dense matrices. This allows to apply some of the extremely efficient techniques developed for parallel dense matrix factorization.

2 Ordering methods

The problem of determining a permutation matrix P so that the Cholesky factor $\tilde{L}$ of PAP^t suffers minimum fill is a NP-complete problem [74]. Therefore, during the last decades much effort has been devoted to the development of powerful heuristic algorithms. All heuristics are based on the observation that a symmetric $n \times n$ matrix can be interpreted as the adjacency matrix of an undirected graph with n vertices. More formally, let $G = (X, E)$ denote an undirected graph with vertex set X and edge set E. Two vertices $x, y \in X$ are said to be adjacent in G, if $\{u, v\} \in E$. The set of vertices adjacent to x is denoted by $\mathrm{adj}_G(x)$. We extend the notation to $\mathrm{adj}_G(U)$ for the adjacent set of a subset U of nodes. The degree of a node x is defined by $\deg_G(x) = |\mathrm{adj}_G(x)|$. The undirected graph associated with A, denoted by $G^A = (X^A, E^A)$, has n vertices $x_1, \ldots, x_n$ and an edge between x_i and x_j if and only if $a_{ij} \neq 0$. An ordering (or labelling)

π is a bijection $X^A \mapsto \{1, \ldots, n\}$. Any ordering π of the nodes in G^A induces a permutation matrix P. Therefore, a fill-reducing permutation matrix P can be obtained by finding a "good" ordering of the nodes in G^A. It should be noticed that the definition of G^A induces an ordering π on the nodes in X^A with $\pi(x_i) = i$. The permutation matrix associated with this ordering is the identity matrix I.

Eq. 1 defines a transformation of $A_0 = A$ to $A_1 = B - vv^t/d$. As observed by Parter [61] and Rose [69] the graph G^{A_1} associated with A_1 can be obtained by applying the following modifications to G^{A_0}:

(1) delete node x_1 and its incident edges
(2) add edges so that all nodes adjacent to x_1 are pairwise connected

G^{A_1} is called the elimination graph of G^{A_0}. Therefore, symmetric Gaussian elimination can be interpreted as generating a sequence of elimination graphs G^{A_i}, $i = 1, \ldots, N$, where G^{A_i} is obtained from $G^{A_{i-1}}$ by deleting node x_i as described above ($G^{A_N} = \emptyset$). To ease the presentation we will use G^i instead of G^{A_i}. Each edge added in (2) corresponds to a fill element in $L + L^t$. Define the filled graph of G^A to be the graph $G^F = (X^F, E^F)$, $X^F = X^A$, whose edge set E^F consists of all edges in E^A and those added in the elimination graphs. It is obvious that G^F is the undirected graph associated with the matrix $F = L + L^t$. The edge sets E^F and E^A are related by the following theorem due to Rose et al. [70].

Theorem 1. *Let π be an ordering of the nodes in $X^A = \{x_1, \ldots, x_n\}$. Then $\{x_i, x_j\}$ is an edge in the corresponding filled graph if and only if $\{x_i, x_j\} \in E^A$ or there exists a path $x_i, x_{p_1}, \ldots, x_{p_t}, x_j$ in G^A such that $\pi(x_{p_k}) < \min\{\pi(x_i), \pi(x_j)\}$ for all $1 \le k \le t$.*

The simplest methods to determine an ordering π are band and profile schemes [22, 29]. Roughly speaking, the objective is to cluster all nonzeros in PAP^t near the main diagonal. Since this property is retained in the corresponding Cholesky factor all zeros outside the cluster can be ignored during factorization. The main disadvantage of band and profile schemes is that the factorization process cannot benefit from zeros located inside the cluster. In the following we present more advanced ordering methods that allow to exploit all zero elements during numerical factorization.

2.1 Minimum degree ordering

The minimum degree ordering algorithm determines a new ordering π of the nodes in $G^0 = G^A$ by generating a special sequence of elimination graphs. In each iteration G^i is constructed from G^{i-1} by deleting a node x with minimum degree. This strategy locally minimizes the number of fill produced when applying Eq. 1 to A_{i-1}. Fig. 1 presents the basic minimum degree ordering algorithm. It is apparent that the performance of the algorithm crucially depends on how the transformation from G^{i-1} to G^i is implemented. Much effort has been devoted in designing appropriate data structures that enable an efficient integration of additional fill-edges [14, 23, 24].

```
G⁰ := Gᴬ;
for i := 1 to N do
    select a node x of minimum degree in Gⁱ⁻¹;
    π(x) := i;
    transform Gⁱ⁻¹ to Gⁱ as described above;
od
```

Fig. 1. Basic minimum degree algorithm.

Each time a node x with minimum degree has been deleted, the degrees of all nodes adjacent to x must be updated. This "degree update" is one of the most time consuming steps of the entire algorithm. Various techniques (incomplete degree update, delayed degree update, mass elimination) have been developed to reduce the number of degree updates performed by the basic algorithm. A detailed description of these techniques can be found in [25]. In the following we only introduce the concept of mass elimination.

Two nodes y, z are said to be indistinguishable in G^i if $\mathrm{adj}_{G^i}(z) \cup \{z\} = \mathrm{adj}_{G^i}(y) \cup \{y\}$. The property of being indistinguishable defines an equivalence relation on the nodes of G^i that is preserved under elimination graph transformation. Let $Y = \{z \in \mathrm{adj}_{G^i}(y);\ \mathrm{adj}_{G^i}(z) \cup \{z\} = \mathrm{adj}_{G^i}(y) \cup \{y\}\}$. The set $Y \cup \{y\}$ defines an equivalence class represented by y. The following theorem provides the basis for the concept of mass elimination.

Theorem 2. *If y is selected as the minimum degree node in G^i, then the nodes in $Y = \{z \in adj_{G^i}(y);\ adj_{G^i}(z) \cup \{z\} = adj_{G^i}(y) \cup \{y\}\}$ can be selected (in any order) as minimum degree nodes during the next $|Y|$ iterations of the algorithm.*

Therefore, all nodes in $Y \cup \{y\}$ can be merged together and treated as one supernode [12]. Instead of performing $|Y \cup \{y\}|$ iterations of the for-loop in Fig. 1, the whole supernode is eliminated in one iteration. This significantly reduces the number of graph transformations and degree update steps. Another approach to speed up the minimum degree algorithm is to compute upper bounds on the degrees rather than exact degrees [2, 30]. In this approximate minimum degree algorithm (AMD) a node with minimum upper bound is selected in each iteration.

The quality of the orderings produced by the minimum degree algorithm is quite sensitive to the way ties are broken when there is more than one node of minimum degree [9]. The multiple elimination technique proposed by Liu [51] provides a limited form of tie-breaking. Here, a maximal independent set T of nodes with minimum degree is selected for elimination in each round of the algorithm. As noted in [51], this modification may not provide a genuine minimum degree ordering, since all nodes in T are eliminated before the new degrees of all other nodes are calculated. This version of the original algorithm is called multiple minimum degree algorithm (MMD).

To summarize, MMD and AMD represent the state-of-the-art in minimum degree based ordering algorithms. Both methods produce high quality orderings for matrices arising in finite-element applications. Unfortunately, the basic minimum degree algorithm is inherently sequential in nature. Parallelism can be exploited by eliminating independent sets of nodes [49, 51]. However, the various enhancements applied to the basic minimum degree heuristic have resulted in an extremely efficient sequential algorithm [25]. The only attempt to perform a minimum degree ordering in parallel we are aware of [28] was not able to outperform the sequential algorithm with its enhancements.

2.2 Nested Dissection Ordering

Another effective algorithm for reducing fill in sparse Cholesky factorization is nested dissection [19, 20], which is based on a divide-and-conquer paradigm. In this scheme vertex-separators play a central role. Let $G = (X, E)$ be an undirected graph. A subset $S \subset X$ is called vertex-separator of G, if the section graph $G(X - S) = (X - S, E(X - S))$, $E(X - S) = \{\{x, y\}; \ x, y \in X - S\}$ is disconnected. The method takes as input the undirected graph G^A associated with A and computes a minimal vertex-separator S^A. All nodes in S^A are numbered after those in the disconnected components. The method is recursively applied to the subgraphs induced by the disconnected components until a component consists of only one node. The ordering guarantees that for two nodes x and y belonging to different components no fill edge $\{x, y\}$ will be added during the elimination process. The reason for this can be seen as follows: Any path connecting x and y in G^A must contain a node $s \in S^A$. Since s has a higher ordering number than x, y and $\{x, y\} \notin E^A$ it follows immediately from Theorem 1 that $\{x, y\} \notin E^F$.

Similar to the argumentation above it can be shown that all nodes in a vertex-separator are pairwise connected in the filled graph G^F. Therefore, the effectiveness of nested dissection in reducing fill is highly dependent on the size of the separators used to split the graph. In the following we describe two different approaches that can be used to construct a vertex-separator. In the first approach an edge-bisection heuristic is applied to G^A to construct an edge-separator. Then, the edge separator is transformed into a small vertex-separator. In the second approach a vertex-separator is constructed directly by partitioning the nodes of G^A in several domains separated from each other by a multisector. Due to Ashcraft and Liu [5] this is called a domain decomposition of G^A.

Using edge-bisection heuristics to construct vertex-separators. The objective of the bisection problem is to find a minimal set of edges $E_c \subset E$, whose removal splits G in two equally sized parts U, V (i.e., $||U| - |V|| \leq 1$). E_c can be used to determine a vertex-separator. To explain this, we require some additional graph terminology. A bipartite graph is an undirected graph $G = (X, E)$ such that $X = X_1 \cup X_2$, $X_1 \cap X_2 = \emptyset$ and every edge connects a vertex from X_1 with a vertex from X_2. A vertex cover is a subset of vertices so that every edge in the graph is incident to at least one vertex in the vertex cover. A matching is a subset M of E such that no two edges in M have a node in common. A maximum matching is a matching with the largest possible size.

An edge-separator E_c induces a bipartite graph $H = (U' \cup V', E_c)$, such that U' (V') consists of all nodes in U (V) incident to an edge in E_c. Any vertex cover S in H defines a vertex-separator in G. An initial vertex cover is easily obtained by setting $S = V'$ (we assume that $|U'| \geq |V'|$). In a bipartite graph the size of a maximum matching is closely related to the size of a minimum vertex cover. This fundamental result is due to König [48].

Theorem 3. *If H is a bipartite graph, then a maximum matching and minimum vertex cover have the same cardinality in H.*

In [53] Liu uses matching techniques to construct a subset Z of S with the property $|adj_H(Z)| < |Z|$. If such a subset exists, we can set $\tilde{S} = (S - Z) \cup adj_H(Z)$ so that $|\tilde{S}| < |S|$. In more advanced algorithms [7, 63] subsets of S are not only used to reduce the size of the separator but also to balance the size of the disconnected components. For this, the Dulmage-Mendelsohn decomposition [13] is applied to a maximum matching in H to identify subsets Z with $|adj_H(Z)| = |Z|$. With the help of these subsets the separator can be moved while preserving its size.

Graph bisection has received a great deal of attention during the recent years. The CHACO [37], METIS [45], PARTY [67], and WGPP [33] software packages provide a variety of different bisection methods. All methods can be broadly classified into two categories: Global (or construction) heuristics that generate a bisection only from the description of the graph and local (or iterative) heuristics that try to improve a given bisection by local search.

The algorithm proposed by Kernighan-Lin [46] is one of the most frequently used local improvement methods. Starting with a given initial bisection, it tries to find a sequence of node-pair exchanges that reduce the number of edges in the separator (i.e., the cut size of the separator). Let U, V denote the disconnected parts of G after removal of E_c. For a pair (u, v), $u \in U$, $v \in V$ of nodes let diff(u, v) denote the change in cut size if u and v are exchanged. The Kernighan-Lin algorithm repeats choosing pairs (u, v) of unmarked nodes with diff(u, v) maximal, exchanges them logically and marks them. The vertices are exchanged even if diff(u, v) is negative. The idea behind this is that a deterioration of the cut size might allow to discover a better bisection during subsequent exchanges. If all nodes are marked, the algorithm will choose the sequence of pair exchanges that leads to a maximal decrease in cut size and it will exchange the corresponding nodes physically. The Kernighan-Lin algorithm as originally described in [46] is not very efficent, because each iteration takes $O(|E| \log |E|)$ time. There are several possible improvements proposed in literature, among which the version of Fiduccia and Mattheyses [15] is the most important. Instead of performing pair exchanges it only moves nodes with maximal diff-value while keeping the balance below a certain constant. In this way, a single search for a better bisection (i.e., a single iteration of the iterative algorithm) can be performed in time $O(|E|)$ when using an appropriate data structure.

Just like the Kernighan-Lin algorithm the Helpful-Set heuristic [11] is based on local search. The algorithm starts on nodes of a certain type to search for k-helpful sets, i.e., subsets of nodes from either U or V that decrease the cut size by k if moved to the other side. If such a set is found in one part of the graph, say U, it will be moved to V. The algorithm then starts to search for a balancing set in V that increases the cut size by not more than $k - 1$ edges if moved to U.

Simple global heuristics like linear-, scattered- or random-bisection are based on node numbering and do not consider the adjacency structure of the graph. The more sophisticated spectral bisection heuristic is based on algebraic graph theory [64]. The bisection is constructed using the eigenvector belonging to the second smallest eigenvalue of the Laplacian matrix (Fiedler vector). Iterative eigensolvers like Lanczos algorithm combined with edge contraction schemes are used to speed up the computation of the Fiedler vector [8, 38, 37]. However, even spectral bisection with all its enhancements takes a large amount of time. In fact, the time required to compute a nested dissection ordering based on spectral methods can be several orders of magnitude higher than the time taken by the parallel factorization algorithm.

Multilevel methods are capable of finding high quality bisections very quickly. These methods are based on graph contraction using certain operations to group nodes together. Two main contraction schemes have been proposed in literature. The first one determines a matching and collapses matched vertices into a multinode [32, 39, 40, 44], while the second identifies highly connected components and melts the corresponding nodes together [10, 35]. In many graphs arising in structual engineering applications the degree of each vertex is fairly close to the average degree of the graph. Therefore, most researchers apply a matching based contraction scheme when the multilevel algorithm is used to obtain a nested dissection ordering for these graphs. The coarsening of nodes is repeated until the graph has shrunken to a certain size. The resulting coarsed graph can be partitioned by using a global method like spectral bisection [39] or a simple graph growing heuristic [44]. In the following uncoarsening steps the bisection is refined using a local improvement heuristic like Kernighan-Lin, Fiduccia-Mattheyses, or Helpful-Set.

Using domain decomposition to construct vertex-separators. Domain decomposition is a well known approach in the solution of partial differential equations (PDE's). In graph theoretic terms a domain decomposition of an undirected graph $G = (X, E)$ is a partition of X in $\Phi \cup \Omega_1 \cup \ldots \cup \Omega_d$ [5, 6]. Each Ω_i denotes a domain, i.e., a large connected subset of vertices from X, and Φ is the set of remaining vertices that seperate the domains from each other. Φ is called the multisector of the domain decomposition. In contrast to the multilevel methods described above domain decomposition schemes use a two-level approach to construct vertex-separators. The entire scheme consists of three major steps: First, construct a domain decomposition of G, second, determine a subset Φ' of the multisector that may serve as an initial vertex-separator, and, finally, improve the vertex-separator using a graph matching scheme as described above.

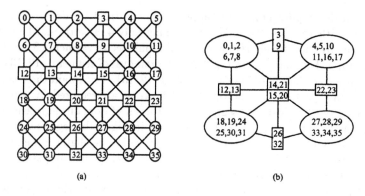

Fig. 2. Domain decomposition for a 6×6 grid graph (a). Vertices belonging to a domain are represented by circles, multisector vertices by squares. The corresponding segmented graph is shown in (b).

Fig. 2 (a) shows a domain decomposition for a 6×6 grid graph. The vertices are partitioned into four domains $\Omega_1 = \{0, 1, 2, 6, 7, 8\}$, $\Omega_2 = \{4, 5, 10, 11, 16, 17\}$, $\Omega_3 = \{18, 19, 24, 25, 30, 31\}$, and $\Omega_4 = \{27, 28, 29, 33, 34, 35\}$. Vertices belonging to a domain are represented by circles, multisector vertices by squares. The initial vertex-separator Φ' can be computed using a simple coloring scheme. For this, each domain Ω_i is assigned a color from $\{\text{black}, \text{white}\}$. The color of a vertex $v \in \Phi$ is determined as follows: If all domains adjacent to v have the same color $c \in \{\text{black}, \text{white}\}$, color v with c; otherwise color v with gray. Φ' consists of all vertices in Φ whose color is gray. Unfortunately, this simple coloring scheme may not always produce a feasible vertex-separator. For example, let us assume that in Fig. 2 (a) domains Ω_1, Ω_2 are colored white and Ω_3, Ω_4 black. When using the above coloring scheme multisector vertex 20 is colored black and multisector vertex 15 is colored white. Since both vertices are connected by an edge, no valid vertex-separator is obtained. As shown by Ashcraft and Liu [5], the problem can be solved by clustering adjacent multisector vertices that have no common adjacent domain. Concerning our example this means that vertices in $\{14, 21\}$ and $\{15, 20\}$ are comprised to one multinode. Further multinodes can be generated by clustering all multisector vertices that are adjacent to the same set of domains. In this way all multisector vertices are blocked into segments. Figure 2 (b) shows the segmented graph obtained from (a).

A given vertex-separator Φ' can be improved by changing the coloring of the domains. For two domains Ω_i, Ω_j with different colors let $\text{diff}(\Omega_i, \Omega_j)$ denote the change in separator size when the domains exchange their colors. With this definition a local search algorithm like Kernighan-Lin can be easily adopted to improve Φ'.

The size of the final separator is highly dependent on the structure of the domain decomposition. In [5] Ashcraft and Liu use a simple graph growing scheme to construct the domains. Their scheme offers only little possibilities to influence the shape and orientation of the generated domains. Currently, one of the authors [72] is developing a multilevel approach to construct the domains. The idea is to start with very small domains containing only 4–8 vertices. The domains are enlarged by successively removing certain multisector segments from the segmented graph. For example, in Fig. 2 (b) the removal of segment $\{12, 13\}$ yields a new domain consisting of nodes $\Omega_1 \cup \Omega_3 \cup \{12, 13\}$. To explain the coarsening process in more detail the following two definitions are required: A segment ϕ_1 is said to be outmatched by a segment ϕ_2 if the set of domains adjacent to ϕ_1 is a subset of those adjacent to ϕ_2. For example, in Fig. 2 (b) segment $\{12, 13\}$ is outmatched by $\{14, 15, 20, 21\}$. A segment ϕ_1 is said to be eligible for elimination if no adjacent segment is outmatched by ϕ_1. In each coarsening step an independent set of eligible segments is removed from the current segmented graph. The independent sets need not to be maximal – they are chosen so that the domains grow equally. Similar to the refinement of edge-separators in multilevel bisection heuristics, the vertex-separator obtained in the coarsed graph can be refined during the uncoarsening steps.

Conclusion: State-of-the-art nested dissection ordering algorithms [5, 33, 41, 45] use a multilevel bisection heuristic or a domain decomposition approach to compute the required vertex-separators. Although algorithms in [33, 41] use a multilevel edge contraction scheme, vertex-separators are computed in each uncoarsening step. It should be noticed that the algorithms do not provide a genuine nested dissection ordering. As soon as the remaining components are smaller than a certain size they are ordered using constrained minimum degree [55]. For many problems arising in structual engineering applications the orderings produced by these "hybrid" algorithms are better than those produced by a minimum degree algorithm. But for highly irregular problems, nested dissection is much less effective. The run time of these nested dissection based ordering algorithms is a small constant factor higher than the run time of minimum degree based heuristics. However, in Section 3.1 we show that nested dissection orderings are more suitable for parallel numerical factorization.

3 Numerical factorization

There are three major schemes to carry out the numerical factorization, namely Column-, Row-, and Submatrix-Cholesky. The schemes differ in the way matrix elements are accessed and updated. Historically, the fan-out algorithm proposed by George et al. [27] was first to be implemented on a distributed memory machine. The fan-out algorithm is a data-driven algorithm based on Submatrix-Cholesky. On condition that the $n \times n$ input matrix A is column-wise partitioned among p processors the total communication volume is $O(np \log n)$ [26]. An improved distributed factorization algorithm based on Column-Cholesky is the fan-in algorithm introduced by Ashcraft et al. [4]. Again, the input matrix A is column-wise partitioned among the p processors, but the total communication volume is reduced to $O(np)$. It has been shown in [26] that a lower bound on the total communication volume of any column-based partitioning scheme is $\Omega(np)$.

This limits the overall scalability of any column-based parallel factorization algorithm [59, 3, 65, 62, 27, 18, 36, 42, 71]. To improve the performance of parallel sparse matrix factorization the update matrix must be be column- and row-wise partitioned [31, 34, 43, 60, 68]

In Section 3.2 we describe a parallel factorization algorithm based on the multifrontal method [12, 57] that has originally been proposed in [34, 43]. As shown in [34, 43] the comunication overhead of this algorithm is only $O(n\sqrt{p})$. It should be noticed that the communivation volume represents a lower bound on the communication overhead. The multifrontal method (a variant of Submatrix-Cholesky) reduces the factorization of a sparse input matrix A to the factorization of several dense matrices. These dense matrices are column- and row-wise splitted among subsets of processors. The whole factorization process is guided by a special data-structure called elimination tree. A formal definition of this tree is given in Section 3.1.

3.1 The elimination tree

The elimination tree T associated with the Colesky factor $L = (l_{ij})$ of A consists of n nodes each corresponding to a column in L. T is defined as follows: Let p, j be two columns in L. Node p is the parent of node j if and only if $p = \min\{i > j;\ l_{ij} \neq 0\}$ (i.e., the first subdiagonal nonzero element in column j can be found in row p). Fig. 3 shows a nested dissection ordering of the nodes of a 3×3 grid graph, the structure of the corresponding factor matrix and the elimination tree.

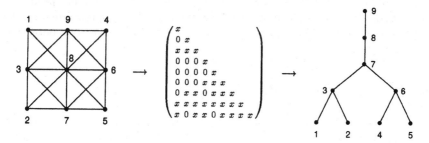

Fig. 3. 3×3 grid graph with nested dissection ordering, structure of the corresponding factor matrix (each x denotes a nonzero element), and the resulting elimination tree.

The elimination tree is a spanning tree of the filled graph G^F and can be constructed in a bottom-up manner using Theorem 1 (for a detailed description the reader is refered to [56]). The structure of the tree is highly dependent on the ordering of the nodes in G^A. For a nested dissection ordering the vertices of the first separator are placed at the top of the elimination tree. The separators obtained for the two disconnected components are placed below the first separator, and so on. For example, in Fig. 3 the first separator is given by $\{7, 8, 9\}$ and the separators $\{3\}$ and $\{6\}$ obtained for the two subgraphs are direct successors of the first separator. In general, the vertices of a separator are indistinguishable and form a chain in T. As described in section 2.1 a set of indistinguishable nodes is comprised to a single supernode. The levels of T are defined with respect to these supernodes. That is, the topmost level of T consists of the supernode associated with the first vertex-separator.

The elimination tree has the important property that columns in different branches can be factorized in parallel. Therefore, an ordering appropriate for parallel Cholesky factorization must generate a balanced elimination tree. Nested dissection algorithms try to achieve this by splitting the graph in two parts containing roughly the same number of vertices. However, it is much more important that the number of floating point operations required to factorize the nodes of a component is balanced (see the mapping algorithm described in Section 3.2). Several tree restructuring algorithm are proposed in literature [50, 52, 54, 58] that take as input a low-fill ordering of the matrix and try to produce an equivalent ordering that is more suitable for parallel factorization. By equivalent we mean an ordering that generates the same fill edges but a new shape of the elimination tree. Since the adjacency structure of the filled graph is preserved the number of nonzeros in the Cholesky factor remains the same. Also, the number of arithmetic operations to perform the factorization is unchanged. The objective of all reordering algorithms is to reduce the height of the elimination tree. But the height is not a very accurate indicator of the actual parallel factorization time. As already concluded in [36], equivalent orderings have the capacity to modify only relatively minor features of the initial fill reducing ordering. The methods are not able to unveil all the parallelism available in the underlying problem. In our opinion the development of new ordering algorithms appropriate for performing the subsequent factorization efficiently in parallel offers a great potential for further research.

3.2 The multifrontal method

The key feature of the multifrontal method is that the update vv^t (cf. Eq. 1) is not applied directly to the submatrix B. Instead, these updates are accummulated in a so called update matrix which is passed on along the edges of the elimination tree in a bottom up manner. Each node j of the elimination tree has associated with it a frontal matrix F_j. This matrix is initialized with the nonzero elements in column j of A. All update matrices "received" from the children of j in T are added to the frontal matrix. Since only nonzero elements are stored in frontal and update matrices (i.e., these matrices are full), the update matrix may have to be extended to conform with the subscript structure of the frontal matrix by introducing a number of zero rows and columns. We shall refer to this extended-add operation by using the symbol "$\oplus$". Column j of L and the update matrix U_j is obtained by applying Eq. 1 to F_j. U_j is passed on to the parent of j. Fig. 4 captures the essence of the multifrontal method. For a more detailed description the reader is refered to [57].

for column $j := 1$ to n do
    Let $j, i_1, \ldots, i_r$ be the locations of nonzeros in column j of L;
    Let $c_1, \ldots, c_s$ be the children of j in the elimination tree;
    Form the frontal matrix F_j using the update matrices of all children of j;

$$
F_j := \begin{pmatrix} a_{j,j} & a_{j,i_1} & \cdots & a_{j,i_r} \\ a_{i_1,j} & & & \\ \vdots & & 0 & \\ a_{i_r,j} & & & \end{pmatrix} \oplus U_{c_1} \oplus \ldots \oplus U_{c_s};
$$

Factor frontal matrix F_j into

$$
\begin{pmatrix} l_{j,j} & 0 & \cdots & 0 \\ l_{i_1,j} & & & \\ \vdots & & I & \\ l_{i_r,j} & & & \end{pmatrix} \begin{pmatrix} 1 & 0 & \cdots & 0 \\ 0 & & & \\ \vdots & & U_j & \\ 0 & & & \end{pmatrix} \begin{pmatrix} l_{j,j} & l_{j,i_1} & \cdots & l_{j,i_r} \\ 0 & & & \\ \vdots & & I & \\ 0 & & & \end{pmatrix};
$$

od

Fig. 4. Core of the multifrontal method.

The efficiency of the multifrontal method can be improved significantly when using the concept of supernodes. In matrix terms a supernode is a group of consecutive columns in L such that the corresponding diagonal block is full triangular and all columns have an identical off-block-diagonal subscript structure. Let us assume that columns $j, j+1, \ldots, j+t$ form a supernode in L. In the supernodal version of the multifrontal method F_j is initialized by

$$
F_j := \begin{pmatrix} a_{j,j} & a_{j,j+1} & \cdots & a_{j,j+t} & a_{j,i_1} & \cdots & a_{j,i_r} \\ a_{j+1,j} & a_{j+1,j+1} & \cdots & a_{j+1,j+t} & a_{j+1,i_1} & \cdots & a_{j+1,i_r} \\ \vdots & \vdots & \cdots & \vdots & \vdots & \cdots & \vdots \\ a_{j+t,j} & a_{j+t,j+1} & \cdots & a_{j+t,j+t} & a_{j+t,i_1} & \cdots & a_{j+t,i_r} \\ a_{i_1,j} & a_{i_1,j+1} & \cdots & a_{i_1,j+t} & & & \\ \vdots & \vdots & \cdots & \vdots & & 0 & \\ a_{i_r,j} & a_{i_r,j+1} & \cdots & a_{i_r,j+t} & & & \end{pmatrix}
$$

After having added the update matrices $U_{c_1}, \ldots, U_{c_s}$ we can perform $t+1$ consecutive factorization steps. This yields columns $j, j+1, \ldots, j+t$ of L and the update matrix U_{j+t}. It should be emphasized that the factorization steps are performed on a full frontal matrix. Therefore, techniques originally developed to speed-up parallel dense matrix factorization can be applied to factor F_j.

The scalability of a parallel algorithm based on the multifrontal method depends quite crucially on how the elements of the dense frontal matrices are partitioned among the processors. In our implementation [73] we use a 2-dimensional partitioning scheme based on the one proposed in [34]. The scheme can be best described using a simple example. Let us assume that a balanced elimination tree is given, and that we want to use 4 processors for the computation of L (cf. Fig. 5). As noted earlier, each supernode forms a chain in T and the levels of T are defined with respect to these supernodes. In Fig. 5 the supernodes are indicated by an oval. The nodes in the 4 subtrees T_0, T_1, T_2, T_3 are completely mapped to the processors $P_{00}, P_{01}, P_{10}, P_{11}$, respectively. The factorization of the associated columns can be done without any communication. In the next level all even columns of the factor matrix F_i (F_j) associated with the supernode i (j) are mapped to processor P_{00} (P_{10}) and all odd columns to processor P_{01} (P_{11}). As shown in Fig. 5 the factor matrix F_k associated with the supernode in the topmost level is distributed over all 4 processors. For example, processor P_{10} stores all elements of F_k with even column and odd row index.

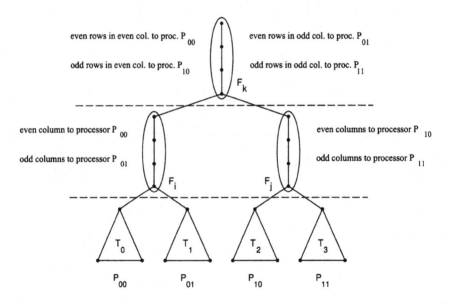

Fig. 5. A simple mapping example. The dashed lines separate the upper levels of the elimination tree. Each supernode forms a chain in T. The oval around a chain indicates the corresponding supernode.

In general the elimination tree is unbalanced so that a more complicated mapping scheme has to be used (cf. [18, 66, 43]). Again, let us assume that the nodes of the elimination tree have to be mapped on four processors $P_{00}, P_{01}, P_{10}, P_{11}$. The scheme used in our implementation works as follows: For each node u of the elimination tree let $T[u]$ denote the subtree rooted at u and weight(u) the number of floating point operations required to factorize all nodes in $T[u]$. The mapping algorithm uses a queue Q initialized with the supernode in the topmost level of T and proceeds from the root of T down to its leaves. Let us assume that Q contains the supernodes $\{u_1, \ldots, u_t\}$. A bin-packing heuristic is is used to split Q in two equally weighted parts Q_1 and Q_2. If the weight difference of Q_1 and Q_2 exceeds a certain threshold the algorithm removes the supernode u_j with the maximal weight from Q and inserts all children of u_j in Q. This is iterated until the bin-packing heuristic is successful. In this case, all subtrees rooted at a node in Q_1 are mapped to the processor group P_{00}, P_{01} and all subtrees rooted at a node in Q_2 are mapped to the group P_{10}, P_{11}. The mapping algorithm is recursively called for the first processor group with a queue initialized to Q_1 and for the second processor group with a queue initialized to Q_2. All frontal matrices corresponding to supernodes that have been removed from Q are distributed among the processors $P_{00}, P_{01}, P_{10}, P_{11}$ in the same way as F_k in Fig. 5.

In this way supernodes in the upper part of the elimination tree are mapped to certain processor groups. In each recursive call the size of the processor group is halfed. If a group consists of only one processor, no further recursive call is initiated. The processor is responsible for the factorization of all subtrees $T[u]$, whose root u is contained in the queue received by the last recursive call. If the elimination tree is balanced with respect to the weights of the supernodes, only one supernode has to be splitted in each recursion level. This may lead to a reduction of the communication overhead, since more work can be exclusively assigned to a single processor. On the other hand, additional overhead can occur if the top-level supernodes in a balanced elimination tree are much larger than in an unbalanced one. This trade-off seems to be an interesting topic for further research.

The frontal matrices associated with the supernodes are column- and row-wise partitioned among the processors of the group. For dense parallel matrix factorization it is well known that this 2-dimensional partitioning scheme is much more efficient than a column-based scheme. The efficiency can be further increased by grouping several consecutive columns and rows into blocks. This reduces the number of communication steps required to perform the factorization at the cost of some added load imbalance. Figure 6 shows a block-cyclic column- and row-wise partitioning of a 12×12 matrix among 4 processors.

To summarize, the parallel algorithm benefits from a careful mapping of supernodes to processor groups and a 2-dimensional block-cyclic partitioning of the corresponding dense frontal matrices among the processors of the group. In [34, 43] it is shown that the algorithm is as scalable as any parallel formulation of dense matrix factorization on mesh and hypercube architectures.

References

1. F.L. Alvarado, A. Pothen, R. Schreiber, *Highly parallel sparse triangular solution*, RIACS Techn. Rep. 92.11, NASA Ames Research Center, 1992.

$$
\begin{pmatrix}
0 & & & & & & & & & & & \\
2 & 3 & & & & & & & & & & \\
0 & 1 & 0 & & & & & & & & & \\
2 & 3 & 2 & 3 & & & & & & & & \\
0 & 1 & 0 & 1 & 0 & & & & & & & \\
2 & 3 & 2 & 3 & 2 & 3 & & & & & & \\
0 & 1 & 0 & 1 & 0 & 1 & 0 & & & & & \\
2 & 3 & 2 & 3 & 2 & 3 & 2 & 3 & & & & \\
0 & 1 & 0 & 1 & 0 & 1 & 0 & 1 & 0 & & & \\
2 & 3 & 2 & 3 & 2 & 3 & 2 & 3 & 2 & 3 & & \\
0 & 1 & 0 & 1 & 0 & 1 & 0 & 1 & 0 & 1 & 0 & \\
2 & 3 & 2 & 3 & 2 & 3 & 2 & 3 & 2 & 3 & 2 & 3
\end{pmatrix}
\qquad
\begin{pmatrix}
0 & & & & & & & & & & & \\
0 & 0 & & & & & & & & & & \\
2 & 2 & 3 & & & & & & & & & \\
2 & 2 & 3 & 3 & & & & & & & & \\
0 & 0 & 1 & 1 & 0 & & & & & & & \\
0 & 0 & 1 & 1 & 0 & 0 & & & & & & \\
2 & 2 & 3 & 3 & 2 & 2 & 3 & & & & & \\
2 & 2 & 3 & 3 & 2 & 2 & 3 & 3 & & & & \\
0 & 0 & 1 & 1 & 0 & 0 & 1 & 1 & 0 & & & \\
0 & 0 & 1 & 1 & 0 & 0 & 1 & 1 & 0 & 0 & & \\
2 & 2 & 3 & 3 & 2 & 2 & 3 & 3 & 2 & 2 & 3 & \\
2 & 2 & 3 & 3 & 2 & 2 & 3 & 3 & 2 & 2 & 3 & 3
\end{pmatrix}
$$

Fig. 6. A 2-dimensional block-cyclic partitoning of a 12×12 matrix among 4 processors. On the left side the block-size is set to one, on the right side it is set to two.

2. P. Amestoy, T.A. Davis, I.S. Duff, *An approximate minimum degree ordering algorithm*, Techn. Rep. TR/PA/95/09, Parallel Algorithm Project, CERFACS, Toulouse, 1995.

3. C. Ashcraft, S.C. Eisenstat, J.W.H. Liu, A. Sherman, *A comparison of three distributed sparse factorization schemes*, SIAM Symposium on Sparse Matrices, 1989.

4. C. Ashcraft, S.C. Eisenstat, J.W.H. Liu, *A fan-in algorithm for distributed sparse numerical factorization*, SIAM J. Sci. Stat. Comput., Vol. 11, No. 3, 593–599, 1990.

5. C. Ashcraft, J.W.H. Liu, *Using domain decomposition to find graph bisectors*, Techn. Rep. ISSTECH-95-024, Boeing Computer Services, Seattle, 1995.

6. C. Ashcraft, J.W.H. Liu, *Robust ordering of sparse matrices using multisection*, Techn. Rep. ISSTECH-96-002, Boeing Computer Services, Seattle, 1996.

7. C. Ashcraft, J.W.H. Liu, *applications of the Dulmage-Mendelsohn decomposition and network flow of graph bisection improvement*, Techn. Rep., Boeing Computer Services, Seattle, 1996.

8. S.T. Barnard, H.D. Simon, *A fast multilevel implementation of recursive spectral bisection*, Proc. of 6th SIAM Conf. Parallel Processing for Scientific Computing, 711–718, 1993.

9. P. Berman, G. Schnitger, *On the performance of the minimum degree ordering for Gaussian elimination*, SIAM J. Matrix Anal. Appl., Vol. 11, No. 1, 83–88, 1990.

10. C.-K. Cheng, Y.-C. Weil, *An improved two-way partitioning algorithm with stable performance*, IEEE Transactions on Computer Aided Design, Vol. 10, No. 12, 1502–1511, 1991.

11. R. Diekmann, B. Monien, R. Preis, *Using helpful sets to improve graph bisections*, DIMACS Series in Discrete Mathematics and Theoretical Computer Science, American Mathematical Society, Volume 21, 1995.

12. I.S. Duff, J.K. Reid, *The multifrontal solution of indefinite sparse symmetric linear equations*, ACM Trans. Math. Software, Vol. 9, 302–325, 1983.

13. A. Dulmage, N. Mendelsohn, *Coverings of bipartite graphs*, Can. J. Math., Vol. 10, 517–534, 1958.

14. S.C. Eisenstat, M.H. Schultz, A.H. Sherman, *Algorithms and data structures for sparse symmetric Gaussian elimination*, SIAM J. Sci. Stat. Comput., Vol. 2, No. 2, 225–237, 1981.

15. C.M. Fiduccia, R.M. Mattheyses, *A linear-time heuristic for improving network partitions*, 19th IEEE Design Automation Conference, 175–181, 1982.

16. K.A. Gallivan, R.J. Plemmons, A.H. Sameh, *Parallel algorithms for dense linear algebra computations*, SIAM Review Vol. 32, No. 1, 54–135, 1990.

17. G.A. Geist, C.H. Romine, *LU factorization algorithms on distributed-memory multiprocessor architectures*, SIAM J. Sci. Stat. Comput., Vol. 9, No. 4, 639–649, 1988.

18. G.A. Geist, E. Ng, *Task scheduling for parallel sparse Cholesky factorization*, International Journal of Parallel Programming, Vol. 18, No. 4, 291–314, 1989.

19. A. George, *Nested dissection of a regular finite element mesh*, SIAM J. Numer. Anal., Vol. 10, No. 2, 345–363, 1973.

20. A. George, J.W.H. Liu, *An automatic nested dissection algorithm for irregular finite element problems*, SIAM J. Numer. Anal., Vol. 15, No. 5, 1053–1069, 1978.

21. A. George, M.T. Heath, E. Ng, J.W.H. Liu, *Symbolic Cholesky factorization on a local-memory multiprocessor*, Parallel Computing, 85–95, 1987.

22. J.A. George, J.W.H. Liu, *Computer Solution of Large Sparse Positive Definite Systems*, Prentice-Hall, Englewood Cliffs, NJ, 1981.

23. A. George, J.W.H. Liu, *A minimal storage implementation of the minimum degree algorithm*, SIAM J. Numer. Anal., Vol. 17, No. 2, 282–299, 1980.

24. A. George, J.W.H. Liu, *A fast implementation of the minimum degree algorithm using quotient graphs*, ACM Trans. Math. Software, Vol. 6, 337–358, 1980.

25. A. George, J.W.H. Liu, *The evolution of the minimum degree ordering algorithm*, SIAM Review, Vol. 31, No. 1, 1–19, 1989.

26. A. George, J.W.H. Liu, E. Ng, *Communication results for parallel sparse Cholesky factorization on a hypercube*, Parallel Computing 10, 287–298, 1989.

27. A. George, M.T. Heath, J.W.H. Liu, E. Ng, *Sparse Cholesky factorization on a local-memory multiprocessor*, SIAM J. Sci. Stat. Comput., Vol. 9, No. 2, 327–340, 1988.

28. M. Ghose, E. Rothberg, *A parallel implementation of the multiple minimum degree ordering heuristic*, Techn. Rep., Old Dominion University, Norfolk, 1994.

29. N.E. Gibbs, W.G. Poole, P.K. Stockmeyer, *An algorithm for reducing the bandwidth and profile of a sparse matrix*, SIAM J. Numer. Anal., Vol. 13, 236–250, 1976.

30. J.R. Gilbert, C. Moler, R. Schreiber, *Sparse matrices in MATLAB: design and implementation*, SIAM J. Matrix Anal. Appl., Vol. 13, 333–356, 1992.

31. J.R. Gilbert, R. Schreiber, *Highly sparse Cholesky factorization*, SIAM J. Sci. Stat. Comput., Vol. 13, No. 5, 1151–1172, 1992.

32. A. Gupta, *Fast and effective algorithms for graph partitioning and sparse matrix ordering*, IBM T.J. Watson Research Center, Research Report RC 20496, New York, 1996.

33. A. Gupta, *WGPP: Watson graph partitioning (and sparse matrix ordering) package, users manual*, IBM T.J. Watson Research Center, Research Report RC 20453, New York, 1996.

34. A. Gupta, V. Kumar, *A Scalable parallel algorithm for sparse matrix factorization*, Tech. Rep. 94-19, CS-Dep., Univ. Minnesota, 1994.

35. L. Hagen, A. Kahng, *A new approach in effective circuit clustering*, Proc. of IEEE International Conference on Computer Aided Design, 422–427, 1992.

36. M.T. Heath, E. Ng, B.W. Peyton, *Parallel algorithms for sparse linear systems*, SIAM Review, Vol. 33, No. 3, 420–460, 1991.

37. B. Hendrickson, R. Leland, *The chaco user's guide*, Tech. Rep. SAND94-2692, Sandia Nat. Lab., 1994.

38. B. Hendrickson, R. Leland, *An improved spectral graph partitioning algorithm for mapping parallel computations*, SIAM J. Sci. Comput., Vol. 16, 1995.

39. B. Hendrickson, R. Leland, *A multilevel algorithm for partitioning graphs*, Proc. of Supercomputing'95, 1995.

40. B. Hendrickson, E. Rothberg, *Improving the runtime and quality of nested dissection ordering*, Techn. Rep., SAND96-0868, Sandia Nat. Lab., 1996.

41. B. Hendrickson, E. Rothberg, *Effective sparse matrix ordering: just around the BEND*, Proc. of 8th SIAM Conf. Parallel Processing for Scientific Computing, 1997.

42. L. Hulbert, E. Zmijewski, *Limiting communication in parallel sparse Cholesky factorization*, SIAM J. Sci. Stat. Comput., Vol. 12, No. 5, 1184–1197, 1991.

43. G. Karypis, V. Kumar, *A high performance sparse Cholesky factorization algorithm for scalable parallel computers*, Tech. Rep. 94-41, CS-Dep., Univ. Minnesota, 1994.

44. G. Karypis, V. Kumar, *A fast and high quality multilevel scheme for partitioning irregular graphs*, Tech. Rep. 95-035, CS-Dep., Univ. Minnesota, 1995.

45. G. Karypis, V. Kumar, *METIS: unstructured graph partitioning and sparse matrix ordering system*, Techn. Rep., CS-Dep., Univ. Minnesota, 1995.

46. B.W. Kernighan, S. Lin, *An effective heuristic procedure for partitioning graphs*, The Bell Systems Technical Journal, 291–308, 1970.

47. V. Kumar, A. Grama, A. Gupta, G. Karypis, *Introduction to Parallel Computing: Design and Analysis of Algorithms*, Benjamin Cummings Publishing Company, Redwood City, CA, 1994.

48. D. König, *Über Graphen und ihre Anwendung auf Determinantentheorie und Mengenlehre*, Math. Ann., 77, 453–465, 1916.

49. M. Leuze, *Independent set orderings for parallel matrix factorization by Gaussian elimination*, Parallel Computing, Vol. 10, 177–191, 1989.

50. J.G. Lewis, B.W. Peyton, A. Pothen, *A fast algorithm for reordering sparse matrices for parallel factorization*, SIAM J. Sci. Stat. Comput., Vol. 10, No. 6, 1146–1173, 1989.

51. J.W.H. Liu, *Modification of the minimum-degree algorithm by multiple elimination*, ACM Trans. Math. Software, Vol. 11, No. 2, 141–153, 1985.

52. J.W.H. Liu, *Equivalent sparse matrix reordering by elimination tree rotations*, SIAM J. Sci. Stat. Comput., Vol. 9, No. 3, 424–444, 1988.

53. J.W.H. Liu, *A graph partitioning algorithm by node separators*, ACM Trans. Math. Software, Vol. 15, No. 3, 198–219, 1989.

54. J.W.H. Liu, *Reordering sparse matrices for parallel elimination*, Parallel Computing 11, 73–91, 1989.

55. J.W.H. Liu, *The minimum degree ordering with constraints*, SIAM J. Sci. Stat. Comput., Vol. 10, No. 6, 1136–1145, 1989.

56. J.W.H. Liu, *The role of elimination trees in sparse factorization*, SIAM J. Matrix Anal. Appl., Vol. 11, No. 1, 134–172, 1990.

57. J.W.H. Liu, *The multifrontal method for sparse matrix solutions: theory and practice*, SIAM Review, Vol. 34, No. 1, 82–109, 1992.

58. J.W.H. Liu, A. Mirzaian, *A linear reordering algorithm for parallel pivoting of chordal graphs*, SIAM J. Disc. Math., Vol. 2, No. 1, 100–107, 1989.

59. R.F. Lucas, T. Blank, J.J. Tiemann, *A parallel solution method for large sparse systems of equations*, IEEE Transactions on Computer Aided Design, Vol. 6, No. 6, 981–991, 1987.

60. M. Mu, J.R. Rice, *A grid-based subtree-subcube assignment strategy for solving partial differential equations on hypercubes*, SIAM J. Sci. Stat. Comput., Vol. 13, No. 3, 826–839, 1992.

61. S.V. Parter, *The use of linear graphs in Gauss elimination*, SIAM Review, Vol. 3, 119–130, 1961.

62. R. Pozo, S.L. Smith, *Performance evaluation of the parallel multifrontal method in a distributed-memory environment*, Proc. of 6th SIAM Conference on Parallel Processing for Scientific Computing, 1993.

63. A. Pothen, C.-J. Fan, *Computing the block triangular form of a sparse matrix*, ACM Trans. Math. Software, Vol. 16, No. 4, 303–324, 1990.

64. A. Pothen, H.D. Simon, K.-P. Liou, *Partitioning sparse matrices with eigenvectors of graphs*, SIAM J. Matrix Anal. Appl., Vol. 11, No. 3, 430–452, 1990.

65. A. Pothen, C. Sun, *Distributed multifrontal factorization using clique trees*, Proc. 5th SIAM Conference on Parallel Processing for Scientific Computing, 34–40, 1991.

66. A. Pothen, C. Sun, *A mapping algorithm for parallel sparse Cholesky factorization*, SIAM J. Sci. Comput., Vol. 14, No. 5, 1253–1257, 1993.

67. R. Preis, R. Diekmann, *The PARTY partitioning library user guide – version 1.1*, Techn. Rep., CS-Dept., Univ. of Paderborn, 1996.

68. E. Rothberg, A. Gupta, *An efficient block-oriented approach to parallel sparse Cholesky factorization*, Proc. of Supercomputing'92, 1992.

69. D.J. Rose, *A graph-theoretic study of the numerical solution of sparse positive definite systems of linear equations*, in *Graph-Teory and Computing*, R. Read (Ed.), Academic Press, New York, 1972.

70. D.J. Rose, R.E. Tarjan, G.S. Luecker, *Algorithmic aspects of vertex elimination on graphs*, SIAM J. Comput., Vol. 5, No. 2, 266–283, 1976.

71. R. Schreiber, *Scalability of sparse direct solvers*, in *Sparse Matrix Computations: Graph Theory Issues and Algorithms*, J.R. Gilbert, J.W.H. Liu (Eds.), Springer Verlag, 1992.

72. J. Schulze, *A new multilevel scheme for constructing vertex separators*, Techn. Rep., CS-Dept., Univ. of Paderborn, 1997.

73. J. Schulze, *Implementation of a parallel algorithm for sparse matrix factorization*, Techn. Rep., CS-Dept., Univ. of Paderborn, 1996.

74. M. Yannakakis, *Computing the minimum fill-in is NP-complete*, SIAM J. Alg. Disc. Meth., Vol. 2, No. 1, 77–79, 1981.

Unstructured Graph Partitioning for Sparse Linear System Solving*

Jean Michel, François Pellegrini and Jean Roman

LaBRI, URA CNRS 1304
Université Bordeaux I
351, cours de la Libération, 33405 TALENCE, FRANCE
{jmichel|pelegrin|roman}@labri.u-bordeaux.fr

Abstract. Solving large sparse linear systems is a key issue, arising in a variety of scientific and engineering applications. This paper presents the work carried out at the LaBRI on graph partitioning and its application to efficient parallel solutions of this problem.

1 Introduction

Many scientific and engineering problems can be modeled by sparse linear systems, which are solved either by iterative or direct methods. In this paper, we summarize the work carried out at the *Laboratoire Bordelais de Recherche en Informatique* (LaBRI) of the Université Bordeaux I, on methods and tools for solving sparse linear systems.

Our research effort is currently focused on two main issues. The first one, presented in section 2, is the SCOTCH project, whose goal is to study the applications of graph theory to scientific computing using a "divide and conquer" approach, and which has resulted in the development of efficient static mapping and graph partitioning algorithms [41]. The second one, described in section 3, is the development of preprocessing techniques for parallel sparse block Cholesky factorization, including the production of low fill-in sparse matrix orderings [44] and efficient block data distributions for MIMD machines. Currently available software tools are presented in section 4.

2 Graph partitioning

2.1 Graph partitioning with external constraints

Many problems arising in the context of parallel computing can be conveniently modeled in terms of graphs. Most often, the solution of these problems amounts to partitioning these graphs according to some problem-specific constraints. The most common constraints consist in minimizing the number of edges that link vertices belonging to distinct blocks, while evenly distributing

---

* This work was supported by the French GDR PRS

the vertices across the blocks of the partition. This problem, referred to as *graph partitioning*, has been addressed by many authors, using numerous techniques such as inertial or geometric [7] methods, graph heuristics [6, 14, 33, 35], spectral partitioning [4, 10, 15, 24], simulated annealing [5], genetic algorithms [9, 38] and neural networks [11]. However, in many cases, additional constraints must be added to capture the true nature of the problem, leading to a more general optimization problem termed *skewed graph partitioning* by some authors [28]. For instance, in a divide-and-conquer context, one must be able to represent the influence of the outer environment on a subproblem that is being solved locally by partitioning of the associated subgraph, so that a globally coherent solution can be achieved.

In the context of static mapping, we have extended graph partitioning algorithms to handle independent external constraints, that is, external constraints that apply independently to every vertex, so that locally moving some vertex from one block of the partition to another does not alter the external constraints applied to other vertices. The main advantage of independent external constraints is that they can be precomputed for every vertex, and thus allow incremental update of partitions, which is extremely important for the efficient execution of many partitioning algorithms such as taboo search, simulated annealing, spectral methods, or graph heuristics [27, 41].

2.2 Graph partitioning techniques

By definition of the goals of the SCOTCH project, we have focused on graph heuristics for computing partitions of graphs.

Improvements of the Fiduccia–Mattheyses heuristic. The Fiduccia–Mattheyses graph bipartitioning heuristic [14] is an almost-linear improvement of the Kernighan-Lin algorithm [35]. Its goal is to minimize the cut between two vertex subsets, while maintaining the balance of their cardinals within a limited user-specified tolerance. Starting from an initially balanced solution of any cut value, it proceeds iteratively by trying, at each stage, to reduce the cut of the current solution. To achieve fast convergence, vertices whose displacement would bring the same *gain* to the cut are linked into lists which are indexed in a *gain array*. This structure is repeatedly searched for the vertex of highest gain, until all vertices have been chosen or until no move of yet unprocessed vertices would keep the load balance within the tolerance. Once the ordering is complete, the new solution is built from the current one by moving as many vertices of the ordering as necessary to get the maximum accumulated gain. Thus, by considering the accumulated gain, the algorithm allows *hill-climbing* from local minima of the cut cost function.

Independent external constraints can easily be incorporated into the above algorithm, by adding to the current gain of every vertex an external contribution equal to the difference between the external costs of its presence in the other part and in the current one. Since we only consider independent external constraints, external gains need be computed only once, during the initial gain calculations, leaving most of the code unchanged [28, 41].

However, this update has important consequences for the efficiency of the algorithm. As a matter of fact, the almost-linearity in time of the original FM algorithm is based on the assumption that the range of gain values is small, so that the search in the gain array for the vertices of best gain takes an almost-constant time. To handle the huge gains generated by the possibly large gains of the vertices, we have implemented a logarithmic indexing of gain lists, which keeps the gain array a reasonable size and so guarantees an almost-constant access time. Results obtained with linear and logarithmic data structures are equivalent in quality, which shows that the approximation induced by logarithmic indexing is of same order of magnitude as the one inherent to the FM algorithm [43].

Multilevel graph partitioning. We have developed a multi-level framework around our graph partitioning algorithms. The multi-level method, which derives from the multi-grid algorithms used in numerical physics and has already been studied by several authors in the context of straight graph partitioning [4, 25, 31], repeatedly reduces the size of the graph to bipartition by finding matchings that collapse vertices and edges, computes a partition for the coarsest graph obtained, and projects the result back to the original graph. Experiments that have been carried out to date show that the multi-level method, used in conjunction with the FM algorithm to compute the initial partitions and refine the projected partitions at every level, yield partitions of better quality than those obtained using FM alone, in less time. By coarsening the graph used by the FM method to compute and project back the initial partition, the multi-level algorithm broadens the scope of the FM algorithm, and makes it possible for it to account for topological structures of the graph that would otherwise be of too high a level for it to encompass in its local optimization process [31, 42].

3 Application to sparse linear system solving

The graph partitioning techniques developed within the SCOTCH project find their use both in iterative and direct linear system solving. Applied to element graphs—that is, graphs whose vertices and edges represent elements and dependences between element nodes, respectively—static mapping allows one to compute partitions that are highly suitable for parallel iterative solving by domain decomposition methods. For direct solving, graph partitioning is used to compute vertex separators for nested dissection orderings.

3.1 Static mapping

The Dual Recursive Bipartitioning algorithm. The efficient execution of a parallel program on a parallel machine requires that the communicating processes of the program be assigned to the processors of the machine so as to minimize the overall running time. When processes are assumed to coexist simultaneously for the duration of the entire execution, this optimization problem is called *mapping*. It amounts to balancing the computational weight of the processes among the processors of the machine, while reducing the communication overhead induced by parallelism by keeping intensively intercommunicating processes on nearby processors. In many such programs, the underlying computational structure can be conveniently modeled as a graph in which vertices correspond to processes that handle distributed pieces of data, and edges reflect data dependencies. The mapping problem can then be addressed by assigning processor labels to the vertices of the graph, so that all processes assigned to some processor are loaded and run on it. A mapping is called *static* if it is computed prior to the execution of the program and is never modified at run-time. Static mapping is NP-complete in the general case [16]. Therefore, many studies have been carried out in order to find suboptimal solutions in reasonable time. Specific algorithms have been proposed for mesh [11, 40] and hypercube [13, 22] topologies.

The static mapping algorithm that has been developed within the SCOTCH project, called *Dual Recursive Bipartitioning* [41], is based on a divide and conquer approach. It starts by considering a set of processors, also called the *domain*, containing all the processors of the target machine, and to which is associated the set of all the processes to map. At each step, the algorithm bipartitions a yet unprocessed domain into two disjoint subdomains, and calls a graph bipartitioning algorithm to split the subset of processes associated with the domain across the two subdomains. Whenever a domain is restricted to a single processor, its associated processes are assigned to it and recursion stops. The association of a subdomain with every process defines a *partial mapping* of the process graph. The *complete mapping* is achieved when successive dual bipartitionings have reduced all subdomain sizes to one.

Domains are recursively bipartitioned by applying our graph partitioning heuristics to the weighted graph that describes the target architecture, such that strongly connected clusters of processors are kept uncut as long as possible and therefore strongly connected clusters of processes tend to be mapped onto them. Process subgraphs are recursively bipartitioned using external constraints that model the dilations of cocycle edges, that is, edges that have exactly one end in the subgraph being bipartitioned and their other end in subgraphs already (partially) mapped onto different subdomains. Thus, when deciding in which subdomain to put a given process, one can account for the communication costs induced by the neighbor processes, whether they are handled by the job itself or not, since one can estimate the dilation of the corresponding edges. This results in an interesting feedback effect: once an edge has been kept in a cut between two subdomains, the distance between its end vertices will be accounted for in the partial communication cost function to be minimized, and following jobs will thus more likely keep these vertices close to each other [43].

Experimental results. The mappings computed by SCOTCH exhibit a great locality of communications with respect to the topology of the target architec-

Name	Vertices	Edges	Description
144	144649	1074393	3D finite element mesh
3ELT	4720	13722	2D finite element mesh
598A	110971	741934	3D finite element mesh
AUTO	448695	3314611	3D finite element mesh
BCSSTK29	13992	302748	3D stiffness matrix
BCSSTK30	28924	1007284	3D stiffness matrix
BCSSTK31	35588	572914	3D stiffness matrix
BCSSTK32	44609	985046	3D stiffness matrix
BODY	45087	163734	3D finite element mesh
BRACK2	62631	366559	3D finite element mesh
BUMP	9800	28989	2D finite element mesh
M14B	214765	3358036	3D finite element mesh
PWT	36519	144793	3D finite element mesh
ROTOR	99617	662431	3D finite element mesh
TOOTH	78136	452591	3D finite element mesh

Table 1. Various matrices used for our tests.

ture. For instance, when mapping the toroidal graph PWT (see table 1) onto a hypercube with 16 vertices, SCOTCH finds a hamiltonian path in the hypercube, such that about 98 percent of the edges are kept local, 2 percent are mapped at distance 1, and less than 0.2 percent are of dilation 2, as illustrated in figure 1. Edges of dilation 2 are only located on the small sphere detailed in figure 1.b, which is mapped onto a sub-hypercube of dimension 2—that is, a square. In this area, SCOTCH minimizes the communication cost function by assigning buffer zones to intermediate processors, so that edges of dilation 2 are replaced by at most twice their number of edges of dilation 1.

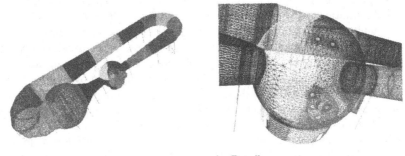

a. Global view.   b. Detail.

Fig. 1. Result of the mapping of graph PWT onto a hypercube with 16 vertices. Vertices with the same grey level are mapped onto the same processor.

Graph	Chaco 1.0			P-Metis 2.0		
	64	128	256	64	128	256
4ELT	2928	**4514**	6869	2965	4600	6929
BCSSTK30	241202	318075	423627	190115	271503	384474
BCSSTK31	65764	98131	141860	**65249**	**97819**	140818
BCSSTK32	106449	153956	223181	106440	**152081**	222789
BRACK2	34172	46835	66944	29983	42625	60608
PWT	9166	12737	18268	**9130**	**12632**	**18108**
ROTOR	53804	75140	104038	53228	75010	103895

Graph	Scotch 3.1		
	64	128	256
4ELT	**2906**	4553	**6809**
BCSSTK30	**188240**	**270165**	**382888**
BCSSTK31	66780	98148	140703
BCSSTK32	**104651**	152082	220476
BRACK2	**29187**	**42169**	**59454**
PWT	9225	13052	18459
ROTOR	**52864**	**73461**	**102697**

Table 2. Edge cuts produced by Chaco 1.0, P-Metis 2.0, and Scotch 3.1 for partitions with 64, 128, and 256 blocks (Chaco and Metis data extracted from [31]).

When mapping onto the complete graph, our program behaves as a standard graph partitioner, since all external constraints are set to zero. Table 2 summarizes edge cuts that we have obtained for classical test graphs, compared to those computed by the recursive graph bisection algorithms of the Chaco 1.0 [25] and Metis 2.0 [32] software packages. Over all the graphs that have been tested, Scotch produces the best partitions of the three in two-thirds of the runs. It can therefore be used as a state-of-the-art graph partitioner. However, not accounting for the target topology generally leads to worse performance results of the mapped applications [22, 26], due to long-distance communication, which makes static mapping more attractive than strict partitioning for most communication-intensive applications.

Tables 3 and 4 summarize some results that have been obtained by Chaco 2.0 using *terminal propagation* (which enables it to handle external constraints [27]) and by Scotch 3.1 when mapping graphs *4ELT* and *BCSSTK32* onto hypercubes and meshes of various sizes; f_C is the communication cost function to minimize, equal to the sum over all edges of the source graph of the product of their weight and their dilation in the resulting mapping. Here again, Scotch outperforms Chaco in most cases.

A complexity analysis of the DRB algorithm shows that, provided that the running time of all graph bipartitioning algorithms is linear in the number of edges of the graphs, the running time of the mapper is linear in the number of edges of the source graph, and logarithmic in the number of vertices of the target

Target	CHACO 2.0-TP		SCOTCH 3.1	
	cut	f_C	cut	f_C
H(1)	168	168	166	**166**
H(2)	412	484	396	**447**
H(3)	769	863	708	**841**
H(4)	1220	1447	1178	**1405**
H(5)	1984	2341	2050	**2332**
H(6)	3244	3811	3194	**3712**
H(7)	5228	6065	5051	**5887**
$M_2(5,5)$	1779	2109	1629	**2039**
$M_2(10,10)$	4565	6167	4561	**6001**

Table 3. Edge cuts and communication costs produced by CHACO 2.0 with Terminal Propagation and by SCOTCH 3.1 for mappings of graph *4ELT* onto hypercube (H) and bidimensional grid (M_2) target architectures (CHACO data extracted from [23]).

Target	CHACO 2.0-TP		SCOTCH 3.1	
	cut	f_C	cut	f_C
H(1)	5562	5562	4797	**4797**
H(2)	15034	15110	10100	**11847**
H(3)	26843	**27871**	24354	28813
H(4)	49988	53067	43078	**49858**
H(5)	79061	89359	71934	**87093**
H(6)	119011	143653	112580	**141516**
H(7)	174505	218318	164532	**211974**
$M_2(5,5)$	64156	**76472**	69202	87737
$M_2(10,10)$	150846	**211672**	150715	223968

Table 4. Edge cuts and communication costs produced by CHACO 2.0 with Terminal Propagation and by SCOTCH 3.1 for mappings of graph *BCSSTK32* onto hypercube (H) and bidimensional grid (M_2) target architectures (CHACO data extracted from [23]).

graph. This is verified in practice for the graph partitioning algorithms that we use [43].

3.2 Sparse matrix ordering

Nested dissection and vertex separators. An efficient way to compute fill reducing orderings for Cholesky factorization of symmetric sparse matrices is to use incomplete recursive nested dissection [19]. It amounts to computing a vertex set S that separates the adjacency graph of the matrix into two parts A and B, ordering S last, and proceeding recursively on parts A and B until their sizes become smaller than some threshold value. This ordering guarantees that no non-zero term can appear in the factorization process between unknowns of A and unknowns of B.

The main issue of the nested dissection ordering algorithm is thus to find small vertex separators that balance the remaining subgraphs as evenly as possible. Most often, vertex separators are computed from edge separators [46, and included references] by minimum cover techniques [12, 30] or heuristics [36, 29], but other techniques such as spectral vertex partitioning have also been used [47]. Provided that good vertex separators are found, the nested dissection algorithm produces orderings which, both in terms of fill-in and operation count, compare very favorably [21, 31] to the ones obtained with the multiple minimum degree algorithm [37]. Moreover, the elimination trees induced by nested dissection are broader, shorter, and better balanced than those issued from multiple minimum degree, and therefore exhibit much more concurrency in the context of parallel Cholesky factorization [2, 17, 18, 20, 51, and included references].

An efficient approach to improve both the computation time and the quality of orderings consists in using a hybrid of the nested dissection and minimum degree methods [29, 44]. Several levels of the nested dissection algorithm are performed on the original graph (to exploit the benefits of this ordering at the top levels, where most of the factorization work is performed), while minimum degree ordering is used on the remaining subgraphs (this latter ordering usually being more efficient for modest-sized graphs).

Our ordering program implements this hybrid ordering algorithm. As do several other authors [12, 32], we compute the vertex separator of a graph by first computing an edge separator, and then turning it into a vertex separator by using the method proposed by Pothen and Fang [46]. This method requires the computation of maximal matchings in the bipartite graphs associated with the edge cuts, which are built using Duff's variant [12] of the Hopcroft and Karp algorithm [30]. The ordering program is completely parameterized by two strategies: the vertex separation strategy, which defines the partitioning methods to apply to build the separator tree, according to parameters of the subgraphs to separate (e.g. size[2] and average degree), and the ordering strategy, which can be used for sequential or parallel [20, 48, 49, 50] block solving to select ordering algorithms that minimize the number of extra-diagonal blocks [8], thus allowing for efficient use of BLAS3 primitives, and for the reduction of inter-processor communication.

Experimental results. The quality of orderings is evaluated with respect to several criteria. The first one, NNZ, is the number of non-zero terms of the factored reordered matrix to store. The second one, OPC, is the operation count, that is, the number of arithmetic operations required to factor the matrix. To comply with existing RISC processor technology, the operation count that we consider accounts for all operations (additions, subtractions, multiplications, divisions) required by Cholesky factorization, except square roots; it is equal to

---

[2] This parameter allows the user to end the recursion, by selecting a void separation method when graph size becomes too small.

Graph	SCOTCH 3.2			O-METIS 2.0		
	NNZ	OPC	Time	NNZ	OPC	Time
144	**5.04302e+07**	**6.30637e+10**	54.18	5.14774e+07	6.63160e+10	32.07
3ELT	**9.34520e+04**	**2.91077e+06**	0.44	9.78490e+04	3.43192e+06	0.27
598A	2.80753e+07	2.15109e+10	32.85	**2.80390e+07**	**2.10550e+10**	22.69
AUTO	**2.42459e+08**	**5.45092e+11**	673.78	–	–	–
BCSSTK29	**1.95601e+06**	**4.98316e+08**	5.05	2.00891e+06	5.22292e+08	2.88
BCSSTK30	**4.70394e+06**	**1.35818e+09**	16.79	4.91682e+06	1.48445e+09	10.26
BCSSTK31	**5.22795e+06**	1.84196e+09	12.64	5.27408e+06	**1.72527e+09**	7.73
BCSSTK32	**6.31493e+06**	**1.69557e+09**	20.12	6.93198e+06	2.23416e+09	12.20
BODY	1.21614e+06	1.08353e+08	8.45	**1.10336e+06**	**7.92548e+07**	4.44
BRACK2	7.38112e+06	**2.54711e+09**	15.07	**7.20380e+06**	2.56784e+09	9.19
BUMP	**2.67353e+05**	**1.48797e+07**	1.02	2.73569e+05	1.55700e+07	0.63
M14B	**6.77465e+07**	**7.08264e+10**	102.05	6.89223e+07	7.18790e+10	52.67
PWT	**1.43140e+06**	**1.22120e+08**	6.05	1.53126e+06	1.40441e+08	3.76
ROTOR	**1.76431e+07**	**1.11101e+10**	28.88	1.81358e+07	1.16120e+10	18.36
TOOTH	**1.31559e+07**	**9.02613e+09**	19.48	1.32536e+07	9.44619e+09	12.34

Graph	MMD		
	NNZ	OPC	Time
144	9.81030e+07	2.62934e+11	28.30
3ELT	**8.98200e+04**	**2.84402e+06**	0.06
598A	4.63319e+07	6.43375e+10	19.96
AUTO	5.16945e+08	2.41920e+12	139.14
BCSSTK29	**1.69480e+06**	**3.930592+08**	0.76
BCSSTK30	**3.84344e+06**	**9.28353e+08**	1.58
BCSSTK31	5.30825e+06	2.55099e+09	1.96
BCSSTK32	**5.24635e+06**	**1.10873e+09**	2.14
BODY	**1.00908e+06**	8.93073e+07	1.25
BRACK2	7.30755e+06	2.99122e+09	3.90
BUMP	2.81482e+05	1.76288e+07	0.12
M14B	1.22359e+08	2.56226e+11	44.07
PWT	1.59060e+06	1.75944e+08	0.70
ROTOR	2.60982e+07	2.86029e+10	11.27
TOOTH	1.56226e+07	1.65121e+10	5.06

Table 5. Orderings produced by SCOTCH 3.2, O-METIS 2.0, and a multiple minimum degree (MMD) algorithm. Bold values in the SCOTCH and METIS sub-arrays indicate best results among these two sub-arrays. Bold values in the MMD sub-array indicate best results over all sub-arrays. Dashes indicate abnormal termination due to lack of memory.

$\sum_c n_c^2$, where n_c is the number of non-zero values in column c of the factored matrix, diagonal included.

Table 5 summarizes the values of NNZ and OPC obtained by SCOTCH, by the O-METIS sparse matrix ordering software [32], and by a multiple minimum degree algorithm (MMD) [45], for all of our test graphs. By default, SCOTCH performs nested dissection until the size of the resulting blocks becomes smaller than 50 vertices, and runs the MMD algorithm of [45] on the remaining subblocks. SCOTCH 3.2 outperforms O-METIS 2.0 in 80 % of the runs. On average, its NNZ is smaller by 2%, and its OPC by 4%. These figures, which may seem small, amount, for large graphs, to gains of millions of terms and billions of operations. For instance, for graph *M14B*, the gain over O-METIS is of about 120 thousand terms and 1 billion operations. Our experiments confirm the efficiency of state-of-the-art nested dissection methods compared to multiple minimum degree algorithms. SCOTCH 3.2 outperforms the MMD algorithm in about 70 % of the runs. On average, its NNZ is smaller by 19%, and its OPC by 47%. For graph *AUTO*, the gains for NNZ and OPC represent savings of about 274 million terms and 1.9 trillion operations. The main advantage of MMD resides in its speed, which is 1.5 to 7 times that of current nested dissection algorithms. However, the elimination trees generated by nested dissection are by far better balanced than the ones computed by MMD (the good balance of elimination trees for nested dissection can be inferred from the pattern of the factored matrices; see for instance figure 2.a), and therefore are much more suitable for parallel Cholesky factorization [20, 49, 50]. Moreover, the MMD algorithm is inherently sequential, so nested dissection seems much more promising for parallelization [34].

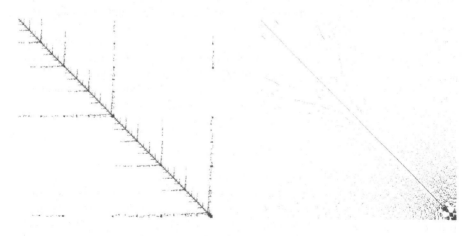

a. SCOTCH 3.2.   b. MMD.

Fig. 2. Patterns of the factored matrices built from orderings of graph *BUMP* computed by SCOTCH 3.2 and the MMD algorithms.

3.3 Data distribution for high performance parallel sparse Cholesky factorization

The hybrid ordering defined in the previous section allows us to compute a block data structure for the factored matrix by block symbolic factorization [8], following a supernodal approach [3]. Then, BLAS3 primitives can be used to achieve better efficiency on vector or RISC cache-based processors.

We are currently working on a block distribution of the factored matrix across the processors of distributed memory machines for a parallel block fan-in (left-looking) Cholesky factorization algorithm [39]. The distribution of the blocks is performed using the block elimination tree, first by subtree-to-subcube mapping of the blocks belonging to the lowest levels, to exploit the concurrency induced by sparsity, and second by distribution of the other blocks, taken in ascending order, according to a load balancing criterion; the most costly blocks in term of operation count are split in a block cyclic manner, to exhibit concurrency from dense computations.

We have carried out implementations of this solver on the Intel Paragon and on the IBM SP2; the first results to date are encouraging, for moderately large distributed memory parallel computers.

4 Software tools

SCOTCH 3.2 is a software package for static mapping, graph partitioning, and sparse matrix ordering, which embodies the algorithms developed within the SCOTCH project. Apart from the static mapper and the sparse matrix orderer themselves, the SCOTCH package contains programs to build and test graphs, compute decompositions of target architectures for mapping, and visualize mapping and ordering results. Advanced command-line interface and vertex labeling capabilities make them easy to interface with other programs.

The SCOTCH 3.2 academic distribution may be obtained from the SCOTCH WWW page at http://www.labri.u-bordeaux.fr/~pelegrin/scotch/, or by ftp at ftp.u-bordeaux.fr in directory /pub/Local/Info/Software/Scotch. It is labeled scotch_3.2A.tar.gz. It contains the executables for several machines and operating systems, along with documentation and sample files.

5 Future work

Work in progress follows four main directions.

The first one is the parallelization of the existing sequential graph partitioning tools, not only to increase partitioning speed, but mainly to process very large problems that do not fit in the memory of centralized architectures. The parallel graph partitioning library will make up the core of parallel static mapping and ordering libraries.

The second one is the development of new (combinations of) vertex separation methods, in a way similar to the work of [29]. We are planning to implement the Approximate Minimum Degree algorithm of Amestoy, Davis, and Duff [1] to compute the orderings of the leaves of the separation tree.

The third one is the definition of ordering strategies that minimize the number of extra-diagonal blocks in the factored matrices, for our parallel sparse block Cholesky solver.

The fourth one is the improvement of block data distribution techniques to achieve scalability on very large distributed memory parallel computers, and improve the overall performance of the solver [49, 50, 51].

References

1. P. Amestoy, T. Davis, and I. Duff. An approximate minimum degree ordering algorithm. Technical Report RT/APO/95/5, ENSEEIHT-IRIT, 1995. To appear in SIAM Journal of Matrix Analysis and Applications.

2. C. Ashcraft, S. Eisenstat, J. W.-H. Liu, and A. Sherman. A comparison of three column based distributed sparse factorization schemes. In *Proc. Fifth SIAM Conf. on Parallel Processing for Scientific Computing*, 1991.

3. C. Ashcraft, R. Grimes, J. Lewis, B. Peyton, and H. Simon. Recent progress in sparse matrix methods for large linear systems. *Intl. J. of Supercomputer Applications*, 1(4):10–30, 1987.

4. S. T. Barnard and H. D. Simon. A fast multilevel implementation of recursive spectral bisection for partitioning unstructured problems. *Concurrency: Practice and Experience*, 6(2):101–117, 1994.

5. S. W. Bollinger and S. F. Midkiff. Processor and link assignment in multicomputers using simulated annealing. In *Proceedings of the 11th Int. Conf. on Parallel Processing*, pages 1–7. The Penn. State Univ. Press, August 1988.

6. T. N. Bui, S. Chaudhuri, F. T. Leighton, and M. Sipser. Graph bisection algorithms with good average case behavior. *Combinatorica*, 7(2):171–191, 1987.

7. F. Cao, J. R. Gilbert, and S.-H. Teng. Partitioning meshes with lines and planes. Technical Report CSL-96-01, XEROX Corporation, Palo Alto, California, 1996.

8. P. Charrier and J. Roman. Algorithmique et calculs de complexité pour un solveur de type dissections emboîtées. *Numerische Mathematik*, 55:463–476, 1989.

9. T. Chockalingam and S. Arunkumar. A randomized heuristics for the mapping problem: The genetic approach. *Parallel Computing*, 18:1157–1165, 1992.

10. W. Donath and A. Hoffman. Algorithms for partitioning of graphs and computer logic based on eigenvectors of connection matrices. *IBM Technical Disclosure Bulletin*, 15:938–944, 1972.

11. M. Dormanns and H.-U. Heiß. Mapping large-scale FEM-graphs to highly parallel computers with grid-like topology by self-organization. Technical Report 5/94, Fakultät für Informatik, Universität Karlsruhe, February 1994.

12. I. Duff. On algorithms for obtaining a maximum transversal. *ACM Trans. Math. Software*, 7(3):315–330, September 1981.

13. F. Ercal, J. Ramanujam, and P. Sadayappan. Task allocation onto a hypercube by recursive mincut bipartitioning. *JPDC*, 10:35–44, 1990.

14. C. M. Fiduccia and R. M. Mattheyses. A linear-time heuristic for improving network partitions. In *Proc. 19th Design Autom. Conf.*, pages 175–181. IEEE, 1982.

15. M. Fiedler. Algebraic connectivity of graphs. *Czech. Math. J.*, 23:298–305, 1973.

16. M. R. Garey and D. S. Johnson. *Computers and Intractablility: A Guide to the Theory of NP-completeness.* W. H. Freeman, San Francisco, 1979.

17. G. A. Geist and E. G.-Y. Ng. Task scheduling for parallel sparse Cholesky factorization. *International Journal of Parallel Programming*, 18(4):291–314, 1989.

18. A. George, M. T. Heath, J. W.-H. Liu, and E. G.-Y. Ng. Sparse Cholesky factorization on a local memory multiprocessor. *SIAM J. Sci. Stat. Comp.*, 9:327–340, 1988.

19. J. A. George and J. W.-H. Liu. *Computer solution of large sparse positive definite systems.* Prentice Hall, 1981.

20. A. Gupta, G. Karypis, and V. Kumar. Highly scalable parallel algorithms for sparse matrix factorization. TR 94-063, University of Minnesota, 1994. To appear in *IEEE Trans. on Parallel and Distributed Systems*, 1997.

21. A. Gupta, G. Karypis, and V. Kumar. Scalable parallel algorithms for sparse linear systems. In *Proc. Stratagem'96, Sophia-Antipolis*, pages 97–110. INRIA, July 1996.

22. S. W. Hammond. *Mapping unstructured grid computations to massively parallel computers.* PhD thesis, Rensselaer Polytechnic Institute, February 1992.

23. B. Hendrickson. Mapping results. Personal communication, July 1996.

24. B. Hendrickson and R. Leland. Multidimensional spectral load balancing. Technical Report SAND93-0074, Sandia National Laboratories, January 1993.

25. B. Hendrickson and R. Leland. The CHACO user's guide – version 2.0. Technical Report SAND94-2692, Sandia National Laboratories, 1994.

26. B. Hendrickson and R. Leland. An empirical study of static load balancing algorithms. In *Proc. SHPCC'94, Knoxville*, pages 682–685. IEEE, May 1994.

27. B. Hendrickson, R. Leland, and R. Van Driessche. Enhancing data locality by using terminal propagation. In *Proceedings of the 29th Hawaii International Conference on System Sciences.* IEEE, January 1996.

28. B. Hendrickson, R. Leland, and R. Van Driessche. Skewed graph partitioning. In *Proceedings of the 8th SIAM Conference on Parallel Processing for Scientific Computing.* IEEE, March 1997.

29. B. Hendrickson and E. Rothberg. Improving the runtime and quality of nested dissection ordering. Technical Report SAND96-0868J, Sandia National Laboratories, March 1996.

30. J. Hopcroft and R. Karp. An $n^{5/2}$ algorithm for maximum matchings in bipartite graphs. *SIAM Journal of Computing*, 2(4):225–231, December 1973.

31. G. Karypis and V. Kumar. A fast and high quality multilevel scheme for partitioning irregular graphs. TR 95-035, University of Minnesota, June 1995.

32. G. Karypis and V. Kumar. METIS – *Unstructured Graph Partitioning and Sparse Matrix Ordering System – Version 2.0.* University of Minnesota, June 1995.

33. G. Karypis and V. Kumar. Multilevel k-way partitioning scheme for irregular graphs. Technical Report 95-064, University of Minnesota, August 1995.

34. G. Karypis and V. Kumar. Parallel multilevel k-way partitioning scheme for irregular graphs. TR 96-036, University of Minnesota, 1996.

35. B. W. Kernighan and S. Lin. An efficient heuristic procedure for partitionning graphs. *BELL System Technical Journal*, pages 291–307, February 1970.

36. C. Leiserson and J. Lewis. Orderings for parallel sparse symmetric factorization. In *Third SIAM Conference on Parallel Processing for Scientific Computing, Tromsø.* SIAM, 1987.

37. J. W.-H. Liu. Modification of the minimum-degree algorithm by multiple elimination. *ACM Trans. Math. Software*, 11(2):141–153, 1985.

38. T. Muntean and E.-G. Talbi. A parallel genetic algorithm for process-processors mapping. *High performance computing*, 2:71–82, 1991.

39. E. Ng and B. Peyton. Block sparse Cholesky algorithms on advanced uniprocessor computers. *SIAM J. on Scientific Computing*, 14:1034–1056, 1993.

40. D. M. Nicol. Rectilinear partitioning of irregular data parallel computations. *Journal of Parallel and Distributed Computing*, 23:119–134, 1994.

41. F. Pellegrini. Static mapping by dual recursive bipartitioning of process and architecture graphs. In *Proc. SHPCC'94, Knoxville*, pages 486–493. IEEE, May 1994.

42. F. Pellegrini. Application of graph partitioning techniques to static mapping and domain decomposition. In *ETPSC 3, Faverges-de-la-Tour*, August 1996. To appear in a special issue of Parallel Computing.

43. F. Pellegrini and J. Roman. Experimental Analysis of the Dual Recursive Bipartitioning Algorithm for Static Mapping. Research Report 1138-96, LaBRI, Université Bordeaux I, September 1996. Available at URL http://www.labri.u-bordeaux.fr/~pelegrin/papers/scotch_expanalysis.ps.gz.

44. F. Pellegrini and J. Roman. Sparse matrix ordering with SCOTCH. In *Proc. HPCN'97, Vienna*, April 1997. To appear.

45. A. Pothen. MMD code. Personal communication, February 1997.

46. A. Pothen and C.-J. Fan. Computing the block triangular form of a sparse matrix. *ACM Trans. Math. Software*, 16(4):303–324, December 1990.

47. A. Pothen, H. D. Simon, and K.-P. Liou. Partitioning sparse matrices with eigenvectors of graphs. *SIAM Journal of Matrix Analysis*, 11(3):430–452, July 1990.

48. E. Rothberg. Performance of panel and block approaches to sparse Cholesky factorization on the iPSC/860 and Paragon multicomputers. In *Proceedings of SHPCC'94, Knoxville*, pages 324–333. IEEE, May 1994.

49. E. Rothberg and A. Gupta. An efficient block-oriented approach to parallel sparse Cholesky factorization. In *Supercomputing'93 Proceedings*. IEEE, 1993.

50. E. Rothberg and R. Schreiber. Improved load distribution in parallel sparse Cholesky factorization. In *Supercomputing'94 Proceedings*. IEEE, 1994.

51. R. Schreiber. Scalability of sparse direct solvers. Technical Report TR 92.13, RIACS, NASA Ames Research Center, May 1992.

Author Index

Springer
and the
environment

At Springer we firmly believe that an
international science publisher has a
special obligation to the environment,
and our corporate policies consistently
reflect this conviction.

We also expect our business partners –
paper mills, printers, packaging
manufacturers, etc. – to commit
themselves to using materials and
production processes that do not harm
the environment. The paper in this
book is made from low- or no-chlorine
pulp and is acid free, in conformance
with international standards for paper
permanency.

Springer

Lecture Notes in Computer Science

For information about Vols. 1–1170

please contact your bookseller or Springer-Verlag